NJA Sloo

Keep.

Jake on a long trip? No.

ALGEBRAIC NUMBER FIELDS

This is Volume 55 in
PURE AND APPLIED MATHEMATICS
A series of Monographs and Textbooks
Editors: PAUL A. SMITH AND SAMUEL EILENBERG
A complete list of titles in this series appears at the end of this volume

ALGEBRAIC NUMBER FIELDS

GERALD J. JANUSZ

University of Illinois at Urbana-Champaign
Urbana, Illinois

ACADEMIC PRESS New York and London 1973

A Subsidiary of Harcourt Brace Jovanovich, Publishers

ACADEMIC PRESS, INC.
111 Fifth Avenue, New York, New York 10003

United Kingdom Edition published by
ACADEMIC PRESS, INC. (LONDON) LTD.
24/28 Oval Road, London NW1

LIBRARY OF CONGRESS CATALOG CARD NUMBER: 72-9330

AMS (MOS) 1970 Subject Classifications: 12A01, 12A02, 12B01, 13F01

PRINTED IN THE UNITED STATES OF AMERICA

Dedicated to my mother, Beatrice Janusz
and to the memory of my father,
Andrew Janusz (1915–1964).

CONTENTS

PREFACE

This book contains an exposition of the main theorems of the class field theory of algebraic number fields along with the necessary introductory material. An attempt is made to keep the exposition self-contained. The only material presupposed is that which would be used in elementary Galois theory. The structure theorem for finitely generated modules over a principal ideal domain is used, but a proof is included in an appendix.

We use the direct approach to the subject by congruence subgroups of the ideal group rather than the more subtle description involving the cohomology of groups. This may be considered the historical approach to the subject, but we have presented it because it seems most useful for mathematicians who are specialists in other areas but wish to use it. From the student's point of view, this approach seems to require less background preparation and so is desirable for them.

The student who is not particularly interested in the theory of class fields can profitably read the first three chapters for an introduction to the study of arithmetic in fields, Dedekind domains, valuations, and general background material necessary for further work in several directions.

The first chapter contains an introduction to the algebra of number theory. The basic properties of Dedekind domains are presented using rather general ring theoretic arguments as much as possible. Emphasis is placed upon local methods and proofs by localization. The results are given for rather general fields except in the last three sections. There we discuss cyclotomic extensions of the rational field and prove the unit theorem and finiteness of the class number for algebraic number fields.

Valuations and complete fields are discussed in the second chapter. This depends partly upon the previous chapter.

Chapter three contains material connecting Galois groups and ramification. The last section introduces the Artin map, which is of fundamental importance for the rest of the book.

In Chapter four the material becomes more specialized. We work exclusively with algebraic number fields and their completions. The analytic theorems proved include the Frobenius density theorem and Dirichlet's theorem on primes in an arithmetic progression.

Chapter five contains the main results on class fields. A small amount of cohomology is developed here. We use only H^0 and H^1 and then only for cyclic groups. In this case a concrete description of the cohomology groups is used so the reader is not carried far from the fields and Galois groups. The approach to the main theorems makes systematic use of the Artin map and the Artin reciprocity theorem. We close the chapter with a discussion of the Hilbert class field and Artin's reduction of the principal ideal theorem.

The last chapter is intended primarily for illustration of the concepts introduced in the earlier chapters. We study mainly quadratic fields and prove a result which goes back to Gauss giving information about the class group of a quadratic number field. A few calculations are made to illustrate the use of the norm residue symbols.

The material in this text was used in a year-long course at the University of Illinois in 1970–1971. The first three chapters were covered in the first semester and the balance in the second semester.

Chapter I

SUBRINGS OF FIELDS

1. LOCALIZATION

Let R be an integral domain. This means R is a commutative ring with identity having no zero divisors.

Let S be a subset of R which does not contain zero and which contains the product of any two elements in S. A set satisfying these conditions is called a *multiplicative set* in R.

1.1 Proposition. There is a ring R_S which contains R as a subring (up to isomorphism) and such that each element of S has a multiplicative inverse in R_S.

PROOF. Let us first consider the collection of all pairs (r, s) in $R \times S$. Call two such pairs (r, s) and (q, t) equivalent if $qs = rt$. We leave as an exercise the verification that this is an equivalence relation. Let r/s denote the equivalence class containing (r, s). By definition of the equivalence class we see that $r/s = rt/st$ for any t in S.

Let $R_S = \{r/s \mid r \in R,\ s \in S\}$. Addition and multiplication are defined for elements of R_S by the rules

$$r/s + r'/s' = rs' + r's/ss'$$

$$r/s \cdot r'/s' = rr'/ss'.$$

One can check that these operations are well defined and R_S is a ring. To imbed

R into R_S we first fix a particular $s \in S$ and use the mapping

$$r \to rs/s.$$

This is a ring homomorphism and is in fact one to one. If we identify r in R with any of the (equal) elements rs/s, then we do have $R \subseteq R_S$.

If t is in S, then $1/t$ is in R_S and $t \cdot 1/t = 1/1$, which is the identity in R_S. This completes the proof. □

The ring R_S constructed in the proof has the following universal property: If ϕ is a homomorphism of R into a ring T such that every element in $\phi(S)$ has an inverse in T, then ϕ has a unique extension to a homomorphism from R_S into T. The extended map is defined by $\phi(r/s) = \phi(r)\phi(s)^{-1}$. This kind of reasoning shows the ring R_S is characterized as the smallest ring containing R and inverses for the elements of S.

Definition. The ring R_S is called the *localization* of R at S.

One may verify that R_S is also an integral domain.

EXERCISE 1. Let R and S be as above and let $S^* = \{1\} \cup S$. Prove S^* is a multiplicative set such that $R_S \cong R_{S^*}$.

This shows we may assume 1 is in S and the mapping of R into R_S may be taken as $r \to r/1$. We shall identify r with r/1 so $R \subseteq R_S$ has this meaning.

EXAMPLE 1. $R = Z =$ integers, $S = \{1, n, n^2, \ldots\}$ for some fixed nonzero integer n. Then Z_S is the collection of all rational numbers a/n^i, $i \in Z$

EXAMPLE 2. $R = Z$, $S = \{$all positive integers not divisible by 3$\}$, $R_S = \{a/b \mid 3 \text{ does not divide } b\}$.

EXAMPLE 3. $R =$ any integral domain

$$S = R - \{0\}.$$

Then R_S is a field (all nonzero elements have an inverse) and we call this the *quotient field* of R. It is the smallest field containing R.

We next look for some relation between ideals in R and ideals in R_S.

Definition. An ideal $\mathfrak{P}$ of R is *prime* if whenever ab belongs to $\mathfrak{P}$ with a, b in R, then either a or b already belongs to $\mathfrak{P}$. We shall *exclude* the case $\mathfrak{P} = R$ always.

EXERCISE 2. $\mathfrak{P}$ is a prime ideal of R if and only if the factor ring $R/\mathfrak{P}$ is an integral domain.

1.2 Proposition. Let R be an integral domain and S a multiplicative set in R. There is a one-to-one correspondence between the prime ideals of R_S and the prime ideals of R which have empty intersection with S. Under the correspondence, a prime $\mathfrak{P}$ of R is associated with the ideal $\mathfrak{P}R_S$ in R_S.

PROOF. Let $\mathfrak{Q}$ be a prime ideal in R_S. From the definition it is immediate that $\mathfrak{P} = \mathfrak{Q} \cap R$ is a prime ideal of R. Then $\mathfrak{P}R_S$ is an ideal of R_S contained in $\mathfrak{Q}$. We show these are equal. Let q/s be any element in $\mathfrak{Q}$ with q in R and s in S. Then $q = (q/s)s$ is in $R \cap \mathfrak{Q} = \mathfrak{P}$. Thus q/s is in $\mathfrak{P}R_S$ since $q(1/s) = q/s$, q in $\mathfrak{P}$, $1/s$ in R_S. So far we have proved that every prime ideal in R_S has the form $\mathfrak{Q} = \mathfrak{P}R_S$ with $\mathfrak{P} = \mathfrak{Q} \cap R_S$ uniquely determined by $\mathfrak{Q}$. Since every element in S has an inverse in R_S we know $\mathfrak{Q} \cap S$ is empty. Thus $\mathfrak{P} \cap S$ is empty.

Now suppose we start with the prime ideal $\mathfrak{P}$ of R which has no elements in S. Let $\mathfrak{Q} = \mathfrak{P}R_S$. This is an ideal of R_S which we shall prove is prime. Suppose a, b are elements in R_S with ab in $\mathfrak{Q}$, then $ab = x/s$ with some x in $\mathfrak{P}$ and some s in S. Suppose $a = r_1/s_1$, $b = r_2/s_2$ with r_1, r_2 in R and s_1, s_2 in S.

We have $r_1 r_2 s = x s_1 s_2$ belongs to $\mathfrak{P}$. Thus one of the elements r_1, r_2 or s in $\mathfrak{P}$ because $\mathfrak{P}$ is prime. Also s is not in $\mathfrak{P}$ by choice of $\mathfrak{P}$. Thus r_1 or r_2 belongs to $\mathfrak{P}$ and so $a = r_1/s_1$ or $b = r_2/s_2$ is in $\mathfrak{Q}$. Thus $\mathfrak{Q}$ is prime. Now finally we prove $\mathfrak{Q} \cap R = \mathfrak{P}$. If u is in $\mathfrak{Q} \cap R$ then $u = x/s$ with x in $\mathfrak{P}$ because $\mathfrak{Q} = \mathfrak{P}R_S$. But u also belongs to R and so $x = us$ implies u or s is in $\mathfrak{P}$. Since s is not, we have u is in $\mathfrak{P}$. Hence the correspondences $\mathfrak{P} \to \mathfrak{P}R_S$ and $\mathfrak{Q} \to \mathfrak{Q} \cap R$ are inverses of one another and the proposition is proved. □

EXAMPLE 4. Let $\mathfrak{P}$ be a prime ideal in the domain R and let $S = \{r \mid r \text{ not in } \mathfrak{P}\} = R - \mathfrak{P}$. The definition of prime ideal is equivalent with the assertion that S is a multiplicative set. Then R_S can be identified with $\{a/b \mid a, b \text{ in } R,\ b \text{ not in } \mathfrak{P}\}$.

This is the most important example of localization. It will be encountered so frequently that we shall use the notation $R_{\mathfrak{P}}$ to denote the localization of R at $S = R - \mathfrak{P}$ when $\mathfrak{P}$ is a prime ideal. Since a prime ideal can *never* be a multiplicative set, this should not cause a conflict.

Observe that the prime ideals of R which have empty intersection with $S = R - \mathfrak{P}$ are those prime ideals contained in $\mathfrak{P}$. Hence the only prime ideals of $R_{\mathfrak{P}}$ are those contained in $\mathfrak{P}R_{\mathfrak{P}}$. Maximal ideals are always prime so $\mathfrak{P}R_{\mathfrak{P}}$ is the only maximal ideal in $R_{\mathfrak{P}}$. We summarize these facts.

1.3 Proposition. If $\mathfrak{P}$ is a prime ideal in R then $R_{\mathfrak{P}}$ has only one maximal ideal; namely $\mathfrak{P}R_{\mathfrak{P}}$.

EXERCISE 3. Let R be a domain and $\mathfrak{P}$ a maximal ideal. Show there is an isomorphism between the fields $R/\mathfrak{P}$ and $R_{\mathfrak{P}}/\mathfrak{P}R_{\mathfrak{P}}$.

EXERCISE 4. If S is a multiplicative set in a noetherian domain R, then R_S is also noetherian.

Rings having only one maximal ideal occur frequently and we give them a special term.

Definition. A ring having only one maximal ideal is called a *local ring*.

A few properties of local rings will be developed here.

Let R be a local ring (always with identity) and $\mathfrak{P}$ be the unique maximal ideal.

Lemma. Every element in R which is not in $\mathfrak{P}$ has a multiplicative inverse. In particular for m in $\mathfrak{P}$, the element $1+m$ is invertible.

PROOF. If x is not in $\mathfrak{P}$ then Rx is not in $\mathfrak{P}$. Since $\mathfrak{P}$ is the only maximal ideal, it follows that $Rx = R$. There exists a y in R such that $yx = 1$. $m \in \mathfrak{P} \Rightarrow 1+m \notin$

1.4 Proposition. Let M be a finitely generated R module such that $\mathfrak{P}M = M$. Then $M = (0)$.

Corollary (Nakayama's Lemma). Let M be a finitely generated R-module, and L a submodule of M such that $L + \mathfrak{P}M = M$. Then $L = M$.

PROOF OF COROLLARY. The equality $L + \mathfrak{P}M = M$ implies that every coset of L in M has a representative in $\mathfrak{P}M$. This means $\mathfrak{P}(M/L) = M/L$. By the Proposition 1.4 it follows $M = L$.

Before giving the proof of Proposition 1.4 we recall some facts about matrices.

Let $A = |a_{ij}|$ be a matrix (size $n \times n$) with all a_{ij} in some commutative ring. The (i,j) cofactor is $(-1)^{i+j}\alpha_{ij} = b_{ji}$, where α_{ij} is the determinant of the $(n-1)\times(n-1)$ matrix obtained from A by removing the ith row and the jth column. The matrix $B = |b_{ij}|$ is called the adjoint matrix of A. Then $BA = AB = \operatorname{diag}\{d, \ldots, d\}$ with $d = \det(A)$.

PROOF OF PROPOSITION 1.4. Let $M = \sum Rm_i$ with $m_1, \ldots, m_n$ in M. Since $\mathfrak{P}M = M$ there exist elements a_{ij} in $\mathfrak{P}$ such that

$$m_i = \sum_j a_{ij} m_j. \tag{$*$}$$

Let A be the matrix $|a_{ij}| - I$. Then $A(m_1, \ldots, m_n)^t = 0$ is just a restatement of $(*)$. Let $B =$ adjoint of A so that by the above remarks $BA(m_1, \ldots, m_n)^t = (dm_1, \ldots, dm_n)^t = 0$ with $d = \det A$. It follows $dM = (0)$. Now consider the expansion of $\det A$. The term d is a sum of a large number of terms each of which is a product involving certain a_{ij} except for one term which is $(-1)^n$. Thus $d = (-1)^n + (\text{sum of elements of } \mathfrak{P})$. By the lemma above, d has an inverse in R so that $dM = 0$ implies $M = 0$ as required.

2. INTEGRAL DEPENDENCE

Let R be a subring of the commutative ring R' and assume the identity of R is the identity of R'.

Definition. An element b in R' is *integral over* R if there is a monic polynomial $f(X)$ in $R[X]$ such that $f(b) = 0$. We say $f(X)$ is the equation of *integral dependence*.

2.1 Proposition. The following statements are equivalent:

(1) The element b of R' is integral over R.

(2) $R[b]$ is a finitely generated R module.

(3) $R[b]$ is contained in a subring B of R' which is a finitely generated R module.

(4) There exists in R' an $R[b]$-module M such that M is finitely generated over R and the only element y in $R[b]$ for which $yM = 0$ is $y = 0$.

PROOF. (1) → (2). If the monic polynomial satisfied by b has degree $n+1$ then $R[b]$ is generated by $1, b, \ldots, b^n$.

(2) → (3). Take $B = R[b]$.

(3) → (4). Take $M = B$. Since 1 is in B, yB always contains y.

(4) → (1). Let $m_1, \ldots, m_n$ be a set of R generators for M. Let r_{ij} be elements of R such that

$$bm_i = \sum_j r_{ij} m_j.$$

This can be rewritten as

$$0 = \sum (r_{ij} - b\delta_{ij}) m_j,$$

where δ_{ij} = Kronecker delta. If A is the matrix of the coefficients in these equations then (by the same method as the proof of Proposition 1.4) $dM = 0$ when $d = \det(A)$. This forces $d = 0$. Consider the polynomial $\det(X \cdot I - |r_{ij}|) = f(X)$. The expansion of this determinant shows $f(X)$ is a monic polynomial with coefficients in R. Moreover $0 = d = f(b)$ so b is integral over R. ☐

2.2 Proposition. Suppose $b_1, \ldots, b_n$ are elements of R' which are integral over R. Then $R[b_1, \ldots, b_n]$ is a finitely generated R module.

PROOF. Use induction on n. $R[b_1]$ is finitely generated by Part (2) of Proposition 2.1. Assume that the ring $R[b_1, \ldots, b_{n-1}] = R''$ is finitely generated over R with generators $a_1, a_2, \ldots, a_t$. Then b_n is integral over R'' so $R''[b_n]$ is generated by $1, b_n, \ldots, b_n{}^k$ (for some k) over R''. Thus the finite set $a_i b_n{}^j$ of $t(k+1)$ elements generates $R[b_1, \ldots, b_n]$ over R.

Corollary. The set of all elements of R' which are integral over R is a subring of R' containing R.

PROOF. Suppose x, y are in R' and are integral over R. Then the proposition just proved says $R[x, y] = B$ is a subring which is finitely generated over R. Since $x \pm y$ and xy are in B we see by Part (3) of Proposition 2.1 that $x \pm y$, xy are integral over R. Thus the set of integral elements forms a subring. It is clear that R is integral over R.

A situation frequently encountered involves an integral domain R contained in its quotient field K. The set of all elements of K integral over R is a subring called the *integral closure of* R. The domain R is said to be *integrally closed* if every element in the quotient field which is integral over R is already in R.

One way to obtain integrally closed domains is to start with any domain R and let R' denote the integral closure of R (in its quotient field). It follows from the next proposition that R' is integrally closed.

2.3 Proposition. If $R \subseteq R' \subseteq R''$ are rings with R' integral over R and R'' integral over R', then R'' is integral over R.

PROOF. Take any b in R''. There is a polynomial $f(X) = X^n + r_1 X^{n-1} + \cdots + r_n$ with r_i in R' such that $f(b) = 0$. Then by Proposition 2.2 $R[r_1, \ldots, r_n] = B$ is a finitely generated R module and so $B[b]$ is also a finitely generated R module. By Part (3) of Proposition 2.1, b is integral over R.

EXAMPLE 1. Let R be any unique factorization domain (UFD). Then R is integrally closed.

To prove this we suppose x, y are in R and x/y is an element in the quotient field which is integral over R. It is necessary to prove x/y is in R. There is a relation

$$(x/y)^n = \sum_{0}^{n-1} r_i (x/y)^i$$

with r_i in R. Because R is a UFD we may suppose at the start that x and y have no common factor apart from units.

Now multiply the equation above by y^n and find

$$x^n = y \sum_{0}^{n-1} r_i x^i y^{n-1-i}$$

This shows y divides x^n and so y must be a unit of R for otherwise a prime divisor of y also divides x, contrary to assumption. So x/y is in R since y^{-1} is in R.

EXAMPLE 2. Any PID (principle ideal domain) is integrally closed because it is a UFD.

EXAMPLE 3. If R is integrally closed in its quotient field and if S is a multiplicative set in R, then R_S is integrally closed.

The proof of this uses computations similar to those above. We suppose u is an element of the quotient field which is integral over R_S. Thus u is the root of a monic polynomial with coefficients in R_S. We find a common denominator for all these coefficients so that u is a root of

$$X^n + (r_1/s)X^{n-1} + \cdots + (r_n/s).$$

Multiply this by s^n and discover that su is the root of a monic polynomial with coefficients in R. Thus su is in R because R is integrally closed. Finally $u = su/s$ is in R_S as required.

EXERCISE 1. Prove that the integral closure of Z in the field $Q(\sqrt{3})$ is just $Z+Z\sqrt{3}$.

EXERCISE 2. Prove the ring $Z[\sqrt{5}]$ is not integrally closed. In fact $\frac{1}{2}(1+\sqrt{5})$ is an integral element in the quotient field.

EXERCISE 3. If K is a field and $\{R_i\}$ is a family of integrally closed subrings of K, then the intersection $\bigcap R_i$ is also integrally closed.

The procedure for determining whether or not an element is integral over R might generally be a lengthy one since there is no clear way to select the polynomial which expresses the integral dependence. The next proposition shows a circumstance where this procedure is simplified.

Suppose R is a domain with quotient field K.

2.4 Proposition. Let b be an element in an extension field of K. Let $f(X)$ be a monic irreducible polynomial in $K[X]$ having b as a root. If b is integral over R then the coefficients of $f(X)$ are integral over R. If R is integrally closed then b is integral over R if and only if $f(X) \in R[X]$.

PROOF. Extend the field from which b is taken to a splitting field of $f(X)$ over K. Let $b_1, \ldots, b_n$ be all the roots with $b = b_1$, say.

Suppose b is integral over R and $g(X)$ is the equation of integral dependence. Since $f(X)$ is irreducible and has a root in common with $g(X)$, it follows that $f(X)$ divides $g(X)$ in $K[X]$. Hence the roots $b_1, \ldots, b_n$ of $f(X)$ are also roots of $g(X)$. So $b_1, \ldots, b_n$ are all integral over R. It follows that the coefficients of $f(X) = (X-b_1)\cdots(X-b_n)$ are integral over R.

These coefficients belong to K so when R is integrally closed we see $f(X)$ is in $R[X]$.

This shows that one can test for integral dependence of an element by looking at its minimum polynomial over the quotient field.

EXERCISE 4. Let d be an integer not divisible by the square of any prime. Show that the integral closure of Z in $Q(\sqrt{d})$ is

$$Z[\sqrt{d}] = Z + Z\sqrt{d} \qquad \text{if} \quad d \equiv 2, 3 \bmod 4,$$

and

$$Z\left[\frac{1+\sqrt{d}}{2}\right] = Z + Z\frac{1+\sqrt{d}}{2} \qquad \text{if} \quad d \equiv 1 \bmod 4.$$

3. DISCRETE VALUATION RINGS AND DEDEKIND RINGS

Definition. A ring is called a *discrete valuation ring* (DVR) if it is a principal ideal domain with only one maximal ideal.

Let R be a DVR and let π be an element such that $\mathfrak{P} = R\pi$ is the unique maximal ideal.

3.1 Elementary Properties

(i) R is a noetherian ring. (PIDs are noetherian.)

(ii) Every nonzero element of R has the form $u\pi^k$ for some nonnegative integer k and some unit u in R.

This follows at once because R is a UFD and hence can contain only one prime element up to unit multiples.

(iii) Every nonzero ideal has the form $R\pi^k$ for some k. This follows from (ii) and the fact that R is a PID.

(iv) R is integrally closed (because it is a UFD).

(v) R has only one nonzero prime ideal [immediate from (iii)].

Definition. A ring R is called a *Dedekind ring* if it is a noetherian integral domain such that $R_{\mathfrak{P}}$ is a DVR for every nonzero prime ideal $\mathfrak{P}$ of R.

EXAMPLE. If R is a PID then R is a Dedekind ring.

3.2 Elementary Properties of a Dedekind ring R

(1) Every nonzero prime ideal of R is a maximal ideal.

Suppose $\mathfrak{P}_1 \subseteq \mathfrak{P}_2$ are nonzero prime ideals, $\mathfrak{P}_1 \neq \mathfrak{P}_2$. Then Proposition 1.2 implies $\mathfrak{P}_1 R_{\mathfrak{P}_2} \subseteq \mathfrak{P}_2 R_{\mathfrak{P}_2}$ are distinct prime ideals. Since $R_{\mathfrak{P}_2}$ is a DVR there can be only one prime ideal. It follows that no chain $\mathfrak{P}_1 \subseteq \mathfrak{P}_2$ can exist in R.

(2) If S is a multiplicative set in R then R_S is a Dedekind ring.

A prime ideal of R_S has the form $\mathfrak{P}R_S$ for some prime ideal $\mathfrak{P}$ of R. One can verify that

$$R_{\mathfrak{P}} \cong (R_S)_{\mathfrak{P}R_S}$$

so the localization of R_S at $\mathfrak{P}R_S$ is also a DVR.

The goal of this section is to obtain a factorization theorem for ideals in a Dedekind ring. It need not happen that elements have unique factorization but we will prove that ideals have unique factorization as a product of prime ideals.

We begin with a very useful tool that can be applied in many situations.

3.3 Chinese Remainder Theorem (CRT). Let B be a ring with identity, $\mathfrak{Q}_1, \ldots, \mathfrak{Q}_n$ a set of ideals such that $B = \mathfrak{Q}_i + \mathfrak{Q}_j$ for $i \neq j$. Let $\mathfrak{J} = \bigcap \mathfrak{Q}_i$. Then

$$B/\mathfrak{J} \cong B/\mathfrak{Q}_1 \oplus \cdots \oplus B/\mathfrak{Q}_n.$$

PROOF. If we replace B and the $\mathfrak{Q}_i$ by $B/\mathfrak{I}$ and $\mathfrak{Q}_i/\mathfrak{I}$ it will be sufficient to prove the theorem under the assumption that $\mathfrak{I} = (0)$. Suppose first $n = 2$ so $\mathfrak{Q}_1 + \mathfrak{Q}_2 = B$ and $\mathfrak{Q}_1 \bigcap \mathfrak{Q}_2 = (0)$. We map B to the direct sum $B/\mathfrak{Q}_1 \oplus B/\mathfrak{Q}_2$ by

$$b \to (b+\mathfrak{Q}_1, b+\mathfrak{Q}_2).$$

This is a ring homomorphism with kernel $\mathfrak{Q}_1 \cap \mathfrak{Q}_2 = (0)$. It must be shown that the map is onto the direct sum. Since $\mathfrak{Q}_1 + \mathfrak{Q}_2 = B$ there exists elements q_i in $\mathfrak{Q}_i$ such that $q_1 + q_2 = 1$. Then

$$q_1 \to (q_1+\mathfrak{Q}_1, q_1+\mathfrak{Q}_2) = (0, 1-q_2+\mathfrak{Q}_2) = (0, 1+\mathfrak{Q}_2).$$

Thus Bq_1 maps onto $(0, B/\mathfrak{Q}_2)$. Similarly Bq_2 maps onto $(B/\mathfrak{Q}_1, 0)$ so the map is onto in the case $n = 2$.

Now suppose $n > 2$. Let $\mathfrak{Q}_{n-1} \cap \mathfrak{Q}_n = \mathfrak{Q}'_{n-1}$ and let $\mathfrak{Q}_j = \mathfrak{Q}_j'$ for $j < n-1$. The first step is to show that induction can be applied to the $n-1$ ideals $\mathfrak{Q}_j'$. Clearly $\bigcap \mathfrak{Q}_j' = \bigcap \mathfrak{Q}_i$ and $\mathfrak{Q}_i' + \mathfrak{Q}_j' = B$ if i, j are both $\neq n-1$. It is necessary to show

$$\mathfrak{Q}'_{n-1} + \mathfrak{Q}_j' = B \qquad \text{for} \quad j \neq n-1.$$

We have $B = B \cdot B = (\mathfrak{Q}_n + \mathfrak{Q}_j)(\mathfrak{Q}_{n-1} + \mathfrak{Q}_j) \subseteq \mathfrak{Q}_n \mathfrak{Q}_{n-1} + \mathfrak{Q}_j$. It is always true that $\mathfrak{Q}_n \mathfrak{Q}_{n-1} \subseteq \mathfrak{Q}_n \cap \mathfrak{Q}_{n-1}$ so

$$B \subseteq \mathfrak{Q}'_{n-1} + \mathfrak{Q}_j'.$$

Equality must follow so by induction we have

$$B \cong B/\mathfrak{Q}_1' \oplus \cdots \oplus B/\mathfrak{Q}'_{n-1}.$$

In the ring $B/\mathfrak{Q}'_{n-1}$ the two ideals $\mathfrak{Q}_{n-1}/\mathfrak{Q}'_{n-1}$ and $\mathfrak{Q}_n/\mathfrak{Q}'_{n-1}$ have sum equal to the whole ring $B/\mathfrak{Q}'_{n-1}$ and intersection equal to zero. By the case $n = 2$ we get $B/\mathfrak{Q}'_{n-1} \cong B/\mathfrak{Q}_{n-1} \oplus B/\mathfrak{Q}_n$ and so the theorem follows.

3.4 Theorem (CRT for modules). Let B, $\mathfrak{Q}_1, \ldots, \mathfrak{Q}_n, \mathfrak{I}$ be as in Theorem 3.3. If M is a B-module then

$$M/\mathfrak{I}M \cong M/\mathfrak{Q}_1 M \oplus \cdots \oplus M/\mathfrak{Q}_n M.$$

PROOF. We may prove the theorem under the assumption that $\mathfrak{I}M = (0)$. Let u_i be some element of B which maps onto $(0, \ldots, 0, 1, 0, \ldots, 0)$ in the isomorphism of Theorem 3.3. Here the 1 is in the ith position so u_i is in $\mathfrak{Q}_j$ for $j \neq i$ and $u_i - 1$ is in $\mathfrak{Q}_i$. Consider the homomorphism from M to $u_i M$ which sends m to $u_i m$. The kernel consists of all m for which $u_i m = 0$. For such an element we have $m = (1-u_i)m$ is in $\mathfrak{Q}_i M$. On the other hand $u_i \mathfrak{Q}_i M \subset \mathfrak{I}M = (0)$. Thus $u_i M \cong M/\mathfrak{Q}_i M$. To complete the proof it is only necessary to show $u_1 M + \cdots + u_n M$ is a direct sum equal to M. One easily sees the sum equals M

because $u_1 + \cdots + u_n = 1$ modulo $\mathfrak{J}$ and since $\mathfrak{J}M = (0)$ we have $m = u_1 m + \cdots + u_n m$ for all m in M.

Now suppose $\sum u_i m_i = 0$. We must show each $u_i m_i = 0$. For $j \neq i$ we have

$$u_j u_i M \subseteq \mathfrak{J}M = (0),$$

so $0 = u_j \sum u_i m_i = u_j u_j m_j$. From this equation it follows that

$$u_j m_j = (1 - u_j) u_j m_j \in \mathfrak{Q}_j u_j M = (0).$$

This completes the proof.

A little more can be said in the context of Theorem 3.3. Namely we have the following.

3.5 Proposition. Let $B, \mathfrak{Q}_1, \ldots, \mathfrak{Q}_n, \mathfrak{J}$ be as in Theorem 3.3. Then $\mathfrak{J} = \mathfrak{Q}_1 \mathfrak{Q}_2 \cdots \mathfrak{Q}_n$.

PROOF. Use induction on n. Suppose $n = 2$. There exists u_i in Q_i with $u_1 + u_2 = 1$. Now for q in $\mathfrak{Q}_1 \cap \mathfrak{Q}_2$ we have $q = qu_1 + qu_2$. The element qu_1 is in $\mathfrak{Q}_2 \mathfrak{Q}_1$ and qu_2 is in $\mathfrak{Q}_1 \mathfrak{Q}_2$ so q is in $\mathfrak{Q}_1 \mathfrak{Q}_2$. Thus $\mathfrak{Q}_1 \cap \mathfrak{Q}_2 \subseteq \mathfrak{Q}_1 \mathfrak{Q}_2$. The reserve inclusion is immediate because the $\mathfrak{Q}_i$ are ideals so the result holds for two ideals. Now suppose $n > 2$. Let $\mathfrak{Q}'_{n-1} = \mathfrak{Q}_{n-1} \cap \mathfrak{Q}_n$. By the case $n = 2$ we have also $\mathfrak{Q}'_{n-1} = \mathfrak{Q}_{n-1} \mathfrak{Q}_n$. We saw in the proof of Theorem 3.3 that induction can be applied with the $n-1$ ideals $\mathfrak{Q}_1, \ldots, \mathfrak{Q}_{n-2}, \mathfrak{Q}'_{n-1}$. Then we have

$$\bigcap_1^n \mathfrak{Q}_i = \mathfrak{Q}_1 \cap \cdots \cap \mathfrak{Q}_{n-2} \cap \mathfrak{Q}'_{n-1} = \mathfrak{Q}_1 \cdots \mathfrak{Q}_{n-2} \mathfrak{Q}'_{n-1} = \prod_1^n \mathfrak{Q}_i$$

and this completes the proof.

Now let R be a Dedekind ring and $\mathfrak{A}$ a nonzero ideal of R. We shall study the factor ring $R/\mathfrak{A}$.

Observe that the prime ideals in $R/\mathfrak{A}$ are quotients $\mathfrak{P}/\mathfrak{A}$ with $\mathfrak{P}$ a prime ideal of R containing $\mathfrak{A}$. Since nonzero primes in R are maximal ideals we see that all prime ideals in $R/\mathfrak{A}$ are maximal. Also R is a noetherian ring so $R/\mathfrak{A}$ is also a noetherian ring. We shall prove some facts about rings satisfying these conditions.

In what follows we let B denote a noetherian ring in which all prime ideals are maximal. The ring $R/\mathfrak{A} = B$ satisfies this condition.

3.6 Lemma. Every ideal in B contains a product of prime ideals.

PROOF. This proof uses only the fact that B is noetherian. If the lemma is false there is an ideal $\mathfrak{J}$ which is maximal with respect to not containing a product of prime ideals. In particular $\mathfrak{J}$ itself is not prime so there exist elements x and y with xy in $\mathfrak{J}$ but x, y both outside $\mathfrak{J}$. Let $\mathfrak{U} = Bx + \mathfrak{J}$, $\mathfrak{B} = By + \mathfrak{J}$. Then $\mathfrak{U}$ and $\mathfrak{B}$ are both ideals larger than $\mathfrak{J}$ so each one contains a product of prime ideals. But $\mathfrak{U}\mathfrak{B} \subseteq \mathfrak{J}$ so $\mathfrak{J}$ contains a product of prime ideals. This contradiction shows no such $\mathfrak{J}$ can exist.

Apply this to the zero ideal of B.

3.7 Corollary. There exist distinct prime ideals $\mathfrak{P}_1, \ldots, \mathfrak{P}_n$ and positive integers $a_1, \ldots, a_n$ such that

$$\mathfrak{P}_1^{a_1} \cdots \mathfrak{P}_n^{a_n} = (0).$$

Our next object is to apply CRT to these $\mathfrak{P}_i^{a_i}$ but first we must verify the hypothesis of Theorem 3.3.

3.8 Lemma. If $\mathfrak{P}_1, \mathfrak{P}_2$ are distinct maximal ideals of B then $\mathfrak{P}_1{}^a + \mathfrak{P}_2{}^b = B$ for any integers $a, b > 0$.

PROOF. $\mathfrak{P}_1{}^a$ cannot be contained in $\mathfrak{P}_2$ unless $\mathfrak{P}_1 = \mathfrak{P}_2$. Thus $\mathfrak{P}_1{}^a + \mathfrak{P}_2 = B$. Suppose for some integer $c \geqslant 1$ we have

$$\mathfrak{P}_1{}^a + \mathfrak{P}_2{}^c = B.$$

Then $\mathfrak{P}_2{}^c = \mathfrak{P}_2{}^c B = \mathfrak{P}_2{}^c(\mathfrak{P}_1{}^a + \mathfrak{P}_2) \subseteq \mathfrak{P}_1{}^a + \mathfrak{P}_2^{c+1}$. So

$$B = \mathfrak{P}_1{}^a + \mathfrak{P}_2{}^c \subseteq \mathfrak{P}_1{}^a + (\mathfrak{P}_1{}^a + \mathfrak{P}_2^{c+1}) = \mathfrak{P}_1{}^a + \mathfrak{P}_2^{c+1}.$$

The result follows.

3.9 Lemma. Let $\mathfrak{P}_1, \ldots, \mathfrak{P}_n$ be distinct prime ideals such that

$$(0) = \mathfrak{P}_1^{a_1} \cdots \mathfrak{P}_n^{a_n}.$$

Then

$$B \cong B/\mathfrak{P}_1^{a_1} \oplus \cdots \oplus B/\mathfrak{P}_n^{a_n}.$$

PROOF. Each $\mathfrak{P}_i$ is maximal so Lemma 3.8 implies that CRT can be applied to give

$$B/\mathfrak{J} \cong B/\mathfrak{P}_1^{a_1} \oplus \cdots \oplus B/\mathfrak{P}_n^{a_n}$$

when $\mathfrak{J} = \bigcap \mathfrak{P}_i^{a_i}$. We will be done if $\mathfrak{J} = (0)$.

By Proposition 3.5 we see $\mathfrak{J} = \prod \mathfrak{P}_i^{a_i} = (0)$ so the proof is complete.

3.10 Corollary. The ideals $\mathfrak{P}_1, \ldots, \mathfrak{P}_n$ are all the prime ideals of B.

PROOF. One checks easily that $\mathfrak{P}_i/\mathfrak{P}_i^{a_i}$ is the only prime ideal in $B/\mathfrak{P}_i^{a_i}$ and the only ideals in a direct sum $B_1 \oplus \cdots \oplus B_n$ are direct sums $\mathfrak{I}_1 \oplus \cdots \oplus \mathfrak{I}_n$ of ideals $\mathfrak{I}_i$ in the ring $B_i = B/\mathfrak{P}_i^{a_i}$. The ideal $\mathfrak{I}_1 \oplus \cdots \oplus \mathfrak{I}_n$ is prime if and only if $B_1/\mathfrak{I}_1 \oplus \cdots \oplus B_n/\mathfrak{I}_n$ is an integral domain. It follows that $\mathfrak{I}_j = B_j$ for all but one index j and $\mathfrak{I}_j = \mathfrak{P}_j$ for the remaining index.

It follows now that there exist only a finite number of prime ideals of R

which contain $\mathfrak{A}$ (namely the primes which map onto the $\mathfrak{P}_i$ in B) and also $\mathfrak{A}$ contains a product of these primes. Our goal is to show in fact that $\mathfrak{A}$ is equal to a product of primes. To get this it is necessary to use some additional information. Up to now we have used only that primes of R are maximal ideals but we have not used that $R_{\mathfrak{P}}$ is a principal ideal domain when $\mathfrak{P}$ is a nonzero prime of R. We shall bring this into consideration shortly.

3.11 Lemma. Let $\mathfrak{P}$ be a prime ideal of R and a be a positive integer. Then $R/\mathfrak{P}^a \cong R_{\mathfrak{P}}/\mathfrak{P}^a R_{\mathfrak{P}}$.

PROOF. The map $f(r+\mathfrak{P}^a) = r+\mathfrak{P}^a R_{\mathfrak{P}}$ from $R/\mathfrak{P}^a$ to $R_{\mathfrak{P}}/\mathfrak{P}R_{\mathfrak{P}}$ is a ring homomorphism which is one to one. We must show f is onto. Take any r/s in $R_{\mathfrak{P}}$ with s not in $\mathfrak{P}$ but r in R. Since $\mathfrak{P}$ is a maximal ideal it follows that $Rs+\mathfrak{P} = R$. By the method of proof used in Lemma 3.8, one finds $Rs+\mathfrak{P}^a = R$. Thus there exists c in R and q in $\mathfrak{P}^a$ with $cs+q = 1$. Then $f(rc+\mathfrak{P}^a) = rc+\mathfrak{P}^a R_{\mathfrak{P}} = r(1/s-q/s)+\mathfrak{P}^a R_{\mathfrak{P}} = r/s+\mathfrak{P}^a R_{\mathfrak{P}}$. Thus f is onto as required.

3.12 Corollary. Every ideal of $R/\mathfrak{P}^a$ is a power of $\mathfrak{P}/\mathfrak{P}^a$. Moreover $\mathfrak{P}/\mathfrak{P}^a$ is a principal ideal.

PROOF. In view of Lemma 3.11, we may replace R by $R_{\mathfrak{P}}$ in order to prove these assertions. But $R_{\mathfrak{P}}$ is a DVR so the statements follow from Property 3.1 (iii).

3.13 Proposition. Let $\mathfrak{A}$ be a nonzero ideal of R and let $\mathfrak{P}_1, \ldots, \mathfrak{P}_n$ be all the prime ideals of R which contain $\mathfrak{A}$. Then $\mathfrak{A} = \mathfrak{P}_1^{a_1} \cdots \mathfrak{P}_n^{a_n}$ for some positive integers a_i.

PROOF. We have seen above that $\mathfrak{P}_1^{b_1} \cdots \mathfrak{P}_n^{b_n} \subseteq \mathfrak{A}$ for some positive integers b_i.

In the factor ring

$$B = R/\mathfrak{P}_1^{b_1} \cdots \mathfrak{P}_n^{b_n} \cong R/\mathfrak{P}_1^{b_1} \oplus \cdots \oplus R/\mathfrak{P}_n^{b_n}$$

the ideal $\mathfrak{A}$ has image which is necessarily of the form

$$\mathfrak{P}_1^{a_1}/\mathfrak{P}_1^{b_1} \oplus \cdots \oplus \mathfrak{P}_n^{a_n}/\mathfrak{P}_n^{b_n}.$$

for some positive integers a_i because of Corollary 3.12. The ideal $\mathfrak{P}_1^{a_1} \cdots \mathfrak{P}_n^{a_n}$ has the same image so $\mathfrak{A} = \mathfrak{P}_1^{a_1} \cdots \mathfrak{P}_n^{a_n}$ because both contain $\mathfrak{P}_1^{b_1} \cdots \mathfrak{P}_n^{b_n}$.

3.14 Theorem. Let $\mathfrak{A}$ be a nonzero ideal in the Dedekind ring R. Then $\mathfrak{A} = \mathfrak{P}_1^{a_1} \cdots \mathfrak{P}_n^{a_n}$ with $\mathfrak{P}_1, \ldots, \mathfrak{P}_n$ distinct prime ideals uniquely determined by $\mathfrak{A}$ and certain positive integers $a_1, \ldots, a_n$ uniquely determined by $\mathfrak{A}$.

PROOF. Every thing except uniqueness has already been done. The primes $\mathfrak{P}_1, \ldots, \mathfrak{P}_n$ are uniquely determined by $\mathfrak{A}$ because they are all the primes of R

which contain $\mathfrak{A}$. The integer a_i is uniquely determined by the condition that a_i is the least power of the maximal ideal of $R_{\mathfrak{P}_i}/\mathfrak{A}R_{\mathfrak{P}_i}$ which is zero. That is $\mathfrak{A}R_{\mathfrak{P}_i} = \mathfrak{P}_i^{a_i}R_{\mathfrak{P}_i}$.

As an application of Theorem 3.14, we can prove the next result.

3.15 Theorem. If R is a Dedekind ring with only a finite number of prime ideals, then R is a principal ideal domain.

PROOF. Let $\mathfrak{P}_1, \ldots, \mathfrak{P}_n$ be all the nonzero primes of R. The first step is to find an element x_i which is in $\mathfrak{P}_i$ but not in $\mathfrak{P}_i^2$ and also not in $\mathfrak{P}_j$ for $j \neq i$. To produce x_1 we let x_1 be an element of R which maps onto $(\pi, 1, 1, \ldots, 1)$ in

$$R/\mathfrak{P}_1^{\,2}\,\mathfrak{P}_2 \cdots \mathfrak{P}_n \cong R/\mathfrak{P}_1^{\,2} \oplus R/\mathfrak{P}_2 \oplus \cdots \oplus R/\mathfrak{P}_n,$$

where π is a generator of $\mathfrak{P}_1/\mathfrak{P}_1^{\,2}$. Such an x_1 exists by CRT. Clearly it satisfies the required conditions. Similarly select x_i. Now the ideal Rx_i is contained in $\mathfrak{P}_i$ but not in any other prime. Moreover Rx_i is not in $\mathfrak{P}_i^{\,2}$. Hence the only factorization possible is $Rx_i = \mathfrak{P}_i$. Thus every prime is principal. Since each ideal $\neq 0$ is a product of primes, each ideal is also principal.

EXERCISE. If $\mathfrak{A}$ is an ideal of R, we write $x \equiv y \bmod \mathfrak{A}$ to mean $x - y$ is in $\mathfrak{A}$. Let $\mathfrak{P}_1, \ldots, \mathfrak{P}_n$ be distinct prime ideals $\neq (0)$ in the Dedekind ring R; $a_1, \ldots, a_n$ are positive integers; $y_1, \ldots, y_n$ are elements of R. Show there exists an element x in R with $x \equiv y_i \bmod \mathfrak{P}_i^{a_i}$ for $i = 1, 2, \ldots, n$.

Now that the structure of ideals is known, it will be practical to have several ways of identifying Dedekind rings. We offer two alternate characterizations of these rings.

3.16 Theorem. Let R be an integral domain which is not a field. The following are equivalent statements.

(a) R is a Dedekind ring.
(b) For each maximal ideal $\mathfrak{P}$, $R_{\mathfrak{P}}$ is a DVR and for each element $a \neq 0$ there exists only a finite number of prime ideals containing a.
(c) R is noetherian, integrally closed and each prime ideal $\neq (0)$ is a maximal ideal.

PROOF (a) $\to$ (b). If R is Dedekind, then $R_{\mathfrak{P}}$ is a DVR and by the remark following Corollary 3.10, there exist only a finite number of prime ideals containing $Ra = \mathfrak{A}$.

(b) $\to$ (c). Let $(0) \neq \mathfrak{Q}$ be a prime ideal. If $\mathfrak{Q}$ is not maximal then $\mathfrak{Q} \subset \mathfrak{P}$ with $\mathfrak{P}$ maximal. It follows that $\mathfrak{Q}R_{\mathfrak{P}}$ is a nonmaximal prime ideal in $R_{\mathfrak{P}}$ so $R_{\mathfrak{P}}$ is not a DVR. Thus $\mathfrak{Q}$ is maximal. Next we show R is integrally closed. We need an elementary fact before proceeding.

3.17 Lemma. Let R be any integral domain. Then

$$R = \bigcap_{\mathfrak{P}\ \max} R_{\mathfrak{P}},$$

where the intersection taken over all maximal ideals.

PROOF. The inclusion of R into the intersection is immediate. Suppose x is in $R_{\mathfrak{P}}$ for every maximal ideal $\mathfrak{P}$. We may write $x = a/b$ with a, b in R. Let

$$\mathfrak{A} = \{y \in R \mid ya \in Rb\}.$$

$\mathfrak{A}$ is an ideal of R. For any maximal ideal $\mathfrak{P}$ we can find r, s in R with s not in $\mathfrak{P}$ such that $x = a/b = r/s$. In particular $sa = rb$ so s is in $\mathfrak{A}$. This means $\mathfrak{A}$ is not contained in $\mathfrak{P}$. Hence $\mathfrak{A}$ is not contained in any maximal ideal and so $\mathfrak{A} = R$. Thus a belongs to Rb and $a/b = x$ belongs to R.

Now we return to the proof of (c). Each $R_{\mathfrak{P}}$ is a DVR, so $R_{\mathfrak{P}}$ is integrally closed. By Exercise 3 on page 7, $R = \bigcap R_{\mathfrak{P}}$ is integrally closed.

The last assertion to be proved is that R is noetherian. To prove this we shall need another lemma.

3.18 Lemma. Let R be any domain and let $\mathfrak{A} \subseteq \mathfrak{B}$ be two ideals of R such that $\mathfrak{A}R_{\mathfrak{P}} = \mathfrak{B}R_{\mathfrak{P}}$ for all maximal ideals $\mathfrak{P}$ of R. Then $\mathfrak{A} = \mathfrak{B}$.

PROOF. Let b be an element of $\mathfrak{B}$. For each maximal ideal $\mathfrak{P}$, we have $bR_{\mathfrak{P}} \subseteq \mathfrak{A}R_{\mathfrak{P}}$. So there is some a in $\mathfrak{A}$ and some s in R with s not in $\mathfrak{P}$ such that $b = a/s$. The ideal of R defined by

$$\{y \in R \mid by \in \mathfrak{A}\}$$

must contain s and so it does not belong to $\mathfrak{P}$. This ideal is not in any maximal ideal so it is all of R. Thus b is in $\mathfrak{A}$ as required.

Now to complete the proof of (c) it is necessary to prove each ideal is finitely generated. We can prove even more. Namely, each ideal requires at most two generators.

3.19 Proposition. Let R be a domain which satisfies Condition (b) of Theorem 3.16, and let $\mathfrak{A}$ be a nonzero ideal of R. For any a in $\mathfrak{A}$ with $a \neq 0$ there exists b in $\mathfrak{A}$ such that $Ra + Rb = \mathfrak{A}$.

PROOF. Let $\mathfrak{P}_1, \ldots, \mathfrak{P}_n$ be all the prime ideals of R which contain a. Each localization $R_{\mathfrak{P}_i}$ is a PID, so there exists c_i with

$$\mathfrak{A}R_{\mathfrak{P}_i} = c_i R_{\mathfrak{P}_i}.$$

We can write $c_i = x/s$ with x in $\mathfrak{A}$ and s in R and observe that $c_i R_{\mathfrak{P}_i} = xR_{\mathfrak{P}_i}$. So we shall assume that c_i is in $\mathfrak{A}$ to start with. We have enough information already to show $\mathfrak{A}$ is finitely generated. Consider the ideal $\mathfrak{C} = Ra + Rc_1 + \cdots$

$+Rc_n$ which is contained in $\mathfrak{A}$. If $\mathfrak{P}$ is a prime ideal and a is not in $\mathfrak{P}$ then $1/a$ is in $R_{\mathfrak{P}}$ so $\mathfrak{C}R_{\mathfrak{P}} = \mathfrak{A}R_{\mathfrak{P}} = R_{\mathfrak{P}}$. If $\mathfrak{P}$ is one of the $\mathfrak{P}_i$ which contain a, then $\mathfrak{C}R_{\mathfrak{P}}$ contains c_i so again $\mathfrak{C}R_{\mathfrak{P}} = \mathfrak{A}R_{\mathfrak{P}}$. By Lemma 3.18 it follows that $\mathfrak{C} = \mathfrak{A}$. To complete the proof we need to find b. Select b in $\mathfrak{A}$ so that under the isomorphism of Theorem 3.4

$$\mathfrak{A}/\mathfrak{P}_1 \cdots \mathfrak{P}_n \mathfrak{A} \cong \mathfrak{A}/\mathfrak{P}_1 \mathfrak{A} \oplus \cdots \oplus \mathfrak{A}/\mathfrak{P}_n \mathfrak{A},$$

the coset $b+\mathfrak{P}_1 \cdots \mathfrak{P}_n \mathfrak{A}$ maps onto the n-tuple $(c_1+\mathfrak{P}_1\mathfrak{A}, \ldots, c_n+\mathfrak{P}_n\mathfrak{A})$. Then $b-c_i$ is in $\mathfrak{P}_i\mathfrak{A}$ and certainly $Ra+Rb \subseteq \mathfrak{A}$. We show equality by the same technique as above. Let $\mathfrak{D} = Ra+Rb$. For a prime $\mathfrak{P}$ not containing a, $\mathfrak{D}R_{\mathfrak{P}} = \mathfrak{A}R_{\mathfrak{P}} = R_{\mathfrak{P}}$. If $\mathfrak{P}$ is the prime $\mathfrak{P}_i$ then $\mathfrak{D}R_{\mathfrak{P}}+\mathfrak{P}\mathfrak{A}R_{\mathfrak{P}}$ contains $b+(c_i-b) = c_i$. Thus this sum contains $c_i R_{\mathfrak{P}} = \mathfrak{A}R_{\mathfrak{P}}$. On the other hand, $\mathfrak{D} \subseteq \mathfrak{A}$ so the sum is contained in $\mathfrak{A}R_{\mathfrak{P}}$. It follows

$$\mathfrak{D}R_{\mathfrak{P}} + \mathfrak{P}\mathfrak{A}R_{\mathfrak{P}} = \mathfrak{A}R_{\mathfrak{P}}.$$

The hypothesis of Nakayama's Lemma (Corollary to Proposition 1.4) is satisfied: $R_{\mathfrak{P}}$ is local, $\mathfrak{A}R_{\mathfrak{P}}$ is a finitely generated module, $\mathfrak{D}R_{\mathfrak{P}}$ is a submodule. The above equation implies $\mathfrak{D}R_{\mathfrak{P}} = \mathfrak{A}R_{\mathfrak{P}}$. By Lemma 3.18, $\mathfrak{A} = \mathfrak{D} = Ra+Rb$.

This completes the proof that (b) → (c) in Theorem 3.16. We begin the proof that (c) → (a).

Let $\mathfrak{P}$ be a maximal ideal of R. Then $R_{\mathfrak{P}}$ is a noetherian local ring with $\mathfrak{P}R_{\mathfrak{P}}$ the only nonzero prime ideal (because primes in R are maximal). Moreover $R_{\mathfrak{P}}$ is integrally closed because R is integrally closed. It remains to show these conditions imply $R_{\mathfrak{P}}$ is a PID.

3.20 Proposition. Let R be a noetherian, local, integrally closed domain with $\mathfrak{P}$ its only nonzero prime ideal. Then R is a DVR.

PROOF. Select any $a \neq 0$ in $\mathfrak{P}$ and let $M = R/Ra$. For each m in M let

$$\operatorname{ann}(m) = \{r \in R \mid rm = 0\}.$$

Since R is noetherian there is a maximal element in the collection $\{\operatorname{ann}(m) \mid m \neq 0, m \in M\}$. Let b be an element of R such that $\mathfrak{Q} = \operatorname{ann}(b+Ra)$ is such a maximal element. $\mathfrak{Q}$ is nonzero because $a \neq 0$ and a is in $\mathfrak{Q}$. We show now $\mathfrak{Q}$ is prime. If $\mathfrak{Q}$ is not prime there exist elements x, y not in $\mathfrak{Q}$ with xy in $\mathfrak{Q}$. Now $y(b+Ra) \neq 0+Ra$ because y is not in $\mathfrak{Q}$. Then $\operatorname{ann}(yb+Ra)$ contains both $\mathfrak{Q}$ and x which is against the maximal choice of $\mathfrak{Q}$.

Thus $\mathfrak{Q}$ is prime and since R has only one prime $\neq 0$ it follows that the maximal ideal $\mathfrak{P}$ is the set of all elements which multiply b into Ra. That is $\mathfrak{P}b \subseteq Ra$ but b is not in Ra.

We now carry through several steps which lead to the conclusion of the proof.

Step 1. b/a is not in R.
If it were then b is in Ra contrary to $b + Ra \neq 0 + Ra$.

Step 2. $\mathfrak{P} = R(a/b)$.
The inclusion $\mathfrak{P}b \subseteq Ra$ means $\mathfrak{P}b/a$ is an ideal in R. If $\mathfrak{P}b/a \subseteq \mathfrak{P}$ then by Part (4) of Proposition 2.1 it follows b/a is integral over R. Since R is integrally closed we get b/a is in R contrary to Step 1. Thus $\mathfrak{P}b/a = R$ and $\mathfrak{P} = Ra/b$.
We now know the maximal ideal is principal. Let us write $\mathfrak{P} = R\pi$ for short.

Step 3. Every ideal is principal.
Let $\mathfrak{A}$ be a nonzero ideal. Consider the chain

$$\mathfrak{A} \subseteq \mathfrak{A}\pi^{-1} \subseteq \mathfrak{A}\pi^{-2} \subseteq \cdots.$$

If $\mathfrak{A}\pi^{-k} = \mathfrak{A}\pi^{-k-1}$ then π^{-1} sends $\mathfrak{A}\pi^{-k}$ into itself so π^{-1} is integral over R. But this is impossible since π^{-1} cannot be in R.

Since R is a noetherian ring, the part of the chain which falls into R must be finite. Let $\mathfrak{A}\pi^{-n} \subseteq R$, $\mathfrak{A}\pi^{-n-1} \not\subseteq R$. If $\mathfrak{A}\pi^{-n} \subseteq \mathfrak{P} = R\pi$ then $\mathfrak{A}\pi^{-n-1} \subseteq R$ so it must be $\mathfrak{A}\pi^{-n} = R$. Thus $\mathfrak{A} = \mathfrak{R}\pi^{n}$ which completes the proof.

4. FRACTIONAL IDEALS AND THE CLASS GROUP

Throughout this section R denotes a Dedekind ring and K its quotient field.

DEFINITIONS. (1) A *fractional ideal* of R is a nonzero finitely generated R-submodule of K.
(2) If $\mathfrak{M}$ is a fractional ideal, $\mathfrak{M}^{-1}$ is the set $\{x \in K \mid \ x\mathfrak{M} \subseteq R\}$.

EXAMPLES. If y is a nonzero element of K, then Ry is a fractional ideal. The inverse $(Ry)^{-1}$ is Ry^{-1}.
Any nonzero ideal of R is a fractional ideal.

REMARK. If $\mathfrak{M}$ is a fractional ideal then so is $\mathfrak{M}^{-1}$. It is clear that $\mathfrak{M}^{-1}$ is an R-submodule of K. It is necessary to show it is finitely generated. Select any m in $\mathfrak{M}$ with $m \neq 0$. Then $\mathfrak{M}^{-1}m \subseteq R$ so $\mathfrak{M}^{-1} \subseteq Rm^{-1}$. Certainly Rm^{-1} is a finitely generated R module and because R is noetherian, the submodule $\mathfrak{M}^{-1}$ is also finitely generated.

Suppose $\mathfrak{M}$ and $\mathfrak{N}$ are fractional ideals. The product $\mathfrak{M}\mathfrak{N}$ is the collection of all elements of the form $\sum m_i n_i$ with m_i in $\mathfrak{M}$, n_i in $\mathfrak{N}$. If $\{x_j\}$ and $\{y_k\}$ are sets of generators for $\mathfrak{M}$ and $\mathfrak{N}$, respectively, then $\mathfrak{M}\mathfrak{N}$ is generated over R by the products $x_j y_k$. Hence $\mathfrak{M}\mathfrak{N}$ is also a fractional ideal.

Definition. A fractional ideal $\mathfrak{M}$ is *invertible* if $\mathfrak{M}\mathfrak{M}^{-1} = R$.

EXAMPLE. The principal fractional ideal Rx is invertible for any x in K because $(Rx)(Rx)^{-1} = Rxx^{-1} = R$.

EXERCISE. Multiplication and inversion behave properly with respect to localization. That is, if $\mathfrak{P}$ is a prime ideal of R and $\mathfrak{M}$ a fractional ideal of R, then $\mathfrak{M}R_{\mathfrak{P}}$ is a fractional ideal of $R_{\mathfrak{P}}$ and $(\mathfrak{M}R_{\mathfrak{P}})^{-1} = \mathfrak{M}^{-1}R_{\mathfrak{P}}$. Also $(\mathfrak{M}\mathfrak{N})R_{\mathfrak{P}} = (\mathfrak{M}R_{\mathfrak{P}})(\mathfrak{N}R_{\mathfrak{P}})$ for $\mathfrak{M}, \mathfrak{N}$ fractional ideals of R.

4.1 Lemma. A prime ideal of R is invertible.

PROOF. Let $\mathfrak{P}$ be a nonzero prime of R. Then $\mathfrak{P}\mathfrak{P}^{-1} = \mathfrak{A}$ is an ideal of R. For any maximal ideal $\mathfrak{Q}$ we know $R_{\mathfrak{Q}}$ is a PID so $\mathfrak{P}R_{\mathfrak{Q}}$ is principal and hence is invertible. Thus $\mathfrak{A}_{\mathfrak{Q}} = (\mathfrak{P}\mathfrak{P}^{-1})_{\mathfrak{Q}} = \mathfrak{P}_{\mathfrak{Q}} \cdot \mathfrak{P}_{\mathfrak{Q}}^{-1} = R_{\mathfrak{Q}}$. This holds for all maximal $\mathfrak{Q}$, so by Lemma 3.18, $\mathfrak{A} = R$ as required. (In this proof we have written $\mathfrak{A}_{\mathfrak{Q}}$ for $\mathfrak{A}R_{\mathfrak{Q}}$.)

When $\mathfrak{M}$ is a fractional ideal and n a positive integer, we shall write $\mathfrak{M}^{-n}$ to mean $(\mathfrak{M}^{-1})^n$.

4.2 Theorem. Any fractional ideal $\mathfrak{M}$ can be uniquely expressed as a product $\mathfrak{P}_1^{a_1} \cdots \mathfrak{P}_n^{a_n}$ with $\mathfrak{P}_1, \ldots, \mathfrak{P}_n$ distinct prime ideals of R and $a_1, \cdots, a_n$ integers (positive or negative).

PROOF. Let $\mathfrak{M}$ be a fractional ideal with generators $m_1, \ldots, m_k$. Each m_i is in K so there is a "common denominator" s in R such that $m_i s$ is also in R. It follows that $\mathfrak{M}s \subseteq R$. There exist factorizations of the ideals Rs and $\mathfrak{M}s$ as

$$Rs = \prod \mathfrak{Q}_j^{b_j}, \qquad \mathfrak{M}s = \prod \mathfrak{P}_i^{a_i},$$

where the $\mathfrak{P}_i$ and the $\mathfrak{Q}_j$ are the prime ideals of R. It follows $\mathfrak{M}\mathfrak{Q}_1^{b_1} \cdots \mathfrak{Q}_t^{b_t} = \mathfrak{P}_1^{a_1} \cdots \mathfrak{P}_k^{a_k}$. We have seen in Lemma 4.1 that prime ideals are invertible so we obtain

$$\mathfrak{M} = \prod \mathfrak{P}_i^{a_i} \cdot \prod \mathfrak{Q}_j^{-b_j}.$$

This establishes the existence of a factorization of $\mathfrak{M}$ as a product of prime ideals with integral exponents. Now we obtain uniqueness as follows. Suppose

$$\mathfrak{M} = \prod \mathfrak{P}_i^{a_i} \prod \mathfrak{Q}_j^{-b_j} = \prod \mathfrak{X}_i^{c_i} \prod \mathfrak{Y}_j^{-d_j}$$

where $\mathfrak{P}, \mathfrak{Q}, \mathfrak{X}, \mathfrak{Y}$ denote prime ideals and the a_i, b_j, c_i, d_j are positive integers. Then we have $\prod \mathfrak{P}_i^{a_i} \prod \mathfrak{Y}_j^{d_j} = \prod \mathfrak{X}_i^{c_i} \prod \mathfrak{Q}_j^{b_j}$. This is a factorization of ideals in R so the uniqueness statement for ideals in R can be used to get the uniqueness of the expression for $\mathfrak{M}$.

The discussion to this point shows that the collection of all fractional ideals forms a group under the rule for multiplication of fractional ideals described at the beginning of the section. We denote this group by $\mathbf{I}(R)$ and call it simply the *ideal group of* R. The uniqueness statement of Theorem 4.2 implies that $\mathbf{I}(R)$ is a free abelian group with the collection of nonzero prime ideals as free generators. Generally, this is an infinitely generated group.

There is a subgroup of particular interest. Namely the collection of all principal fractional ideals Rx with x in K forms a subgroup of $\mathbf{I}(R)$ which is denoted by $\mathbf{P}(R)$. While $\mathbf{I}(R)$ and $\mathbf{P}(R)$ may be very large Abelian groups, the quotient can be very small. We let

$$\mathbf{C}(R) = \mathbf{I}(R)/\mathbf{P}(R)$$

and call $\mathbf{C}(R)$ the *class group* of R.

The class group is an invariant of the ring R. In many cases R itself is canonically selected from its quotient field K in which case $\mathbf{C}(R)$ may be viewed as an invariant of K.

For example, let K be a finite dimensional extension of the rationals and let R denote the integral closure of Z in K. Then $\mathbf{C}(R)$ is an important invariant of K. We shall prove later that $\mathbf{C}(R)$ is a finite group in this case. We refer to $\mathbf{C}(R)$ as the *class group of K* and its order is the *class number of K*.

It is not generally true that $\mathbf{C}(R)$ is finite for arbitrary Dedekind rings.

Exercise 1. For a Dedekind ring R, $\mathbf{C}(R)$ has order one if and only if R is a PID.

Exercise 2. Let $R = Z + Z\sqrt{-5}$ = integral closure of Z in $Q(\sqrt{-5})$. R is a Dedekind ring (by Theorem 6.1). Show that R is not a UFD (and so not a PID) because

$$3 \cdot 7 = (1 + 2\sqrt{-5})(1 - 2\sqrt{-5})$$

gives two essentially different factorizations of 21. Prove this fact and then find an ideal of R which is not principal. (It happens in this case that $\mathbf{C}(R)$ has order 2.)

5. NORMS AND TRACES

Let K be a field and L a finite-dimensional extension of K. Each element x in L gives rise to a function

$$r_x\colon y \to yx$$

sending L into itself. We may regard L as a finite-dimensional vector space over K and then r_x is a linear transformation. If we select a basis $u_1, \ldots, u_n$ for L over K, then r_x has a matrix representation $|a_{ij}|$, where the a_{ij} are in K and satisfy

$$u_i x = \sum a_{ij} u_j.$$

The mapping of L defined by sending x to $|a_{ij}|$ is called the *regular representation* of L over K. This is a monomorphism of L onto a subfield of the K-algebra of $n \times n$ matrices. The regular representation depends upon the choice of basis.

If another basis is selected and if U is the change of basis matrix, then the regular representation determined by the second basis sends x to $U|a_{ij}|U^{-1}$. The two functions trace $|a_{ij}|$ and det $|a_{ij}|$ are independent of the particular basis and depend only upon the linear transformation r_x, and so only upon x. We use this observation to see that the maps in the following definition depend only upon L and K and not upon the particular basis.

Definition. The *trace map* from L to K is the function $\mathrm{T}_{L/K}(x) = \mathrm{trace}(r_x)$.
The *norm map* from L to K is the function $\mathrm{N}_{L/K}(x) = \det(r_x)$.

5.1 Properties of the Norm and Trace. Let $x, y \in L$ and $a \in K$.

(i) $\mathrm{T}_{L/K}(x+y) = \mathrm{T}_{L/K}(x) + \mathrm{T}_{L/K}(y)$;
(ii) $\mathrm{T}_{L/K}(ax) = a\mathrm{T}_{L/K}(x)$;
(iii) $\mathrm{N}_{L/K}(xy) = \mathrm{N}_{L/K}(x)\,\mathrm{N}_{L/K}(y)$;
(iv) $\mathrm{N}_{L/K}(ax) = a^n\,\mathrm{N}_{L/K}(x)$.

These are easily verified if we simply observe $r_{xy} = r_x r_y$ and $r_{x+y} = r_x + r_y$ along with the fact that for a in K, r_a has a scalar matrix with a on the diagonal.

We shall also require the following transitivity property of the trace. Let $K \subset E \subset L$ be a chain of finite-dimensional extensions.

(v) For x in L, $\mathrm{T}_{L/K}(x) = \mathrm{T}_{E/K}(\mathrm{T}_{L/E}(x))$.

PROOF. Let $a_1, \ldots, a_k$ be a basis for E over K and $b_1, \ldots, b_m$ a basis for L over E. For x in L and y in E let

$$xb_i = \sum \beta_{ij}(x)\,b_j, \qquad ya_i = \sum \alpha_{ij}(y)\,a_j.$$

Then

$$\mathrm{T}_{E/K}(y) = \sum \alpha_{ii}(y), \qquad \mathrm{T}_{L/E}(x) = \sum \beta_{ii}(x).$$

It follows that

$$\mathrm{T}_{E/K}(\mathrm{T}_{L/E}(x)) = \sum\sum \alpha_{ii}(\beta_{jj}(x)).$$

The products $a_i b_j$ give a basis for L over K and

$$\begin{aligned} xa_s b_t &= \sum_j a_s \beta_{tj}(x)\,b_j \\ &= \sum_j \sum_i \alpha_{si}(\beta_{tj}(x))\,a_i b_j. \end{aligned}$$

Thus $\mathrm{T}_{L/K}(x) = \sum\sum \alpha_{ii}(\beta_{jj}(x))$ as required.

The corresponding property of the norm is more complicated to verify in this way so we shall postpone it until later when it can be proved using Galois theory.

The *characteristic polynomial* of the element x in L is defined to be

$$f(t) = \det|tI - r_x|.$$

This is a monic polynomial and we know from matrix theory that $f(r_x) = 0$. This implies $f(x) = 0$ because $f(r_x)$ multiplies elements of L by $f(x)$. The coefficient of t^{n-1} is $-\mathrm{T}_{L/K}(x)$ and the constant term is $(-1)^n \mathrm{N}_{L/K}(x)$.

The trace map is used to define a bilinear form from $L \times L$ to K by the rule $(x, y) = T_{L/K}(xy)$. The bilinear properties are easily verified and also the symmetric property $(x, y) = (y, x)$.

Recall that a symmetric bilinear form is called nondegenerate if $(L, x) = 0$ implies $x = 0$.

This idea is important because it characterizes separable extensions.

5.2 Theorem. The finite-dimensional extension L of K is separable if and only if the bilinear form $(x, y) = T_{L/K}(xy)$ is nondegenerate.

PROOF. Assume L/K is separable. There is an element θ in L such that $L = K(\theta)$. Then $1, \theta, \ldots, \theta^{n-1}$ is a basis for L over K and

$$(x, \textstyle\sum a_i \theta^i) = \textstyle\sum a_i (x, \theta^i).$$

So $(x, L) = 0$ if and only if $(x, \theta^i) = 0$ for $i = 0, 1, \ldots, n-1$. Our problem is to prove that $x = 0$ under these conditions. Let $x = \sum b_i \theta^i$. Then

$$(*) \qquad (x, \theta^j) = \textstyle\sum b_i (\theta^i, \theta^j).$$

Let D denote the matrix $|d_{ij}|$ with $d_{ij} = (\theta^{i-1}, \theta^{j-1})$. The equation $*$ and the assumption $(x, L) = 0$ implies

$$(b_0, b_1, \ldots, b_{n-1})\, D = (0, 0, \ldots, 0).$$

We shall prove D is nonsingular so after one multiplies by D^{-1} it follows that each $b_i = 0$. Thus $x = 0$ as required.

The separability of L is crucial in the proof that D^{-1} exists.

Let $f(t)$ be the (monic) minimum polynomial of θ in K. Let E be any field containing L in which $f(t)$ splits. Then

$$f(t) = (t - \theta_1)(t - \theta_2) \cdots (t - \theta_n)$$

with $\theta = \theta_1$ and all θ_i in E. The separability of L implies $\theta_i \neq \theta_j$ for $i \neq j$.

The characteristic polynomial of θ over K has degree $n = [L:K]$ and is divisible by $f(t)$ because θ is a root of its characteristic polynomial. It follows that $f(t)$ is both the characteristic and minimum polynomial of θ. From the product decomposition of $f(t)$ we find the coefficient of t^{n-1}. It follows that

$$\mathrm{T}_{L/K}(\theta) = \theta_1 + \theta_2 + \cdots + \theta_n.$$

We want a similar formula for $\mathrm{T}_{L/K}(\theta^k)$. The linear transformation r_θ has the distinct characteristic roots $\theta_1, \ldots, \theta_n$ in E. Hence the matrix for r_θ (with respect to some basis of L over K) can be diagonalized over E. So there exists U such that

$$U^{-1} r_\theta U = \operatorname{diag}\{\theta_1, \theta_2, \ldots, \theta_n\}.$$

Raise this to the kth power and it follows that

$$U^{-1} r_{\theta^k} U = \operatorname{diag}\{\theta_1{}^k, \ldots, \theta_n{}^k\}.$$

Take traces to get $T_{L/K}(\theta^k) = \theta_1{}^k + \cdots + \theta_n{}^k$. We shall return to this formula.

Now let $V = V(\theta_1, \ldots, \theta_n)$ denote the matrix

$$\begin{vmatrix} 1 & 1 & \cdots & 1 \\ \theta_1 & \theta_2 & \cdots & \theta_n \\ \theta_1{}^2 & \theta_2{}^2 & \cdots & \theta_n{}^2 \\ \vdots & \vdots & & \vdots \\ \theta_1^{n-1} & \theta_2^{n-1} & \cdots & \theta_n^{n-1} \end{vmatrix}$$

and V^t the transposed matrix. The i,j entry of VV^t is

$$\sum_k \theta_k^{i-1}\theta_k^{j-1} = \sum_k \theta_k^{i+j-2} = T_{L/K}(\theta^{i-1}\theta^{j-1}) = d_{ij}.$$

We have thus $VV^t = D$. Since V and V^t have the same determinant, it follows

$$\det D = (\det V)^2.$$

The matrix V is a van der Monde matrix and

$$\det V = \prod_{i>j} (\theta_i - \theta_j).$$

The θ_i are distinct so $\det V \neq 0$ and this proves D^{-1} exists and consequently shows the form is nondegenerate.

For the converse we suppose L/K is not a separable extension. Then the characteristic of K is a prime $p \neq 0$ and there is a subfield F of L containing K such that

(a) $(L:F) = p^m \neq 1$,
(b) for each $x \in L$, x^p is in F.

We prove $(L, x) = 0$ when x is an element in L but not in F. Start with any element y in L. Suppose first $xy \notin F$. Then the minimum polynomial of xy over F is $t^p - a$ for some a in F. The characteristic polynomial of xy over F must be

$$(t^p - a)^{p^{m-1}}.$$

This means $\mathrm{T}_{L/F}(xy) = 0$ and by transitivity of the trace, $(y, x) = \mathrm{T}_{L/K}(xy) = \mathrm{T}_{F/K}(\mathrm{T}_{L/F}(xy)) = 0$.

Now suppose $xy \in F$. Then

$$\mathrm{T}_{L/F}(xy) = xy\mathrm{T}_{L/F}(1) = p^m xy = 0$$

and just as above $(y, x) = 0$. In all cases $(y, x) = 0$ and the form is degenerate, as we had to show.

We turn next to the situation most frequently encountered: the case where a Galois group is used to describe the norm and trace.

Assume L/K is separable and F is a normal, separable extension of K containing L. Let G denote the Galois group of F/K and H the subgroup which leaves L fixed element-wise. Suppose

$$\sigma_1 H, \ldots, \sigma_n H$$

are the distinct cosets of H in G. Notice that $n = (L:K)$ and the σ_i are the distinct imbeddings over K of L into a normal extension of K.

5.3 Theorem. For each $\theta \in L$ we have

(a) $\mathrm{T}_{L/K}(\theta) = \sigma_1(\theta) + \cdots + \sigma_n(\theta)$,
(b) $\mathrm{N}_{L/K}(\theta) = \sigma_1(\theta)\sigma_2(\theta) \cdots \sigma_n(\theta)$.

PROOF. Let $q(t)$ be the minimum polynomial of θ over K. Then $q(t)$ is the characteristic polynomial of θ acting upon $K(\theta)$. Let

$$(L:K(\theta)) = d \qquad \text{and} \qquad (K(\theta):K) = m.$$

Then L is the vector space direct sum of d copies of $K(\theta)$ and the characteristic polynomial of θ acting on L must be $q(t)^d$.

Let $\theta = \theta_1, \ldots, \theta_m$ be the distinct roots of $q(t)$ in F so that

$$q(t) = \prod (t - \theta_i).$$

Since $\mathrm{T}_{L/K}(\theta)$ is the sum of the roots of the characteristic polynomial of θ on L, and $\mathrm{N}_{L/K}(\theta)$ is the product of these roots, we have

$$(1) \qquad \begin{cases} \mathrm{T}_{L/K}(\theta) = d(\theta_1 + \cdots + \theta_m), \\ \mathrm{N}_{L/K}(\theta) = (\theta_1 \theta_2 \cdots \theta_m)^d. \end{cases}$$

Now let H_1 be the subgroup of G fixing θ. Then $H \subseteq H_1$ and $d = [H_1 : H]$, $m = [G : H_1]$. Make a choice of coset representatives so that

$$\tau_1 H_1 \cup \cdots \cup \tau_m H_1 = G,$$

$$\gamma_1 H \cup \cdots \cup \gamma_d H = H_1.$$

Then the products $\tau_i \gamma_j$ represent the cosets of H in G and we may use these representatives in place of the σs because they differ only by elements which leave θ fixed. Notice that $\gamma_j(\theta) = \theta$ for all j and with suitable numbering we may assume $\tau_i(\theta) = \theta_i$. So finally

$$\sum_1^n \sigma_k(\theta) = \sum_1^m \sum_1^d \tau_i \gamma_j(\theta) = d \sum_1^m \tau_i(\theta) = d(\theta_1 + \cdots + \theta_m),$$

$$\prod \sigma_k(\theta) = (\theta_1 \theta_2 \cdots \theta_m)^d.$$

When these are combined with Eqs. (1), Eqs. (a) and (b) are proved.

Now the transitivity of the norm is easy to prove.

5.4 Corollary. If $K \subset E \subset L$ are finite-dimensional separable extensions of K then $N_{L/K}(\theta) = N_{E/K}(N_{L/E}(\theta))$ for all $\theta \in L$.

PROOF. Let F be a normal extension of K containing L. Let $\gamma_1, \ldots, \gamma_k$ be the distinct imbeddings of L into F which are the identity on E and let $\tau_1, \ldots, \tau_n$ be imbeddings of L into F which give all the distinct imbeddings of E into F over K. Then $N_{L/E}(\theta) = \gamma_1(\theta) \cdots \gamma_k(\theta)$

$$N_{L/K}(\theta) = \prod \tau_i \gamma_k(\theta) = \prod \tau_i(N_{L/E}(\theta)) = N_{E/K}(N_{L/E}(\theta)).$$

We record for future use one consequence of Theorem 5.2.

5.5 Theorem. Let L be a separable finite-dimensional extension of K and let $u_1, \ldots, u_n$ be a basis of L over K. Then there exists a second basis $v_1, \ldots, v_n$ with the property $T_{L/K}(u_i v_j) = \delta_{ij}$ where $\delta_{ij} = 1$ if $i = j$ and $\delta_{ij} = 0$ if $i \neq j$.

PROOF. This is a standard result about vector spaces. Every K-linear function from L to K has the form $x \to (x, y)$ for some unique y in L. We let v_i be the element with the property $x \to (x, v_i)$ is the function equal to zero at $u_j, j \neq i$ and 1 at u_i. It is easy to verify the v_i give a basis of L over K.

6. EXTENSIONS OF DEDEKIND RINGS

The main theorem of this section shows the reason for concentrating on Dedekind rings instead of other kinds of integral domains.

6.1 Theorem. Let R be a Dedekind ring with quotient field K and let L be a finite dimensional extension of K. Then the integral closure of R in L is a Dedekind ring.

PROOF. The extension L can be viewed as a chain of extensions $K \subset E \subset L$ with E separable over K and L purely inseparable over E. Let R' be the integral closure of R in E and R'' the integral closure of R' in L. Then R'' is the integral closure of R in L so the proof of the theorem can be accomplished in two steps. We show R' is a Dedekind ring by using the separability of E over K and then show R'' is a Dedekind ring by using facts about the purely inseparable extension L over E.

We shall prove that R' satisfies Condition (c) of Theorem 3.16.

R' is integrally closed by choice and the transitivity of integral dependence.

To show R' is noetherian we begin by selecting a basis $a_1, \ldots, a_n$ of E over K. If necessary we may multiply this basis by a suitable common denominator to insure each a_i is in R'. Let $b_1, \ldots, b_n$ be a dual basis for E over K which satisfies $(a_i, b_j) = \mathrm{T}_{E/K}(a_i b_j) = \delta_{ij}$ (see Section 5). Let y be any element in R'. There exist elements c_j in K with $y = \sum c_j b_j$. We compute the c_j by using the inner product; that is

$$(y, a_j) = \sum_k c_k (b_k, a_j) = c_j.$$

This element c_j is $\mathrm{T}_{E/K}(ya_j)$ which in turn is a coefficient in the minimum polynomial of ya_j. It follows that c_j is in R because ya_j is in R'. This proves

$$R' \subset \sum_j Rb_j.$$

This means R' is contained in a finitely generated R module. Any ideal of the ring R' is also an R submodule of this finitely generated R module and so it is finitely generated over R. It follows then that every ideal of R' is finitely generated over R'. This proves R' is noetherian.

It remains to prove that prime ideals in R' are maximal ideals or equal to zero. To do this we need the following lemma.

6.2 Lemma. Let $A \subset B$ be integral domains with B integral over A and A integrally closed. If $\mathfrak{P}$ is a nonzero prime ideal of B, then $\mathfrak{P} \cap A$ is a nonzero prime ideal of A.

PROOF. Take an element x in $\mathfrak{P}$, $x \neq 0$ and let $f(t) = \sum a_i t^i$ denote the minimum polynomial of x over the quotient field of A. By Proposition 2.4 we know the coefficients a_i belong to A. Since $f(t)$ is irreducible we find $a_0 \neq 0$ and

$$a_0 = \sum_1^n a_i x^i \in \mathfrak{P} \cap A,$$

which proves the lemma.

6.3 Corollary. If A is a field and B a domain which is integral over A, then B is a field.

PROOF. If B were not a field, there would exist a prime ideal $\mathfrak{P}$ which is nonzero (and not equal to B). By the lemma, $\mathfrak{P} \cap A$ would be a nonzero prime ideal. Since A is a field, $\mathfrak{P} \cap A = A$ so 1 is in P, an impossibility.

Now return to the proof of the theorem. Let $\mathfrak{P}$ denote a nonzero prime ideal in R' so by the lemma, $\mathfrak{P} \cap R = \mathfrak{p}$ is a nonzero prime ideal of R. Thus $R/\mathfrak{p}$ is a field and $R'/\mathfrak{P}$ is a domain containing (an isomorphic copy of) $R/\mathfrak{p}$. Let $\bar{x} = x + \mathfrak{P}$ be any element in $R'/\mathfrak{P}$. There is a monic polynomial

$$f(t) = t^m + r_1 t^{m-1} + \cdots + r_m$$

such that the r_i come from R and $f(x) = 0$. Let $\bar{f}(t)$ denote the polynomial obtained by reducing the coefficients of $f(t)$ modulo $\mathfrak{p} = R \cap \mathfrak{P}$. Then $\bar{f}(\bar{x}) = 0$ so $\bar{x}$ is integral over $R/\mathfrak{p}$. Now the hypothesis of Corollary 6.3 hold with $A = R/\mathfrak{p}$ and $B = R'/\mathfrak{P}$ so $R'/\mathfrak{P}$ is a field. That is $\mathfrak{P}$ is a maximal ideal of R'. This completes the proof in the separable case.

We now make a change of notation. Let R be a Dedekind ring with quotient field E and let R' denote the integral closure of R in the purely inseparable extension L of E.

One of the reasons why the proof is given in two steps is this time we cannot prove R' is contained in a finitely generated R-module. There exist examples where this fails. So the noetherian condition is more difficult to prove. We shall verify that R' satisfies Condition (b) of Theorem 3.16.

Since L is purely inseparable over E and finite dimensional, L must have characteristic p and for some power $p^r = q$ it happens that x^q is in E for all x in L. If x is in R' then x^q is in $E \cap R' = R$ (because R is integrally closed). Conversely if x is in L and x^q is in R, then x is integral over R so x is in R'.

Now let $\mathfrak{P} \neq (0)$ be a prime ideal in R'. Then $\mathfrak{P} \cap R = \mathfrak{Q}$ is a nonzero prime ideal in R so $\mathfrak{Q}$ is a maximal ideal. Now for $x \in \mathfrak{P}$, $x^q \in \mathfrak{Q}$ because x^q is in both R and $\mathfrak{P}$. Suppose x is an element of L with x^q in $\mathfrak{Q}$. Then first of all x is in R' and also x^q is in $\mathfrak{P}$. This implies x is in $\mathfrak{P}$ because $\mathfrak{P}$ is prime. Thus x is in $\mathfrak{P}$ if and only if x^q is in $\mathfrak{Q}$. This sets up a one-to-one correspondence $\mathfrak{P} \leftrightarrow \mathfrak{Q}$ between nonzero primes of R' and of R. Now for $a \neq 0$ in R' we know there exist only a finite number of primes in R which contain a^q and hence only a finite number of primes of R' which contain a. This is half the proof. It remains to show $R_{\mathfrak{P}}'$ is a DVR for each nonzero prime $\mathfrak{P}$ of R'.

We make a simplification by reducing to the case where R is itself a DVR. Let $\mathfrak{Q} = R \cap \mathfrak{P}$ and $S = R - \mathfrak{Q}$, so $R_S = R_{\mathfrak{Q}}$ is a DVR. S is a multiplicative set in R' so R_S' has meaning. We want to assert that $R_S' = R_{\mathfrak{P}}'$. We clearly have $R_S' \subset R_{\mathfrak{P}}'$. Suppose x/y is in $R_{\mathfrak{P}}'$ with y not in $\mathfrak{P}$. Then y^q is in R but not in $\mathfrak{P} \cap R = \mathfrak{Q}$. Thus y^q is in S so $x/y = xy^{q-1}/y^q$ is in R_S' proving equality. We know also that R_S' is the integral closure of R_S so the proof of the theorem reduces to the following situation.

6.4 Lemma. Let R be a DVR with quotient field E, and R' the integral closure of R in an extension field L which satisfies $L^q \subseteq E$. Then R' is a DVR.

PROOF. Let $R\pi$ denote the maximal ideal of R and $\mathfrak{M}$ the maximal ideal of R'. Then $\mathfrak{M}^q$ is an ideal of R which is $\neq R$ so $\mathfrak{M}^q = R\pi^n$ for some positive integer n. Let α be an element of $\mathfrak{M}$ such that $\alpha^q = \pi^n$. We shall prove $\mathfrak{M} = R'\alpha$. Let x be any element of R' and let $x^q = u\pi^d$ with $u =$ unit of R and $d =$ integer. Write $d = nt + r$ with $0 \leqslant r < n$ and observe $(x\alpha^{-t})^q = u\pi^d\pi^{-nt} = u\pi^r$. This is in R so $x\alpha^{-t}$ is in R'. The choice of n insures that $x\alpha^{-t}$ is not in $\mathfrak{M}$ so $x\alpha^{-t}$ is a unit in R'. Let $w = x\alpha^{-t}$ and observe $x = w\alpha^t =$ unit times a power of α. Now let $\mathfrak{A}$ be any ideal $\neq 0$ in R' and let m be the least positive integer with α^m in $\mathfrak{A}$. If x is in $\mathfrak{A}$ then $x = w\alpha^t$ for some t and $t \geqslant m$ implies x is in $R'\alpha^m$. Thus $\mathfrak{A} = R'\alpha^m$ so R' is a PID, which proves the result. ∎

Now we return to the more general situation of $R \subset R'$, two Dedekind rings with quotient fields K and L, respectively. We shall study the relation between the prime ideals in R and the prime ideals in R'.

Let $\mathfrak{p}$ be a nonzero prime ideal of R. Then the ideal $R'\mathfrak{p}$ of R' has a factorization,

$$(*) \qquad R'\mathfrak{p} = \mathfrak{P}_1^{e_1} \cdots \mathfrak{P}_g^{e_g}$$

with $\mathfrak{P}_1, \ldots, \mathfrak{P}_g$ distinct prime ideals in R' and $e_1, \ldots, e_g$ positive integers. Notice that the exponent e_i of $\mathfrak{P}_i$ is completely determined by the prime ideal $\mathfrak{P}_i$ because $\mathfrak{P}_i$ determines the ideal $\mathfrak{p}$ in R. That is $\mathfrak{p} = R \cap \mathfrak{P}_i$.

Definition. The integer e_i is called the *ramification index* of $\mathfrak{P}_i$ with respect to R. We shall sometimes write $e(\mathfrak{P}/R)$ or $e(\mathfrak{P}/\mathfrak{p})$ for the ramification index of $\mathfrak{P}$ over R when $\mathfrak{P} \cap R = \mathfrak{p}$.

EXERCISE 1. Let $R \subset R' \subset R''$ be Dedekind rings and $\mathfrak{P}$ a nonzero prime ideal in R''. Prove

$$e(\mathfrak{P}/R) = e(\mathfrak{P}/R')\,e(\mathfrak{P} \cap R'/R).$$

It is sometimes useful to compare various factor rings of R and R'. Notice that $R'/\mathfrak{P}_i$ is a field which contains an isomorphic copy of $R/\mathfrak{p}$. The following result can be used to insure that $R'/\mathfrak{P}_i$ is finite dimensional over $R/\mathfrak{p}$.

6.5 Lemma. Suppose $(L:K)$ is finite. Let $\mathfrak{A}$ be an ideal of R' such that $\mathfrak{A} \cap R = \mathfrak{p}$ is prime and $\neq (0)$. Then

$$(R'/\mathfrak{A} : R/\mathfrak{p}) \leqslant (L:K).$$

PROOF. The proof can be given most simply if we first reduce to the case where $\mathfrak{p}$ is a principal ideal. Let $S =$ complement of $\mathfrak{p}$ in R so that R_S is a DVR. Then $\mathfrak{A}R_S' \cap R_S = \mathfrak{p}R_S$ and $R_S'/\mathfrak{A}R_S' \cong R'/\mathfrak{A}$. Hence we may prove the lemma with R_S, R_S', and so forth, in place of R, R', and so forth. In particular we may suppose $\mathfrak{p} = R\pi$ is principal.

Let $\{x_i\}$ be a finite set of elements of R' whose cosets $x_i + \mathfrak{A}$ are linearly independent over $R/\mathfrak{p}$. Suppose there is a relation $\sum a_i x_i = 0$ with certain elements a_i in K. We may multiply the a_i by a suitable common denominator to obtain such a relation in which the a_i are in R. Suppose not all the a_i are zero. Then there is a highest power of π which divides all the a_i. After we cancel this highest power we obtain a relation in which not all the a_i are in $R\pi$. It follows that

$$\sum \bar{a}_i \bar{x}_i = \bar{0}$$

is a relation of dependence in $R'/\mathfrak{A}$ contrary to the assumed linear independence of the cosets $\bar{x}_i = x_i + \mathfrak{A}$. Consequently the x_i are linearly independent over K and the inequality of the lemma follows. ▯

Definition. The dimension $f_i = (R'/\mathfrak{P}_i : R/\mathfrak{p})$ is called the *relative degree* of $\mathfrak{P}_i$ over $\mathfrak{p}$. We shall sometimes write $f(\mathfrak{P}_i/R)$ or $f(\mathfrak{P}/\mathfrak{p})$ for this relative degree.

EXERCISE 2. Let $R \subset R' \subset R''$ be Dedekind rings and $\mathfrak{P}$ a prime ideal $\neq (0)$ in R''. Then $f(\mathfrak{P}/R) = f(\mathfrak{P}/R')f(\mathfrak{P} \cap R'/R)$.

We shall now make a connection between the ramification indices, relative degrees, and the dimensions of the quotient field.

6.6 Theorem. The integer $\sum e_i f_i$ is the dimension of $R'/\mathfrak{p}R'$ over $R/\mathfrak{p}$. If the quotient field L of R' has finite dimension over the quotient field K of R, then $\sum e_i f_i \leqslant (L:K)$. If $S =$ the complement of $\mathfrak{p}$ in R and if R_S' is finitely generated over R_S, then $\sum e_i f_i = (L:K)$.

PROOF. We use the factorization (*) of $\mathfrak{p}R'$ and CRT to obtain

$$R'/\mathfrak{p}R' \cong \sum \oplus R'/\mathfrak{P}_i^{e_i}.$$

The first statement will follow if $R'/\mathfrak{P}_i^{e_i}$ has dimension $e_i f_i$ over $R/\mathfrak{p}$. This can be proved as follows. The ring $R'/\mathfrak{P}_i^{e_i}$ is not a vector space over $R'/\mathfrak{P}_i$ (unless $e_i = 1$) but the quotients $\mathfrak{P}_i^a/\mathfrak{P}_i^{a+1}$ are vector spaces over $R'/\mathfrak{P}_i$. If we show this space has dimension one over $R'/\mathfrak{P}_i$ then it will have dimension f_i over $R/\mathfrak{p}$ and the result will follow. To show $\mathfrak{P}_i^a/\mathfrak{P}_i^{a+1}$ is one dimensional over $R'/\mathfrak{P}_i$ it is enough to show the space has one generator. By Proposition 3.19 the ideal $\mathfrak{P}_i^a$ can be generated by two elements x and y and y may be selected as nonzero element in $\mathfrak{P}_i^{a+1}$. Thus $\mathfrak{P}_i^a/\mathfrak{P}_i^{a+1}$ requires only one generator x over R' and so also over $R'/\mathfrak{P}_i$. This proves the first assertion. The second follows from Lemma 6.5.

Now suppose R_S' is a finitely generated module over R_S. R_S is PID with maximal ideal πR_S for some π in R. Let $x_1, \ldots, x_n$ be a minimal generating set of R_S' over R_S. Let us first show these elements are linearly independent over K. If there is a relation

$$\sum a_i x_i = 0$$

with a_i not all zero then the a_i may be multiplied by a common denominator to obtain such a relation with all a_i in R. Since not all a_i are zero, there is a highest power of π which divides all the a_i. After removing this highest power we find at least one of the as, say a_1, which is not in πR_S. Thus a_1 has an inverse in R_S and

$$x_1 = -1/a_1 \sum_{i \neq 1} a_i x_i.$$

This contradicts the choice of the x_i as a minimal set of generators.

Next we prove the x_i in fact give a basis for L over K. If this were not so, there would exist in L an element y such that

$$Ky \cap \sum Kx_i = 0.$$

However there is some element $s \neq 0$ in R such that sy is integral over R. (This is so because y satisfies a monic polynomial over K and with the right choice of s, sy satisfies a polynomial over R). Thus sy belongs to R_S' and to Ky which cannot be consistent with the above intersection unless $y = 0$. Hence R_S' is generated by exactly $n = (L:K)$ elements.

Now we have

$$R_S'/\mathfrak{p}R_S' \cong \sum (R/\mathfrak{p})\,\bar{x}_i$$

and the right-hand side is a vector space direct sum. From above we know the left-hand side has dimension $\sum e_i f_i$ and the right-hand side has dimension $(L:K)$. This completes the proof.

6.7 Corollary. Let L be a finite-dimensional, separable extension of K. Then $\sum e_i f_i = (L:K)$.

PROOF. In case L is separable over K we have proved Theorem 6.1 (proof) that R' is finitely generated over R. In particular then R_S' is finitely generated over R_S so the last result applies.

EXERCISE 3. Let $R = Z$ and R' = integral closure of R in $Q(\sqrt{d})$ with d a square free integer. Let p denote a prime integer. Prove that pR' can have only the following factorizations:

$$\text{(a)}\quad pR' = \mathfrak{P}, \qquad \text{(b)}\quad pR' = \mathfrak{P}^2, \qquad \text{(c)}\quad pR' = \mathfrak{P}\mathfrak{Q},$$

where $\mathfrak{P}$ and $\mathfrak{Q}$ are distinct primes of R'. In each case compute $(R'/\mathfrak{P} : R/p)$.

If it happens that L/K is normal as well as finite dimensional and separable, a little more precise information is available about the e_i and f_i. Suppose G is the Galois group of L over K. For each σ in G, $\sigma(\mathfrak{p}) = \mathfrak{p}$ and $\sigma(R') = R'$ so $\sigma(\mathfrak{p}R') = \mathfrak{p}R'$. It follows from the factorization (*) of $\mathfrak{p}R'$ that $\sigma(\mathfrak{P}_i)$ must be one of the $\mathfrak{P}_j$ for each $i = 1, \ldots, g$. We shall give an argument to show that every $\mathfrak{P}_j$ is the image under G of $\mathfrak{P}_1$.

Suppose this is not the case. Let $\mathfrak{P}_1, \ldots, \mathfrak{P}_r$, $r < g$, be all the images of $\mathfrak{P}_1$ under G. Then G must also permute the set $\mathfrak{P}_{r+1}, \ldots, \mathfrak{P}_g$. The product

$\mathfrak{P}_1 \cdots \mathfrak{P}_r$ is not contained in any $\mathfrak{P}_j$ with $r+1 \leqslant j \leqslant g$. There must exist an element a in $\mathfrak{P}_1 \cdots \mathfrak{P}_r$ which is not in $\mathfrak{P}_g$. Then $\sigma(a)$ is in $\mathfrak{P}_1 \cdots \mathfrak{P}_r$ for each σ in G and

$$\prod_{\sigma \in G} \sigma(a) \in R \cap \mathfrak{P}_1 \cdots \mathfrak{P}_r \subseteq \mathfrak{p}.$$

It follows that $\prod \sigma(a)$ is in $\mathfrak{P}_g$ since $\mathfrak{p}$ is in $\mathfrak{P}_g$. But $\mathfrak{P}_g$ is prime so some $\sigma(a)$ is in $\mathfrak{P}_g$ and thus a belongs to $\sigma^{-1}(\mathfrak{P}_g)$ which means $\sigma^{-1}(\mathfrak{P}_g)$ is one of $\mathfrak{P}_1, \ldots, \mathfrak{P}_r$ contrary to assumption. Hence the assumption cannot stand and G is transitive on the $\mathfrak{P}_i$.

6.8 Proposition. Let L be a normal, separable, and finite dimensional extension of K. Then the factorization (*) of $\mathfrak{p}R'$ has the form $(\mathfrak{P}_1 \cdots \mathfrak{P}_g)^e$. Moreover, all the relative degrees are equal (to f say) and $efg = (L:K)$. Also the Galois group permutes the $\mathfrak{P}_i$ transitively.

PROOF. Let σ be the automorphism of L which maps $\mathfrak{P}_1$ onto $\mathfrak{P}_i$. Then $\sigma(\mathfrak{p}R') = \mathfrak{p}R'$ and the uniqueness of the factorization implies $\sigma(\mathfrak{P}_1^{e_1}) = \mathfrak{P}_i^{e_1}$ so $e_1 = e_i$. It also follows that σ induces an isomorphism of $R'/\mathfrak{P}_1$ onto $R'/\mathfrak{P}_i$ so that $f_i = f_1$. The statement that $efg = (L:K)$ follows from Corollary 6.7.

Definition. The prime $\mathfrak{P}$ of R' is *ramified* with respect to R if $\mathfrak{P}$ has ramification index >1 or if the field $R'/\mathfrak{P}$ fails to be separable over $R/R \cap \mathfrak{P}$.

We say the prime $\mathfrak{p}$ of R is *ramified* in R' if $\mathfrak{p}R'$ is divisible by some ramified prime of R'.

Our next goal is to determine which primes of R ramify in R'. We shall see below there are only a finite number of them.

7. DISCRIMINANT

In this section R is a Dedekind ring with quotient field K; L is a finite-dimensional, separable extension of K; R' is the integral closure of R in L. Let T denote the trace map from L to K (see Section 5).

Let $x_1, \ldots, x_n$ be a basis of L over K. The determinant $\Delta(x_1, \ldots, x_n) = \det|\mathrm{T}(x_i x_j)|$ is called the *discriminant* of the basis $x_1, \ldots, x_n$. If we select the x_i in R', then $x_i x_j$ is in R' so $\mathrm{T}(x_i x_j)$ is in R. As we let $x_1, \ldots, x_n$ range over all possible bases of L/K which lie in R', the discriminants generate an ideal of R which we shall call the *discriminant ideal* of R' over R. We denote this ideal by Δ or $\Delta(R'/R)$.

We begin the study of the discriminant by showing it can be determined by localization.

7.1 Lemma. Let S be a multiplicative set in R. Then $\Delta(R_S'/R_S) = \Delta(R'/R)_S$.

PROOF. If $x_1, \ldots, x_n$ is a basis of L over K contained in R' then the x_i are also in R_S' and of course are still a basis for L. Thus $\Delta(R'/R)$ is contained in $\Delta(R_S'/R_S)$. It follows that

$$\Delta(R'/R)_S \subseteq \Delta(R_S'/R_S).$$

Now let $y_1, \ldots, y_n$ be a K-basis for L with each y_i in R_S'. There exists some s in S with sy_i in R'. Since $\mathrm{T}(sy_i sy_j) = s^2\mathrm{T}(y_i y_j)$ one can compute

$$\Delta(sy_1, \ldots, sy_n) = s^{2n}\Delta(y_1, \ldots, y_n).$$

It follows that $\Delta(sy_1, \ldots, sy_n)$ is in $\Delta(R'/R)$ and so $\Delta(y_1, \ldots, y_n)$ is in $\Delta(R'/R)_S$. This proves both inclusions.

One further computation will be made before we get an application.

7.2 Lemma. If R' is a free R-module on the generators $x_1, \ldots, x_n$, then $\Delta(R'/R) = R\,\Delta(x_1, \ldots, x_n)$.

PROOF. Let $y_1, \ldots, y_n$ be a K-basis of L in R'. Let

$$y_i = \sum_j r_{ij} x_j, \qquad r_{ij} \in R.$$

The existence of such equations is a consequence of the freeness of R' on the x_i. A simple matrix calculation yields

$$|\mathrm{T}(y_i y_j)| = |r_{ij}||\mathrm{T}(x_i x_j)||r_{ij}|^{\mathrm{t}}$$

and so

$$\Delta(y_1, \ldots, y_n) = \det|r_{ij}|^2\,\Delta(x_1, \ldots, x_n).$$

Thus every discriminant of a basis is in the principal ideal generated by $\Delta(x_1, \ldots, x_m)$ which proves the lemma.

Now the connection can be made between the discriminant and the ramified primes of R.

7.3 Theorem. The primes of R which ramify in R' are those which contain $\Delta(R'/R)$.

PROOF. Let $\mathfrak{p}$ be a nonzero prime ideal in R and S the complement of $\mathfrak{p}$ in R. Then $\mathfrak{p}$ contains $\Delta(R'/R)$ if and only if $\mathfrak{p}R_S$ contains $\Delta(R'/R)_S$. Moreover $\mathfrak{p}$ is ramified in R' if and only if $\mathfrak{p}R_S$ is ramified in R_S'. The proof of the theorem will follow then if we can prove it for R_S. Thus since R_S is a DVR we may as well assume at the start that R is a DVR. With this additional information it follows that R' is free over R. Let $x_1, \ldots, x_n$ be free generators of R' over R. Then this set is also a basis for L over K. Then $\bar{x}_1, \ldots, \bar{x}_n$ is a basis for $R'/\mathfrak{p}R'$ over $R/\mathfrak{p}$ (by the freeness of R' over R).

It will be necessary to compare the regular representation of R' over R with that of $R'/\mathfrak{p}R'$ over $R/\mathfrak{p}$. For each y in R' the linear transformation

$$r_y : x \to xy$$

sends R' into itself and r_y has a matrix $|a_{ij}|$ with respect to the basis $x_1, \ldots, x_n$ with a_{ij} in R. The equations which define the a_{ij} are

$$x_i y = \sum a_{ij} x_j.$$

We reduce this modulo $\mathfrak{p}R'$ and obtain

$$\bar{x}_i \bar{y} = \sum \bar{a}_{ij} \bar{x}_j.$$

This means the linear transformation $r_{\bar{y}}$ of $R'/\mathfrak{p}R'$ over $R/\mathfrak{p}$ has matrix $|\bar{a}_{ij}|$. Let tr denote the linear function from $R'/\mathfrak{p}R'$ to $R'/\mathfrak{p}$ defined by $\operatorname{tr}(\bar{y}) = \operatorname{trace}(r_{\bar{y}})$. The above computation proves

(7.4) For y in R', $\overline{\mathrm{T}_{L/K}(y)} = \operatorname{tr}(\bar{y})$.

Now we proceed to the proof of the theorem. By Lemma 7.2 we know the discriminant ideal is generated by $\Delta(x_1, \ldots, x_n)$. Thus $\mathfrak{p} \supset \Delta(R'/R)$ if and only if $\Delta(x_1, \ldots, x_n)$ is in $\mathfrak{p}$. This holds if and only if

(7.5) $$\Delta(\bar{x}_1, \ldots, \bar{x}_n) = \det|\operatorname{tr}(\bar{x}_i \bar{x}_j)| = 0$$

in $R/\mathfrak{p}$. [For this we have used (7.4) and the definition of $\Delta(x_1, \ldots, x_n)$.] It remains to examine the structure of $R'/\mathfrak{p}R'$ under the assumption that Equation (7.5) holds for the basis of $R'/\mathfrak{p}R'$ over $R/\mathfrak{p}$.

Let

$$\mathfrak{p}R' = \mathfrak{P}_1^{e_1} \cdots \mathfrak{P}_g^{e_g},$$

so that by CRT it follows that

$$R'/\mathfrak{p}R' \cong R'/\mathfrak{P}_1^{e_1} \oplus \cdots \oplus R'/\mathfrak{P}_g^{e_g}$$

Consider first the case with $\mathfrak{p}$ not ramified in R'. Then each $e_i = 1$ and $R'/\mathfrak{P}_i$ is a separable extension of $R/\mathfrak{p}$. Let t_i denote the trace map from $R'/\mathfrak{P}_i$ to $R/\mathfrak{p}$ Select a new basis for $R'/\mathfrak{p}R'$ which is compatible with the direct sum decomposition. That is select $u_1, \ldots, u_k$ a basis for $R'/\mathfrak{P}_1$; $u_{k+1}, \ldots, u_{k+l}$ a basis for $R'/\mathfrak{P}_2$, and so forth. Then for $\bar{y}$ in $R'/\mathfrak{p}R'$ we can write $\bar{y} = y_1 + \cdots + y_g$ with y_i in $R'/\mathfrak{P}_i$. The matrix for $r_{\bar{y}}$ has the block decomposition

$$r_{\bar{y}} \to \begin{vmatrix} A_1 & & & 0 \\ & A_2 & & \\ & & \ddots & \\ 0 & & & A_g \end{vmatrix}$$

where A_i is the matrix for r_{y_i} acting on $R'/\mathfrak{P}_i$. It follows that

$$\operatorname{tr}(\bar{y}) = t_1(y_1) + \cdots + t_g(y_g).$$

More importantly the discriminant matrix has the block form

$$\begin{bmatrix} \Delta_1 & & & \\ & \Delta_2 & & \\ & & \ddots & \\ & & & \Delta_g \end{bmatrix}$$

where Δ_i = discriminant matrix of the basis of $R'/\mathfrak{P}_i$ over $R/\mathfrak{p}$. We know from Section 5 that $R'/\mathfrak{P}_i$ separable over $R/\mathfrak{p}$ implies $\det\Delta_i \neq 0$. Thus

$$\Delta(\bar{x}_1, \ldots, \bar{x}_n) = u^2(\det\Delta_1)\cdots(\det\Delta_g) \neq 0,$$

where u is the determinant of the matrix which represents the change of basis.

This proves that $\mathfrak{p}$ not ramified implies $\mathfrak{p}$ does not contain $\Delta(x_1, \ldots, x_n)$, by 7.5.

To complete the proof we must show that whenever some $e_i > 1$ or some $R'/\mathfrak{P}_i$ is not separable over $R/\mathfrak{p}$, then $\Delta(\bar{x}_1, \ldots, \bar{x}_n) = 0$.

Suppose $e_i > 1$. Select a basis $u_1, \ldots, u_k$ for $R'/\mathfrak{P}_i^{e_i}$ such that u_1 is in $\mathfrak{P}_i/\mathfrak{P}_i^{e_i}$. Then $(u_1)^{e_i} = 0$ so that r_{u_1} is a nilpotent linear transformation. Moreover $u_1 u_j$ is also nilpotent so the characteristic polynomial of $r_{u_1 u_i}$ has only zeros for its characteristic roots. Thus

$$\mathrm{t}_i(u_1 u_j) = \operatorname{trace} r_{u_1 u_j} = 0.$$

It follows that the discriminant matrix for $R'/\mathfrak{P}_i$ over $R/\mathfrak{p}$ has a row of zeros and so $\det\Delta_i = 0$. Since $\Delta(\bar{x}_1, \ldots, \bar{x}_n)$ is a product of the $\det\Delta_i$ we get $\Delta(\bar{x}_1, \ldots, \bar{x}_n) = 0$.

Finally suppose all the $e_i = 1$ but $R'/\mathfrak{P}_i$ is not separable over $R/\mathfrak{p}$. By Theorem 5.2 (proof) we know the discriminant of $R'/\mathfrak{P}_i$ over $R/\mathfrak{p}$ is zero so again $\Delta(\bar{x}_1, \ldots, \bar{x}_n) = 0$.

In both of these cases it follows $\mathfrak{p}$ must contain the discriminant ideal.

Next we consider some means by which a factorization of $\mathfrak{p}R'$ can be computed. This procedure will not cover all possible cases but is still rather general.

7.6 Theorem. Let R' denote the integral closure of the Dedekind ring R in a finite-dimensional extension L of the quotient field K of R. Let $\mathfrak{p}$ be a nonzero prime ideal of R. Suppose there is an element θ such that the integral closure of $R_\mathfrak{p}$ in L is $R_\mathfrak{p}[\theta]$. Let $f(X)$ be the minimal polynomial of θ over K. Let $\bar{f}(X)$ denote the polynomial obtained by reducing the coefficients of $f(X)$ modulo $\mathfrak{p}$. Suppose

$$\bar{f}(X) = g_1(X)^{a_1} \cdots g_t(X)^{a_t}$$

is the factorization of $\bar{f}(X)$ as a product of the distinct irreducible polynomials

$g_i(X)$ over $R/\mathfrak{p}$. Then

$$\mathfrak{p}R' = \mathfrak{P}_1^{a_1} \cdots \mathfrak{P}_t^{a_t}$$

for certain primes $\mathfrak{P}_i$ of R' and the relative degree f_i equals the degree of $g_i(X)$.

PROOF. The factorization of $\mathfrak{p}R'$ is completely determined by local information so we may replace R by $R_\mathfrak{p}$ and R' by R_S', $S = R - \mathfrak{p}$. In particular we assume R is a DVR. Then we have $R' = R[\theta]$ and this is isomorphic to $R[X]/(f(X))$. Hence $R'/\mathfrak{p}R'$ is isomorphic to $R[X]$ modulo the ideal generated by $\mathfrak{p}$ and $(f(X))$. If we first divide out by $\mathfrak{p}$ we have finally

$$R'/\mathfrak{p}R' \cong \bar{R}[X]/(\bar{f}(X)),$$

where $\bar{R} = R/\mathfrak{p}$. The factorization of $\bar{f}(X)$ and CRT now yields

$$R'/\mathfrak{p}R' \cong \sum \oplus \bar{R}[X]/(g_i(X)^{a_i}).$$

The prime ideals in this ring are in one-to-one correspondence with the $g_i(X)$ and so it follows that

$$\mathfrak{p}R' = \mathfrak{P}_1^{a_1} \cdots \mathfrak{P}_t^{a_t}$$

with $R'/\mathfrak{P}_i \cong \bar{R}[X]/(g_i(X))$. Thus $\mathfrak{P}_i$ has a relative degree equal to the degree of $g_i(X)$ and the proof is done.

This theorem is limited by the necessity that $R' = R[\theta]$ (locally). It need not happen that such a θ exists. We can give a criterion for this in case R is a DVR.

7.7 Proposition. Let R be a DVR with maximal ideal $\mathfrak{p}$ and let θ be an element of R' such that $L = K(\theta)$. If $\Delta(1, \theta, \ldots, \theta^{n-1})$ is not in $\mathfrak{p}$ then $R' = R[\theta]$.

PROOF. Since R is a PID, R' has a free basis $\alpha_0, \ldots, \alpha_{n-1}$ over R. We have $R[\theta] \subseteq R'$ so each power of θ can be expressed in terms of the basis.

$$\theta^i = \sum_j r_{ij} \alpha_j, \qquad r_{ij} \in R.$$

Then

$$\begin{aligned} \Delta(1, \theta, \ldots, \theta^{n-1}) &= \det |\mathrm{T}_{L|K}(\theta^i \theta^j)| \\ &= \det |r_{ij}|^2 \det |\mathrm{T}_{L/K}(\alpha_i \alpha_j)|. \end{aligned}$$

The elements here are all in R and the element on the left is not in $\mathfrak{p}$. Thus $\det |r_{ij}|$ is not in $\mathfrak{p}$ and since R is a DVR, $\det |r_{ij}|$ has an inverse in R. This means each α_j can be expressed as an R-linear combination of the θ^i. Hence $R' \subseteq R[\theta]$ so equality must hold.

EXAMPLE. We consider $K = Q$ = rationals and $L = Q(\theta)$ with $\theta^3 = 2$. One computes the discriminant first. We know $\mathrm{T}_{L/K}(\theta)$ = sum of the roots of

X^3-2 so $\mathrm{T}_{L/K}(\theta)=\theta+\omega\theta+\omega^2\theta=0$ where $\omega^3=1$ and $\omega\neq 1$. Also $\mathrm{T}_{L/K}(\theta^2)=\theta^2+\omega^2\theta^2+\omega\theta^2=0$. It follows that $\mathrm{T}(\theta^3)=2\mathrm{T}(1)=6$ and $\mathrm{T}(\theta^4)=0$. Then $\Delta(1,\theta,\theta^2)=-2^2\cdot 3^3=\det|\mathrm{T}(\theta^i\theta^j)|$. Let R' be the integral closure of Z in L. The factorization of $\mathfrak{p}R'$ is determined by the factorization of $\mathfrak{p}R_S{}'$ with $S=Z-(\mathfrak{p})$. That is if

$$\mathfrak{p}R' = \prod \mathfrak{P}_i^{e_i}$$

then

$$\mathfrak{p}R_S{}' = (\prod \mathfrak{P}_i^{e_i})_S.$$

Let p denote any prime $\neq 2,3$. By Proposition 7.7 the integral closure of Z_p is $Z_p[\theta]$. By Theorem 7.6 the factorization of p in $Z_p[\theta]$ is determined by the factorization of X^3-2 in Z/p. We consider a few cases.

$p=7$ X^3-2 is irreducible modulo 7 so $7R'=\mathfrak{P}_1$ is prime and we find $R'/P_1 \cong \mathrm{GF}(7^3)$.

$p=29$ $X^3-2=(X+3)(X^2-3X+9)$ modulo 29 and the second factor is irreducible. Thus

$$29R' = \mathfrak{P}_1\mathfrak{P}_2 \qquad \text{with} \quad R'/\mathfrak{P}_1 \cong \mathrm{GF}(29)$$

and $R'/\mathfrak{P}_2 \cong \mathrm{GF}(29^2)$.

$p=31$ $X^3-2=(X-4)(X-7)(X+11)$ modulo 31 and $31R'=\mathfrak{P}_1\mathfrak{P}_2\mathfrak{P}_3$ with $R'/\mathfrak{P}_i \cong \mathrm{GF}(31)$.

Notice that these computations are possible without actually knowing R' explicitly. After a rather lengthy computation, it does follow that $R'=Z[\theta]$. Hence Theorem 7.6 can be applied also to the cases $p=2,3$.

$p=2$ $X^3-2=X^3$ modulo 2 so $2R'=\mathfrak{P}^3$ and $R'/\mathfrak{P}\cong \mathrm{GF}(2)$.

$p=3$ $X^3-2=(X+1)^3$ modulo 3 so $3R'=\mathfrak{P}^3$ and $R'/\mathfrak{P}\cong \mathrm{GF}(3)$.

We will be able to do these calculations in Section 8 without first proving $R'=Z[\theta]$.

EXERCISE. Let d be a square free integer and R' the integral closure of Z in $Q(\sqrt{d})$. As a continuation of Exercise 3 in Section 6 determine which of the three possible factorizations of pR' actually occurs. Prove the following.

(a) Suppose p divides $\Delta(R'/Z)$. Then $\mathfrak{p}R'=\mathfrak{P}^2$.
(b) Suppose p is odd and does not divide $\Delta(R'/Z)$. Then $pR'=\mathfrak{P}\mathfrak{Q}$ with $\mathfrak{P}\neq\mathfrak{Q}$ if and only if d is a quadratic residue modulo p. That is X^2-d has a root in Z/p.
(c) Suppose $p=2$ and does not divide $\Delta(R'/Z)$. Then necessarily $d\equiv 1 \bmod 4$. Show $2R'=\mathfrak{P}\mathfrak{Q}$ if $d\equiv 1 \bmod 8$ and $2R=\mathfrak{P}$ is prime if $d\equiv 5 \bmod 8$.

8. NORMS OF IDEALS

In this section R is a Dedekind ring with quotient field K; L is a finite-dimensional, separable extension field of K, and R' is the integral closure of R in L.

Let N denote the norm function, $N(x) = \det(r_x)$ where r_x is the K-linear transformation on L given by $r_x(y) = yx$ (see Section 5).

For any element x of L, the characteristic polynomial of x is a power of its minimum polynomial. When x is in R', the coefficients of the minimum polynomial are in R (by Proposition 2.4) and so also the coefficients of the characteristic polynomial are also in R. In particular $N(x)$ belongs to R.

Let $\mathfrak{A}$ be an ideal in R'.

Definition. The *norm of* $\mathfrak{A}$, $N(\mathfrak{A})$, is the ideal in R generated by all $N(a)$ with a in $\mathfrak{A}$.

8.1 Properties of the Norm:

(i) $\mathrm{N}(ab) = \mathrm{N}(a)\mathrm{N}(b)$. for elements $a, b \in L$
(ii) $\mathrm{N}(R'a) = R\mathrm{N}(a)$.
(iii) If S is a multiplicative set in R, then $\mathrm{N}(\mathfrak{A})_S = \mathrm{N}(\mathfrak{A}_S)$ for any ideal $\mathfrak{A}$ in R'.
(iv) $\mathrm{N}(\mathfrak{A}\mathfrak{B}) = \mathrm{N}(\mathfrak{A})\mathrm{N}(\mathfrak{B})$ for ideals $\mathfrak{A}, \mathfrak{B}$ in R'.

PROOF. (i) is just a statement about determinants.

(ii) Since 1 is in R' and $\mathrm{N}(1) = 1$ it follows that $\mathrm{N}(R') = R$ and $\mathrm{N}(R'a) = R\mathrm{N}(a)$.

(iii) Any element in $\mathfrak{A}_S$ has the form a/s with a in $\mathfrak{A}$ and s in S. Thus $\mathrm{N}(a/s) = \mathrm{N}(a)/s^n$ if $n = (L:K)$. Thus $\mathrm{N}(\mathfrak{A}_S) \subseteq \mathrm{N}(\mathfrak{A})_S$.

Conversely the ideal $\mathrm{N}(\mathfrak{A})_S$ is generated over R_S by elements $\mathrm{N}(a)$ with a in $\mathfrak{A}$. All such elements are in $\mathrm{N}(\mathfrak{A}_S)$ so the other inclusion follows also.

(iv) We shall use Lemma 3.18 to get the equality. It is necessary to prove equality at the localizations at each maximal ideal. For any maximal ideal $\mathfrak{p}$ of R, let $S = R - \mathfrak{p}$. By Part (3) we know

$$\mathrm{N}(\mathfrak{A})_S = \mathrm{N}(\mathfrak{A}_S), \qquad \mathrm{N}(\mathfrak{B})_S = \mathrm{N}(\mathfrak{B}_S), \qquad \mathrm{N}(\mathfrak{A}\mathfrak{B})_S = \mathrm{N}(\mathfrak{A}_S\mathfrak{B}_S).$$

The ring R_S is a DVR and R_S' has only a finite number of prime ideals. By Theorem 3.15 we obtain R_S' is a PID. Thus $\mathfrak{A}_S = aR_S'$, $\mathfrak{B}_S = bR_S'$ for some a, b in R'. Then

$$\begin{aligned} \mathrm{N}(\mathfrak{A}_S\mathfrak{B}_S) &= \mathrm{N}(abR_S') = \mathrm{N}(ab)\,R_S = \mathrm{N}(a)\,R_S \cdot \mathrm{N}(b)\,R_S \\ &= \mathrm{N}(\mathfrak{A}_S)\,\mathrm{N}(\mathfrak{B}_S). \end{aligned}$$

This Property (iv) shows that we can determine $\mathrm{N}(\mathfrak{A})$ if we can determine $\mathrm{N}(\mathfrak{P})$ for each prime ideal $\mathfrak{P}$ of R'. That is when

$$\mathfrak{A} = \prod \mathfrak{P}_j^{a_j} \qquad \text{then} \quad \mathrm{N}(\mathfrak{A}) = \prod \mathrm{N}(\mathfrak{P}_j)^{a_j}.$$

The computation of $\mathrm{N}(\mathfrak{P})$ is simplified if we first work in the situation where L is normal over K with Galois group G. We assume this is the case.

Now let $\mathfrak{P}$ denote a prime ideal of R' and let $\mathfrak{p} = \mathfrak{P} \cap R$. For an element a in $\mathfrak{P}$ the product of all $\sigma(a)$ must also fall in $\mathfrak{P}$ (because $\mathfrak{P}$ is an ideal) so $\mathrm{N}(a) \in \mathfrak{P} \cap R = \mathfrak{p}$. We argue that $\mathrm{N}(\mathfrak{P})$ must be a power of $\mathfrak{p}$. It is enough to show that no other prime of R can enter into the factorization of $\mathrm{N}(\mathfrak{P})$. Clearly $\mathfrak{P} \supset \mathfrak{p}R'$ so $\mathfrak{P}$ divides $\mathfrak{p}R'$. Thus $\mathrm{N}(\mathfrak{P})$ divides $\mathrm{N}(\mathfrak{p}R') = \mathrm{N}(\mathfrak{p})R = \mathfrak{p}^n$, $n = (L:K)$. So it follows that $\mathrm{N}(\mathfrak{P})$ is a power of $\mathfrak{p}$. We shall now determine the exact power. This power will not be changed if we localize at $S = R - \mathfrak{p}$. Thus we may work with R_S, R_S' in place of R, R'. Both of these rings are now PIDs so let $\mathfrak{P}R_S' = \pi R_S'$ and $\mathfrak{p}R_S = \tau R_S$.

The ramification numbers for the primes of R_S' which divide τ are all the same by Proposition 6.8. We may assume

(8.2) $$\mathfrak{p}R_S' = \tau R_S' = (\mathfrak{P}_1 \cdots \mathfrak{P}_g)^e$$

for certain primes $\mathfrak{P}_i$ of R_S'. We may assume $\mathfrak{P}_1 = \pi R_S' = \mathfrak{P}R_S'$. Now we know the Galois group permutes the primes $\mathfrak{P}_i$ transitively and $|G| = efg$ with $f =$ relative degree of $\mathfrak{P}$. Thus as σ ranges over G, $\sigma(\mathfrak{P}_1)$ ranges over $\mathfrak{P}_1, \ldots, \mathfrak{P}_g$ with each $\mathfrak{P}_i$ counted ef times. It follows that

$$\mathrm{N}(\pi)R_S' = \prod_{\sigma \in G} \sigma(\pi)R_S' = \prod \sigma(\mathfrak{P}_1) = (\mathfrak{P}_1 \cdots \mathfrak{P}_g)^{ef}.$$

But also $\mathrm{N}(\pi R_S') = \mathfrak{p}_S{}^m$ for some m so $\mathrm{N}(\pi)R_S' = \mathfrak{p}^m R_S' = (\mathfrak{P}_1 \cdots \mathfrak{P}_g)^{em}$ in view of the factorization Eq. (8.2). It follows that $m = f =$ relative degree of $\mathfrak{P}$ over R and

(8.3) $$\mathrm{N}(\mathfrak{P}) = \mathfrak{p}^f, \qquad f = \text{ relative degree of } \mathfrak{P} \text{ over } R.$$

Now we drop the assumption that L/K is normal. Let E be a field containing L which is normal, separable, and finite dimensional over K. Let R'' be the integral closure of R in E and let $\mathfrak{Q}$ be some prime of R'' which appears in the factorization of $\mathfrak{P}R''$. Then E is normal over L, so by Eq. (8.3) we find

$$\mathrm{N}_{E/L}(\mathfrak{Q}) = \mathfrak{P}^{f_1}, \qquad f_1 = f(\mathfrak{Q}/R').$$

Also

$$\mathrm{N}_{E/K}(\mathfrak{Q}) = \mathfrak{p}^{f_2}, \qquad f_2 = f(\mathfrak{Q}/R).$$

By the transitivity of the norm (Corollary 5.4), $\mathrm{N}_{E/K}(x) = \mathrm{N}_{L/K}(\mathrm{N}_{E/L}(x))$, and so

$$\mathfrak{p}^{f_2} = \mathrm{N}_{L/K}(\mathrm{N}_{E/L}(\mathfrak{Q})) = \mathrm{N}_{L/K}(\mathfrak{P})^{f_1}.$$

Exercise 2 of Section 6 states in this context that $f_2 = f_1 f(\mathfrak{P}/R)$. Hence Eq, (8.3) holds again.

8.4 Proposition. For any nonzero prime $\mathfrak{P}$ of R', the norm $\mathrm{N}(\mathfrak{P})$ equals $\mathfrak{p}^f$ with $\mathfrak{p} = R \cap \mathfrak{P}$ and f the relative degree of $\mathfrak{P}$ over R.

8.5 Corollary. Let $\mathfrak{A} = \prod \mathfrak{P}_i^{a_i}$ be an ideal of R' and f_i the relative degree of $\mathfrak{P}_i$ over R. Let $\mathfrak{p}_i = \mathfrak{P}_i \cap R$. Then $\mathrm{N}(\mathfrak{A}) = \prod \mathfrak{p}_i^{a_i f_i}$.

Consider now the case with $K = Q$ = rational field and $R = Z$ = rational integers. L and R' have the same meanings as above. For any ideal $\mathfrak{A}$ of R', $\mathrm{N}(\mathfrak{A})$ is an ideal in Z which is necessarily a principal ideal, say $\mathrm{N}(\mathfrak{A}) = Zm = (m)$ for some integer m. If we require that $m \geqslant 0$ then m is uniquely determined. Let us denote the integer m by $\mathcal{N}(\mathfrak{A})$ so that the norm of an ideal $\mathfrak{A} \neq 0$ is now a positive integer. We call $\mathcal{N}(\mathfrak{A})$ the *absolute norm* of $\mathfrak{A}$.

8.6 Proposition. For any nonzero ideal $\mathfrak{A}$ in R' the integer $\mathcal{N}(\mathfrak{A})$ is equal to the number of elements in the ring $R'/\mathfrak{A}$.

PROOF. Let $\mathfrak{A} = \prod \mathfrak{P}_i^{a_i}$ and p_i the prime number such that $(p_i) = \mathfrak{P}_i \cap Z$. Let f_i denote the relative degree of $\mathfrak{P}_i$ over Z. By CRT we know

$$R'/\mathfrak{A} \cong R'/\mathfrak{P}_1^{a_1} \oplus \cdots \oplus R'/\mathfrak{P}_r^{a_r}.$$

We shall first compute the number of elements in each of these summands.

In the proof of Theorem 6.6 we observed that each quotient $\mathfrak{P}_i^b/\mathfrak{P}_i^{b+1}$ is a one-dimensional vector space over $R'/\mathfrak{P}_i$. Thus $R'/\mathfrak{P}_i^{a_i}$ has $|R'/\mathfrak{P}_i|^{a_i}$ elements. Since $R'/\mathfrak{P}_i$ has dimension f_i over $Z/(p_i)$ it follows that $|R'/\mathfrak{P}_i| = p_i^{f_i}$. Thus the order of $R'/\mathfrak{P}_i^{a_i}$ is $p_i^{a_i f_i}$. Consequently

$$|R'/\mathfrak{A}| = \prod p_i^{a_i f_i}$$

By Corollary 8.5 this number is $\mathcal{N}(\mathfrak{A})$.

In this same context we shall make a few remarks about computations.

Suppose $\mathfrak{A}$ is an ideal with $\mathcal{N}(\mathfrak{A}) = p$ = prime integer. It follows that $R'/\mathfrak{A}$ has p elements so $\mathfrak{A}$ is a prime ideal of relative degree $f = 1$. In particular, if x is an element of R and $\mathrm{N}(x) = p$ is prime, then Rx is a prime ideal with relative degree equal to one.

EXAMPLE. In the last section we considered factorization of primes in the ring of integers in $Q(\theta)$ where $\theta^3 = 2$. The factorization of p could be easily obtained by Theorem 7.6 when $p \neq 2, 3$. For $p = 2, 3$ it was necessary to know that $R = Z[\theta]$. We can avoid that last computation. Clearly $N(\theta) = 2$ because $X^3 - 2$ is the minimum polynomial of θ. Thus θR is a prime with relative degree equal to one. Moreover, 2 is in $(\theta R)^3$ so $(\theta R)^3 \supseteq 2R$. The sum of the $e_i f_i$ must equal three so $2R = (\theta R)^3 = \mathfrak{P}^3$ gives the factorization of $2R$.

For the prime $p = 3$ we proceed in a similar way. $f(X) = X^3 - 2$ is the minimum polynomial for θ so $f(X-1) = (X-1)^3 - 2$ is the minimum polynomial for $\theta+1$. It follows that $N(\theta+1) = 3$ so $(\theta+1)R$ is prime. It is slightly more difficult this time to show 3 is in $(\theta+1)^3 R$.

Let $\alpha = \theta+1$ and observe that the minimum equation for α is $\alpha^3 - 3\alpha^2 + 3\alpha - 3 = 0$.

Suppose $3R = (\alpha R)\mathfrak{B}$. Then $3 = \alpha\beta$ and $\beta = 3/\alpha = \alpha^2 - 3\alpha + 3$. This belongs to αR so 3 is in $\alpha^2 R$. Now suppose $3 = \alpha^2 \beta'$. Then

$$\beta' = 3/\alpha^2 = \alpha - 3 + 3/\alpha = \alpha - 3 + \alpha^2 - 3\alpha + 3$$

which also belongs to αR. Hence 3 is in $\alpha^3 R$ so $3R = (\alpha R)^3$ by the same reasoning as before for $p = 2$.

Application of Preceding Results. We shall use the information obtained about norms to prove the following theorem which tells precisely which integers can be expressed as the sum of two squares.

Theorem. The positive integer n can be expressed in the form $n = a^2 + b^2$ with a, b integers if and only if no prime of the form $4k+3$ appears in the factorization of n with an odd exponent.

PROOF. Let i be a root of $x^2 + 1 = 0$. The integral closure of Z in $Q(i)$ is just $Z[i] = R$. If $a+ib$ is in R then $\mathrm{N}(a+ib) = a^2 + b^2$.

Conversely if $n = a^2 + b^2$ then $n = \mathrm{N}(a+ib)$ and $a+ib$ is in R. So the integers we are trying to characterize are precisely the norms of elements in R. Consider an element x in R and let

$$xR = \prod \mathfrak{P}_i^{c_i} \prod \mathfrak{Q}_j^{d_j}$$

be the factorization of xR as a product of primes in R. We select the notation so that

$$\mathfrak{P}_i \cap Z = (p_i), \qquad f(\mathfrak{P}_i/Z) = 2;$$

$$\mathfrak{Q}_j \cap Z = (q_j), \qquad f(\mathfrak{Q}_j/Z) = 1.$$

By Proposition 8.4 we see $\mathcal{N}(\mathfrak{P}_i) = p_i^2$ and $\mathcal{N}(\mathfrak{Q}_j) = q_j$ and so

$$\mathrm{N}(x) = \prod p_i^{2c_i} \prod q_j^{d_j}.$$

We see from this that if a prime factor of $\mathrm{N}(x)$ has an odd exponent in the factorization, then the prime must be one of the q_j. The integral primes q which are divisible in R by some prime with relative degree 1 are precisely those primes q for which $X^2 + 1$ is reducible modulo q. These in turn are the primes of the form $4k+1$ or $q = 2$. This proves half the result. Namely the norm of an element of R cannot have a prime of the type $4k+3$ appear with an odd exponent in the factorization.

For the converse suppose n is a positive integer and

$$n = m^2 p_1 \cdots p_s$$

with $p_1, \ldots, p_s$ distinct primes equal to 2 or numbers $4k+1$. Then $p_i R$ is divisible by a prime $\mathfrak{P}_i$ with $\mathrm{N}(\mathfrak{P}_i) = (p_i)$. We shall leave as an exercise the fact that R is a PID. Thus $\mathfrak{P}_i = w_i R$ for some w_i in $\mathfrak{P}_i$. It follows that $\mathrm{N}(w_i)$ generates (p_i) so $\mathrm{N}(w_i) = \pm p_i$. However, for any w in R we have $\mathrm{N}(w) \geqslant 0$ so $\mathrm{N}(w_i) = p_i$. Now then

$$\mathrm{N}(m w_1 \cdots w_s) = m^2 p_1 \cdots p_s = n$$

which proves n is a sum of two squares.

Exercise. Let $R = Z[\sqrt{-1}]$. Let a, b be nonzero elements in R. Show that there exist q, r in R such that

$$a = bq + r \qquad \text{and} \qquad 0 \leqslant \mathrm{N}(r) < \mathrm{N}(b).$$

Conclude that R is a PID.

9. CYCLOTOMIC FIELDS

For a positive integer m, the splitting field of the polynomial $X^m - 1$ over the rationals is called the *cyclotomic field* of mth roots of unity. If θ is a root of $X^m - 1$ but not a root of $X^n - 1$ for any $n < m$, then θ is a *primitive* mth root of unity. If θ is one primitive mth root of unity, then any other has the form θ^k with k and m relatively prime. From this it follows that $Q(\theta)$ is the splitting field of $X^m - 1$. We shall study this field in some special cases first.

We first fix some notation for later use.

Notation. Let p be a prime and $q = p^a$. The number $p^{a-1}(p-1)$ will be denoted by $\phi(q)$, the Euler function at q.

$$f(X) = \frac{X^{p^a} - 1}{X^{p^{a-1}} - 1} = t^{p-1} + t^{p-2} + \cdots + t + 1, \quad t = X^{p^{a-1}}.$$

$\theta =$ primitive qth root of unity.
$R =$ algebraic integers in $Q(\theta)$.

9.1 Theorem. (a) $(Q(\theta) : Q) = p^{a-1}(p-1)$.

(b) The polynomial $f(X)$ is irreducible over Q and it is the minimum polynomial of θ.

(c) The element $\alpha = 1 - \theta$ is a prime element, αR is a prime ideal and $pR = (\alpha R)^{\phi(q)}$.

(d) The prime p is the only ramified prime.

(e) $R = Z[\theta]$.

PROOF. The element θ is a root of $X^{p^a}-1$ and not a root of $X^{p^{a-1}}-1$ so it follows that $f(\theta)=0$. The other roots of $f(X)$ are the other primitive qth roots of unity, namely θ^k for $(p,k)=1$. We see that θ^k is an algebraic integer for all k. Notice also that

$$\frac{1-\theta^k}{1-\theta} = 1+\theta+\cdots+\theta^{k-1} = u_k$$

belongs to R. If $(p,k)=1$, then θ is a power of θ^k so the same method implies

$$\frac{1-\theta}{1-\theta^k} = u_k^{-1} \qquad \text{is in} \quad R.$$

This u_k is a unit in R and $1-\theta^k=(1-\theta)\,u_k$. Since there are $\phi(q)$ distinct θ^k with $(p,k)=1$ and $f(X)$ has degree $\phi(q)$ it follows that

$$f(X) = \prod_{(k,p)=1} (X-\theta^k).$$

From the definition of $f(X)$ it follows that $f(1)=p$ so we find

$$(1) \qquad p = \prod_{(k,p)=1} (1-\theta^k) = (1-\theta)^{\phi(q)}(\text{unit of } R).$$

Next we compute the norm $\mathrm{N}(1-\theta)$. The field $Q(\theta)$ equals $Q(1-\theta)$ so the minimum polynomial and the characteristic polynomial of $1-\theta$ are the same. Thus $N(1-\theta)$ is the product of the distinct roots of the minimum polynomial. These roots are among the elements $1-\theta^k$, $(k,p)=1$ so $\mathrm{N}(1-\theta)$ divides the product of the $1-\theta^k$. So $\mathrm{N}(1-\theta)=\pm 1$ or $\pm p$. If $\mathrm{N}(1-\theta)=\pm 1$ then $1-\theta$ has an inverse in R and so by (1) p has an inverse in R—impossible. Thus $\mathrm{N}(1-\theta)=\pm p$. This proves $1-\theta$ is a prime element and $(1-\theta)R$ is a prime ideal with relative degree equal to one. Let $\alpha=1-\theta$. Equation (1) implies $pR \subseteq (\alpha R)^{\phi(q)}$. By the general equality $\sum e_i f_i = (Q(\theta):Q)$ we find

$$\phi(q)f = \phi(q) \leqslant (Q(\theta):Q).$$

(We have just seen the relative degree $f=1$.) On the other hand, since θ is a root of $f(X)$ we obtain

$$(Q(\theta):Q) \leqslant \text{degree } f(X) = \phi(q).$$

It follows that $(Q(\theta):Q)=\phi(q)$ and this implies that $f(X)$ is irreducible over Q [otherwise the dimension of $Q(\theta)$ would have to be smaller].

This completes the proof of (a), (b), and (c). The remaining parts require more calculation. It will be convenient to number the roots of $f(X)$ as $\theta = \theta_1, \theta_2, \ldots, \theta_{\phi(q)}$ so that

$$f(X) = \prod (X-\theta_i).$$

Differentiate this by the product rule to obtain

(2) $$f'(\theta_i) = \prod_{j\neq i} (\theta_i - \theta_j).$$

We shall make use of this formula to compute the discriminant

$$\Delta(1, \theta, \ldots, \theta^{\phi(q)-1}) = \Delta.$$

By the results in Section 5 we know

$$\Delta = \prod_{i>j} (\theta_i - \theta_j)^2.$$

This is an integer which divides

(3) $$\prod_{\substack{i\neq j \\ \text{all } i,j}} (\theta_i - \theta_j)^2 = \prod_i f'(\theta_i)^2.$$

Now compute $f'(X)$ from the quotient rule using the definition of $f(X)$ given at the beginning of this section. One obtains

$$f'(\theta_i) = p^a \theta_i^{-1}/(\theta_i^{p^{a-1}} - 1).$$

Use this to evaluate the expansion (3). Observe that $\theta_i^{p^{a-1}}$ is a primitive pth root of unity and as θ_i ranges over the $p^{a-1}(p-1)$ p^ath roots of unity, each pth root of unity will arise p^{a-1} times. If ζ is a primitive pth root of 1 then $\mathrm{N}(1-\zeta) = p$ by the results already obtained. This means that the denominators of the fractions for $f'(\theta_i)^2$ will contribute $p^{2p^{a-1}}$ to the product (3). Also $\mathrm{N}(\theta) = 1$ so it turns out that

$$\prod_{\text{all } i\neq j} (\theta_i - \theta_j) = p^{2(a\phi(q) - p^{a-1})}.$$

We knew this expression was divisible by Δ so we finally obtain

(4) $$\Delta = p^s \qquad \text{for some positive } s.$$

Because Z is a PID, we know R has a free Z-basis $x_1, \ldots, x_{\phi(q)}$. Let $U = |u_{ij}|$ be the matrix with integral entries such that

$$\theta^i = \sum_j u_{ij} x_j.$$

By a (now familiar) matrix calculation one finds

(5) $$\Delta(1, \theta, \ldots, \theta^{d-1}) = (\det U)^2 \Delta(x_1, \ldots, x_{\phi(q)}).$$

The three quantities here are integers and the one on the left is a power of p. It follows that $\Delta(x_1, \ldots, x_{\phi(q)})$ is also a power of p. By Lemma 7.2 one finds

$$\Delta(R/Z) = Z\Delta(x_1, \ldots, x_{\phi(q)}) = (p)^t.$$

So the only ramified prime is p since no other primes divide the discriminant ideal. This proves (d).

Equations (5) and (4) also yield the fact that $\det U$ is a power of p so that U^{-1} is a rational matrix with only powers of p appearing in the denominators. We can express the elements x_j as rational combinations of the θ^i using coefficients with only powers of p in the denominators. Since all the θ^i belong to $Z[\theta]$, there is some positive integer r such that $p^r x_i$ is in $Z[\theta]$ for all i. This implies

$$p^r R \subseteq Z[\theta] \subseteq R. \tag{6}$$

Our goal is to prove $Z[\theta] = R$. The crucial part of the argument is the observation that $R/\alpha R$ is isomorphic to the field $Z/(p)$ of p elements. This follows because we have seen above that αR is prime with relative degree one. Now $Z[\theta]/Z[\theta] \cap \alpha R$ must be isomorphic to a nonzero subring of $R/\alpha R$. The only possibility is that $Z[\theta]/Z[\theta] \cap \alpha R$ also has p elements. Thus every coset of αR in R contains an element in $Z[\theta]$. This means

$$Z[\theta] + \alpha R = R. \tag{7}$$

Multiply by α to get

$$Z[\theta]\alpha + \alpha^2 R = \alpha R.$$

Since $\alpha = 1-\theta$ is in $Z[\theta]$ we may substitute this expression for αR into Eq. (7) to get

$$Z[\theta] + \alpha^2 R = R.$$

Continue this way and by induction one obtains

$$Z[\theta] + \alpha^t R = R, \qquad \text{for all} \quad t \geqslant 1. \tag{8}$$

By Eq. (1) we see $pR = (\alpha R)^{\phi(q)}$ and by expression (6)

$$(pR)^r = (\alpha R)^{r\phi(q)} \subseteq Z[\theta].$$

So in Eq. (8) use $t = r\phi(q)$ to get $\alpha^t R \subseteq Z[\theta]$ and finally $Z[\theta] = R$. This completes the proof of Theorem 9.1.

Exercise. If θ is a primitive p^mth root of unity, then the discriminant $\Delta(1, \theta, \ldots, \theta^{\phi(p^m)-1})$ is $\pm p^c$ with $c = p^{m-1}(mp-m-1)$.

Now we consider the cyclotomic field $Q(\theta)$ with θ a primitive mth root of unity and m not necessarily a prime power.

9.2 Theorem. (a) $(Q(\theta) : Q) = \phi(m)$,

(b) If p is a prime integer which ramifies in $Q(\theta)$ then p divides m,

(c) If $m = p^a m_0$ with p a prime not dividing m_0, then p has ramification number $\phi(p^a)$ in $Q(\theta)$.

Proof. Use induction on m—the theorem being true when m is a prime power. Assume $m = p^a n$ with p a prime not dividing n. Let L_n and L_{p^a} denote, respectively, the fields obtained by adjoining to Q a primitive nth root and a primitive p^ath root of unity. Then $Q(\theta) = L_n L_{p^a}$. We argue next that

$$L_n \cap L_{p^a} = Q.$$

For any subfield K of L_{p^a}, the ramification number of p in K equals $(K:Q)$ because this holds in L_{p^a}. In any subfield of L_n, p has ramification number one because p does not divide n and the induction hypothesis can be applied. Thus Q is the only common subfield of L_n and L_{p^a}. From Galois theory it follows that

$$\begin{aligned}\operatorname{Gal}(Q(\theta)/Q) &= \operatorname{Gal}(Q(\theta)/L_n) \times \operatorname{Gal}(Q(\theta)/L_{p^a})\\ &\cong \operatorname{Gal}(L_{p^a}/Q) \times \operatorname{Gal}(L_n/Q).\end{aligned}$$

By induction, the order of $\operatorname{Gal}(Q(\theta)/Q)$ is $\phi(p^a)\,\phi(n) = \phi(p^a n) = \phi(m)$, which proves (a).

Now let R denote the ring of algebraic integers in L_n, S the algebraic integers in $Q(\theta)$ and ε a primitive p^ath root of unity. Then

$$R[\varepsilon] \subseteq S$$

and the discriminant ideal $\Delta(S/R)$ contains the discriminant $\Delta(1, \varepsilon, \ldots, \varepsilon^{\phi(p^a)-1})$.

Observe that for $x \in L_{p^a}$ we have

$$\begin{aligned}T_{Q(\theta)/L_n}(x) &= \sum_{\sigma \in \operatorname{Gal}(Q(\theta)/L_n)} \sigma(x) = \sum_{\sigma \in \operatorname{Gal}(L_{p^a}/Q)} \sigma(x)\\ &= T_{L_{p^a}/Q}(x).\end{aligned}$$

From the computations in the proof of Theorem 9.1 we see

$$\Delta(1, \varepsilon, \ldots, \varepsilon^{\phi(p^a)-1}) = \text{power of } (p).$$

Hence $\Delta(S/R) \supseteq$ power of pR. The only primes of R which can ramify in S are the divisors of pR. The transitivity of the Galois group implies every prime divisor of pR ramifies in S if any one of them does. Since p has ramification number $\phi(p^a)$ in L_{p^a}, a prime $\mathfrak{P}$ dividing pR ramifies in S with ramification number of $\mathfrak{P}$ cannot exceed the dimension $(Q(\theta):L_n) = \phi(p^a)$. Hence the ramification number of $\mathfrak{P}$ and p in S is exactly $\phi(p^a)$. This proves both (b) and (c).

EXERCISE 1. If m is an odd integer then $Q(\varepsilon_{2m}) = Q(\varepsilon_m)$ so 2 does not ramify in $Q(\varepsilon_{2m})$ even though 2 divides $2m$. Show this is the only exception to the assertion "p ramifies in $Q(\varepsilon_n)$ whenever p is a prime dividing n."

The following exercises are given to show how Theorem 7.6 can be used to obtain the factorization of prime ideals of Z when extended to the ring of algebraic integers in a cyclotomic field. Let m be a fixed positive integer and θ a primitive mth root of unity.

EXERCISE 2. Show $Z[\theta]$ is the full ring of algebraic integers in $Q(\theta)$. {Using the notation of the proof of Theorem 9.2 it is only necessary to show $R[\varepsilon] = S$. This can be done by slightly modifying the proof of Theorem 9.1 (e).}

EXERCISE 3. Let q be a prime integer not dividing m and let $\mathfrak{q}$ be a prime ideal in $Z[\theta]$ containing q. If $\theta^k - 1$ is in $\mathfrak{q}$ then $\theta^k - 1 = 0$. Conclude that $\bar{\theta}$ in $Z[\theta]/\mathfrak{q}$ is still a primitive mth root of unity.

EXERCISE 4. Let $\Phi_m(x)$ be the minimum (monic) polynomial of θ over Q. Then $\Phi_m(x)$ has integer coefficients and we let $\overline{\Phi}_m(x)$ denote the polynomial after reduction mod q, q as in Exercise 3. Show the splitting field of $\overline{\Phi}_m$ over $\mathrm{GF}(q) = Z/q$ is the field $\mathrm{GF}(q^r)$ where r is the least positive integer for which $\mathrm{GF}(q^r)$ contains a primitive mth root of unity. This is the least r such that m divides $q^r - 1$. Conclude every prime factor of $\overline{\Phi}_m$ has degree r.

EXERCISE 5. If q is an integral prime not dividing m then the ideal generated by q in the ring of algebraic integers in $Q(\theta)$ has the factorization $(q) = \mathfrak{P}_1 \cdots \mathfrak{P}_g$, where the $\mathfrak{P}_i$ are distinct primes, $gr = \phi(m)$ and r is the least positive integer such that m divides $q^r - 1$.

EXERCISE 6. If $m = p^a n$ with p a prime not dividing n then the factorization of p in $Z[\theta]$ has the form

$$(p) = (\mathfrak{P}_1 \cdots \mathfrak{P}_g)^{\phi(p^a)}$$

where $gr = \phi(n)$ and r is the least positive integer such that n divides $p^r - 1$.

EXERCISE 7. (Galois groups of cyclotomic fields). Let G_m denote the Galois group of $Q(\theta)/Q$.

(a) If m has the factorization $m = p_1^{a_1} \cdots p_t^{a_t}$, then

$$G_m = G_{p_1^{a_1}} \times \cdots \times G_{p_t^{a_t}}.$$

(b) For p an odd prime G_{p^a} is cyclic of order $(p-1)p^{a-1}$.

Procedure. G_{p^a} is isomorphic to the multiplicative group of units in Z/p^a and is also (by Sylow decomposition) isomorphic to a direct product of a group of order $(p-1)$ with a group of order p^{a-1}. The group of order $p-1$ is isomorphic to the multiplicative group in Z/p so it is cyclic. The element corresponding to $1+p$ in Z/p^a has multiplicative order p^{a-1} so G_{p^a} is the direct product of two cyclic groups of relatively prime orders.

(c) The group G_{2^a} (for $2^a \geqslant 8$) is the direct product of a group of order 2 and a cyclic group of order 2^{a-2}.

Procedure. Show the group of units in $Z/2^a$ is generated as a direct product by the images of -1 and 5.

The groups, G_2 and G_{2^2} have orders 1 and 2, respectively.

Quadratic Reciprocity

Now we turn to some special questions. Assume p is an odd prime and θ a primitive pth root of unity. The Galois group of $Q(\theta)$ over Q is cyclic of order $p-1$. This number is even so there is precisely one subgroup of index two. Accordingly there is a unique quadratic extension of Q contained in $Q(\theta)$. We can describe this quadratic subfield by studying the ramification of primes.

9.3 Theorem. When p is an odd prime, the cyclotomic field of pth roots of unity contains exactly one quadratic subfield over Q and it is $Q([\varepsilon(p)p]^{1/2})$ where $\varepsilon(p) = (-1)^{(p-1)/2}$.

PROOF. We have indicated above why the quadratic subfield is unique. Suppose $Q(\sqrt{d}) \subseteq Q(\theta)$, d square free. Any prime integer q which ramifies in $Q(\sqrt{d})$ also ramifies in $Q(\theta)$. Since p is the only prime which ramifies in $Q(\theta)$ it follows that p is the only prime divisor of the discriminant of $Q(\sqrt{d})$. This discriminant is either d or $4d$ depending upon $d \equiv 1 \bmod 4$ or not. Since p is odd the discriminant is not $4d$ and so $d \equiv 1 \bmod 4$ and p is the only prime divisor of d. Thus $d = \pm p$ and the sign is uniquely determined by the congruence modulo 4. An examination of the cases shows $d = \varepsilon(p)p$ as required.

This computation can be made the basis of one of the many proofs of the law of quadratic reciprocity. We shall present the details.

For an odd prime p, let U_p denote the multiplicative group of the field $Z/(p)$. Then U_p is a cyclic group of order $p-1$. The collection of all squares of elements in U_p forms a subgroup, $U_p{}^2$, of index two. Let $\{\pm 1\} = T$ denote the multiplicative group of order two. There is a unique homomorphism of the group U_p onto T which has kernel $U_p{}^2$. This homomorphism will be denoted by $(\cdot/p)$ and its value at u is written (u/p). We call (u/p) the *Legendre symbol*. It is usually convenient to define (a/p) for a in Z to mean the value of $(\cdot/p)$ at the image of a in U_p when $(p, a) = 1$. If p divides a, then (a/p) is not defined.

9.4 Elementary Properties. Let a, b be integers relatively prime to p.

(1) $(ab/p) = (a/p)(b/p)$.
(2) $(a/p) = 1$ if and only if $a \equiv x^2 \bmod p$ for some x in Z.
(3) $(a/p) = 1$ if and only if $X^2 - a$ is reducible modulo p.

PROOFS. (1) Immediate because $(\cdot/p)$ is a homomorphism.
(2) $(a/p) = 1$ if and only if the image of a in U_p falls into $U_p{}^2$.
(3) $X^2 - a$ is reducible if and only if $a \equiv x^2 \bmod p$ for some x in Z.

In view of Property (3) and the exercise at the end of Section 7, one easily proves the following.

9.5 Lemma. Let a be a square free integer. The odd prime p splits as a product of two distinct primes in $Q(\sqrt{a})$ if and only if p and a are relatively prime and $(a/p) = 1$.

The law of quadratic reciprocity gives a relation between (p/q) and (q/p) for distinct odd primes p and q. We shall obtain this law by examining the decomposition of q in the cyclotomic field of pth roots of unity.

Fix the odd prime p; let θ = primitive pth root of unity; $E = Q([\varepsilon(p)p]^{1/2})$ is the unique quadratic subfield over Q contained in $Q(\theta)$ as in Theorem 9.3. Let R = algebraic integers in E, R' = algebraic integers in $Q(\theta)$.

9.6 Lemma. The prime q splits as a product of two distinct primes in R if and only if q splits as a product of an even number of primes in R'.

PROOF. Let $qR = \mathfrak{P}_1 \mathfrak{P}_2$ for distinct primes in R. These prime ideals must be conjugate within the Galois group of E over Q (by Proposition 6.8) and so there is an automorphism σ in the Galois group of $Q(\theta)$ over Q such that $\sigma(\mathfrak{P}_1) = \mathfrak{P}_2$. Now let $\mathfrak{P}_1 R' = \mathfrak{S}_1 \cdots \mathfrak{S}_k$ with $\mathfrak{S}_i$ primes in R'. It follows that

$$qR' = \mathfrak{P}_1 \mathfrak{P}_2 R' = \mathfrak{S}_1 \cdots \mathfrak{S}_k \sigma(\mathfrak{S}_1) \cdots \sigma(\mathfrak{S}_k)$$

and these must be distinct primes because q is not ramified. Thus q has an even number of factors in R'.

Conversely suppose $qR' = \mathfrak{S}_1 \cdots \mathfrak{S}_{2k}$, $\mathfrak{S}_i$ distinct primes in R'. Let G denote the Galois group of $Q(\theta)$ over Q and H the subgroup of elements σ for which $\sigma(\mathfrak{S}_1) = \mathfrak{S}_1$. Then $|G : H| = 2k$. Let G_1 be the subgroup of G fixing E elementwise so that $|G : G_1| = (E : Q) = 2$. Since G is cyclic, there can be only one subgroup with index 2. Since $|G : H|$ is even, it follows that $H \subseteq G_1$ and $|G_1 : H| = k$. Let $\mathfrak{P} = \mathfrak{S}_1 \cap R$. Then $\mathfrak{P}R'$ is divisible by $\mathfrak{S}_1$ and moreover for σ in G_1 we see $\sigma(\mathfrak{P}R') = \mathfrak{P}R'$ so $\sigma(\mathfrak{S}_1)$ also divides $\mathfrak{P}R'$. This accounts for exactly k distinct primes of R' in the factorization of $\mathfrak{P}R'$. Since G_1 = Galois group of $Q(\theta)$ over E, G_1 is transitive on the primes of R' which divide $\mathfrak{P}R'$. Thus $\mathfrak{P}R'$ has exactly k prime divisors. Now qR' has $2k$ prime divisors so $qR = \mathfrak{P}$ is impossible. The only alternative is $qR = \mathfrak{P}_1 \mathfrak{P}_2$ for some pair of primes in R.

In the proof of the next lemma it is necessary to know the relative degree of a prime divisor of q in R' over q. This has been computed in the exercises just above. An alternate method based on the Frobenius automorphism is described in the exercises following Section 3 of Chapter III. The relative degree of a prime $\mathfrak{S}$ in R' dividing q is the least integer f such that $q^f \equiv 1 \bmod p$.

9.7 Lemma. The prime $q \neq p$ splits as a product of two primes in $Q([\varepsilon(p)p]^{1/2})$ if and only if $(q/p) = 1$.

PROOF. Let $qR' = \mathfrak{S}_1 \cdots \mathfrak{S}_g$ be the factorization of q in R'. Then qR has two prime factors in R if and only if g is even (Lemma 9.6). By Proposition

6.8 we have $efg = fg = p-1 = (Q(\theta):Q)$. Thus g is even if and only if f divides $(p-1)/2$. Because of the characterization of the relative degree mentioned just above, this holds if and only if

$$q^{(p-1)/2} \equiv 1 \bmod p.$$

Now in the cyclic group U_p, an element has order dividing $(p-1)/2$ if and only if that element lies in $U_p{}^2$. Then q has two prime factors in R if and only if $(q/p) = 1$.

One last computation before we reach our goal.

9.8 Lemma. $(-1/p) = (-1)^{(p-1)/2}$ for any odd prime p.

PROOF. $(-1/p) = 1$ if and only if $-1 = u^2$ for some u in U_p. Necessarily this element u has order 4 so 4 divides $p-1$. Conversely if 4 divides $p-1$ then there is an element u of order 4 and $u^2 = -1$ because -1 is the only element in U_p with order 2. Thus $(-1/p) = 1$ if and only if 4 divides $(p-1)$ which is equivalent to $(-1)^{(p-1)/2} = 1$.

9.9 Proposition. Let p and q be distinct odd primes. Then

$$(p/q)(q/p) = (-1)^{\frac{p-1}{2}\frac{q-1}{2}}.$$

PROOF. $(q/p) = 1$ if and only if q splits into two factors in $Q([\varepsilon(p)p]^{1/2})$ by (9.7) and this holds if and only if

$$(\varepsilon(p)p/q) = 1$$

by Lemma 9.5. It follows that

$$(q/p) = (\varepsilon(p)p/q) = (\varepsilon(p)/q)(p/q) = (-1/q)^{\frac{p-1}{2}}(p/q)$$
$$= (p/q)(-1)^{\frac{p-1}{2}\frac{q-1}{2}}.$$

For completeness in this matter we shall also evaluate $(2/p)$. The preceding arguments still apply to obtain the following.

9.10 Lemma. $(2/p) = 1$ if and only if $2R'$ has an even number of prime divisors in R'. This holds if and only if $2R$ has two distinct prime divisors in $Q([\varepsilon(p)p]^{1/2})$.

Now we are unable to proceed as in the odd case because the factorization of $2R$ in E is not determined by the polynomial $X^2-\varepsilon(p)p$. Instead we have $R = Z[w]$ with $2w = 1+(\varepsilon(p)p)^{1/2}$. (For odd q we see $R_{(q)} = Z_{(q)}[\varepsilon(p)p]^{1/2}$ so the factorization of q was determined by $X^2-\varepsilon(p)p$.) The minimum polynomial of w is

$$g(X) = X^2 - X + \frac{1-\varepsilon(p)p}{4},$$

and so by Theorem 7.6, $2R$ has two prime divisors if and only if $g(X)$ is reducible modulo 2. This occurs if and only if $(1-\varepsilon(p)p)/4$ is even. This is equivalent to $\varepsilon(p)p = 8m+1$ for some m. By an examination of the cases it follows that this is equivalent to the assertion $(p^2-1)/8$ is even. Thus $(2/p)=1$ if and only if $(p^2-1)/8$ is even. It follows then

$$(2/p) = (-1)^{(p^2-1)/8}.$$

SUMMARY. Let p and q be distinct odd primes:

(1) $(-1/p) = (-1)^{(p-1)/2}$
(2) $(2/p) = (-1)^{(p^2-1)/8}$
(3) $(q/p) = (p/q)(-1)^{\frac{p-1}{2}\frac{q-1}{2}}$

EXERCISE 1. Let m be an odd positive integer and θ_m a primitive mth root of unity. Describe the quadratic subfields of $Q(\theta_m)$.

EXERCISE 2. Let $m = 2^a \geqslant 8$. Show $Q(\theta_m)$ has exactly three quadratic subfields, $Q(\sqrt{-1})$, $Q(\sqrt{2})$ and $Q(\sqrt{-2})$.

EXERCISE 3. Let d be a square free integer. Show $Q(\sqrt{d})$ is contained in $Q(\theta_m)$ for some primitive mth root of unity. Also determine the least m that will do for a given d. (This is a special case of the theorem of Kronecker–Weber which is proved in Chapter V.)

10. LATTICES IN REAL VECTOR SPACES

In this section $\mathscr{R}$ denotes the real field and Z the ring of integers. Let V denote an n-dimensional $\mathscr{R}$ vector space.

Definition. If $v_1, \ldots, v_r$ are linearly independent vectors in V, the abelian group $Zv_1 + \cdots + Zv_r = \mathscr{L}$ is called an *r-dimensional lattice* in V.

In case $r = n$, we then say $\mathscr{L}$ is a *full lattice* in V.

We shall refer to $v_1, \ldots, v_n$ as a basis of the (full) lattice $\mathscr{L}$. Of course a lattice may have many different bases but any two of them compare in a nice way. That is, a second basis can be carried into the first by a matrix with integer coefficients and determinant equal to ± 1.

Let $\mathscr{L}$ be a full lattice with basis $v_1, \ldots, v_n$. The set

$$T = \{r_1 v_1 + \cdots + r_n v_n \mid r_i \in \mathscr{R},\ 0 \leqslant r_i < 1\}$$

is called a *fundamental parallelopiped* of $\mathscr{L}$. Of course T depends upon the choice of basis for $\mathscr{L}$.

10.1 Lemma. The translates $\lambda+T$, $\lambda \in \mathscr{L}$, cover all of V and they are pairwise disjoint.

PROOF. For any $v = \sum s_i v_i$ with $s_i \in \mathscr{R}$ we write $s_i = n_i + r_i$ with n_i an integer and $0 \leqslant r_i < 1$. Then

$$v = \sum n_i v_i + \sum r_i v_i$$

expresses v as an element in $\lambda+T$ with $\lambda \in \mathscr{L}$.

Now suppose λ_1+T and λ_2+T have a common point for $\lambda_1, \lambda_2 \in \mathscr{L}$. Then T and $(\lambda_2-\lambda_1)+T$ have a common point. Examine the coefficients of the v_i in $\lambda_2-\lambda_1$ to get $\lambda_2 = \lambda_1$.

Definition. A sphere of radius m in V is a set

$$U(m) = \{r_1 v_1 + \cdots + r_n v_n | r_1{}^2 + \cdots + r_n{}^2 \leqslant m^2\}.$$

This depends upon the particular basis.

We can now describe a criterion that applies to subgroups of V to determine whether or not the subgroup is a lattice.

10.2 Theorem. An additive subgroup $\mathscr{L}$ of V is a lattice if and only if every sphere contains only a finite number of points of $\mathscr{L}$.

PROOF. Suppose $\mathscr{L}$ is a lattice with basis $v_1, \ldots, v_r$. Extend this set (if $r \neq n$) to a basis $v_1, \ldots, v_n$ of V. Any sphere of radius m' with respect to some basis is contained in a sphere of radius m with respect to $v_1, \ldots, v_n$, for some m. Now if $\sum n_i v_i$ is in $\mathscr{L}$ and in $U(m)$ then $|n_i| \leqslant m$ so there exist at most a finite number of points in $\mathscr{L}$ and $U(m)$.

Now suppose $\mathscr{L}$ is an additive subgroup of V with only finitely many points in any sphere. Use induction on n. Suppose $V = \mathscr{R}v_1$ has dimension one. Let rv_1 be an element of $\mathscr{L}$ with r positive but as small as possible. Such an r exists because there are only a finite number of points in $\{sv_1 | s^2 \leqslant m^2\}$ for any m. Let $v = rv_1$. Then $\mathscr{L}$ contains Zv and in fact $\mathscr{L} = Zv$. For we may select any sv in $\mathscr{L}$ and write $sv = nv + rv$ with n in Z and $0 \leqslant r < 1$. By choice of v we see $r = 0$.

Now suppose $n > 1$. Because of the induction, we may suppose $\mathscr{L}$ is not contained in any proper subspace of V. Select a basis $v_1, \ldots, v_n$ of V in $\mathscr{L}$ and let V_0 be the subspace of V with basis $v_1, \ldots, v_{n-1}$. By induction, $\mathscr{L}_0 = \mathscr{L} \cap V_0$ is a lattice of rank $n-1$. Let $u_1, \ldots, u_{n-1}$ be a basis of $\mathscr{L}_0$. Now any element of $\mathscr{L}$ can be expressed as

$$\lambda = \sum_1^{n-1} r_i u_i + r_n v_n.$$

If $r_n = 0$ then the r_i are integers. There are only a finite number of λ having the r_i bounded so it is possible to select some λ, having $r_n > 0$ and minimal subject to $|r_i| < 1$, for $i \neq n$. Let u_n denote this particular λ. Since $r_n \neq 0$, $u_1, \ldots, u_n$ is a basis for V. Now for any element λ' in $\mathscr{L}$ we may write

$$\lambda' = \sum a_j u_j, \qquad a_j \in \mathscr{R}.$$

From this subtract an integer multiple of u_n to insure either $a_n = 0$ (in which case $\lambda' \in \mathscr{L}_0$) or $0 < |a_n| < 1$. Then subtract a suitable element of $\mathscr{L}_0$ to insure $|a_i| < 1$ for all $i \neq n$. Now in this case return to the expression involving v_n to see that the coefficient of v_n in λ' is $a_n r_n$ which is smaller in absolute value than r_n. This is against the choice of u_n (and r_n) so in fact this case does not occur. In the expression for λ', then, a_n must be an integer. It follows that

$$\mathscr{L} \subseteq \mathscr{L}_0 + Zu_n \subseteq \mathscr{L}.$$

This completes the proof.

It will be necessary to consider volumes of certain sets in V. We shall consider only "nice" sets for which there will be no question about the existence of volume.

The next theorem gives a means of testing whether or not certain sets X will contain a nonzero point of some given lattice.

10.3 Theorem. (Minkowski). Let $\mathscr{L}$ be a full lattice in V and let Δ denote the volume of a fundamental parallelopiped of $\mathscr{L}$. Let X be a set in V which contains the point $(x_1 - x_2)/2$ whenever x_1, x_2 are in X. If $\operatorname{vol}(X) > 2^n \Delta$, then X contains a nonzero point in $\mathscr{L}$.

PROOF. Let T denote a fundamental parallelopiped of $\mathscr{L}$. We begin by proving the following assertion:

10.4 Assertion. If Y is a bounded subset of V such that the translates $\lambda + Y$, $\lambda \in \mathscr{L}$ are pairwise disjoint, then $\operatorname{vol}(T) \geqslant \operatorname{vol}(Y)$.

To prove this first observe that there can be only a finite number of λ in $\mathscr{L}$ such that $(\lambda + T) \cap Y$ is nonempty. This follows from Theorem 10.2 and the fact that Y is contained in some sphere. By Lemma 10.1 the intersections $(\lambda + T) \cap Y$ are pairwise disjoint and cover Y. Thus

$$\operatorname{vol}(Y) = \sum_{\lambda \in \mathscr{L}} \operatorname{vol}((\lambda + T) \cap Y).$$

It is easy to check that

$$(\lambda + T) \cap Y = [T \cap (Y - \lambda)] + \lambda,$$

and since volume is not changed by translation we find

$$\operatorname{vol}((\lambda + T) \cap Y) = \operatorname{vol}(T \cap (Y - \lambda)).$$

The translates of Y are disjoint so the sets $T \cap (Y - \lambda)$ are disjoint for $\lambda \in \mathscr{L}$.

These sets may not cover T so we obtain

$$\mathrm{vol}(T) \geqslant \sum_{\lambda} \mathrm{vol}(T \cap (Y-\lambda)) = \mathrm{vol}(Y)$$

as required for Assertion 10.4.

Now return to the set X. Consider the set

$$\tfrac{1}{2}X = \{\tfrac{1}{2}x \mid x \in X\}.$$

We have $\mathrm{vol}(\frac{1}{2}X) = 2^{-n}\,\mathrm{vol}(X) > \mathrm{vol}(T) = \Delta$. Thus the translates of $\frac{1}{2}X$ by elements in $\mathscr{L}$ must not be pairwise disjoint in view of Assertion 10.4. There exists $\lambda_1 \neq \lambda_2$ in $\mathscr{L}$ such that

$$\tfrac{1}{2}x + \lambda_1 = \tfrac{1}{2}y + \lambda_2, \qquad x, y \in X.$$

Then $\frac{1}{2}(x-y) = \lambda_2 - \lambda_1$ is in X and in $\mathscr{L}$ and is nonzero.

The Minkowski theorem will be applied to prove that certain lattices must contain points satisfying various conditions. We will select various sets X having sufficiently large volume to force these conditions. We shall describe now two types of sets of X that will appear later.

Let us fix a coordinate system in V so that points are represented by the usual n-tuple of real numbers. Write $n = r+2s$ for some nonnegative integers r, s and let $c_1, \ldots, c_{r+s}$ be positive real numbers. Consider the set

(10.5)
$$X = \{(x_1, \ldots, x_r, y_1, z_1, \ldots, y_s, z_s) \mid |x_i| < c_i,\ 1 \leqslant i \leqslant r,\ y_j^2 + z_j^2 < c_{r+j},\ 1 \leqslant j \leqslant s\}.$$

This set X satisfies the condition, $\frac{1}{2}(x-y)$ is in X whenever x, y are in X. The volume is easily computed since X is just a product of intervals of length $2c_i$ with two-dimensional spheres of radius $(c_{r+j})^{1/2}$. It turns out that

(10.6)
$$\mathrm{vol}(X) = 2^r \pi^s (c_1 c_2 \cdots c_{r+s}).$$

Another set that will be useful is the set

(10.7)
$$X_t = \left\{(x_1, \ldots, x_r, y_1, z_1, \ldots, y_s, z_s) : \sum |x_i| + 2\sum (y_j^2 + z_j^2)^{1/2} < t\right\}.$$

Here t can be any positive real number.

It follows (by the triangle inequality) that $\frac{1}{2}(x-y)$ is in X whenever x, y are in X. The volume of X is somewhat more difficult to compute in this case. The reader may see the calculation in Artin [1] or Lang [7]. We shall simply state the result.

(10.8)
$$\mathrm{vol}(X_t) = 2^r (\pi/2)^s t^n / n!.$$

The sets described here will play a role in the next section.

11. THE UNIT THEOREM AND FINITENESS OF THE CLASS NUMBER

In this section let K denote an *algebraic number field*, that is, a finite-dimensional extension of the rational field Q, and R the ring of algebraic integers in K. Our object is to determine the group of units in R and prove the class group of R, $\mathbf{C}(R)$ defined in Section 4, is a finite group. The proofs of the two results are similar in spirit and so they are both discussed here.

We must fix the notation very carefully. Let E be a normal extension of Q which contains K, G is the Galois group of E over Q, and H the subgroup of G which leaves K fixed elementwise. We shall regard E as a subfield of the field of complex numbers which we denote by $\mathscr{C}$.

Select representatives $\sigma_1, \ldots, \sigma_n$ of the distinct cosets of H in G. Then $n = (K:Q)$ and the σ_i are all of the possible imbeddings of K into $\mathscr{C}$. Some of the fields $\sigma_i(K)$ may actually lie in the field of reals $\mathscr{R}$. Select the numbering so that $\sigma_1, \ldots, \sigma_r$ map K into $\mathscr{R}$. (We allow $r = 0$ in case there are no imbeddings of K into $\mathscr{R}$.) Of the remaining σs we can assert that no one of them is equal to its complex conjugate. [The conjugate of σ is the map $\bar{\sigma}$ which sends x to $\overline{\sigma(x)}$ = complex conjugate of $\sigma(x)$.] Hence there must be an even number of σs remaining. Number them so that

$$\sigma_{r+1}, \ldots, \sigma_{r+s}, \bar{\sigma}_{r+1}, \ldots, \bar{\sigma}_{r+s}$$

are all of the remaining imbeddings of K into $\mathscr{C}$. Then we have $n = r + 2s = (K:Q)$.

Consider the function v defined on K by

$$v(x) = (\sigma_1(x), \ldots, \sigma_r(x), \ldots, \sigma_{r+s}(x)). \tag{11.1}$$

The values of v lie in the space of $(r+s)$-tuples having the first r coordinates real and the last s coordinate complex. There is a natural identification of $\mathscr{C}$ as a two-dimensional vector space over $\mathscr{R}$ and so we may regard $v(x)$ as an element of an $(r+2s)$-dimensional space over $\mathscr{R}$. Let $V = \mathscr{R}^r \times \mathscr{C}^s$ denote this space.

11.2 Lemma. The map v of K into V is an additive monomorphism.

PROOF. Obvious.

The importance of this procedure can now be illustrated.

11.3 Theorem. Let $\mathfrak{A}$ be a nonzero ideal in R = algebraic integers of K. Then $v(\mathfrak{A})$ is a full lattice in V.

PROOF. Since Z is a PID there exists a free basis $a_1, \ldots, a_n$ of $\mathfrak{A}$ over Z. Necessarily these give a Q-basis for K. The image $v(\mathfrak{A})$ has the elements $v(a_i)$ as a free basis over Z so the result will be proved if the elements $v(a_1), \ldots, v(a_n)$

are linearly independent over $\mathscr{R}$. This in turn will be the case if and only if the matrix with row i equal to $v(a_i)$ has a nonzero determinant. Let M denote this matrix with i,j entry equal to the j entry of $v(a_i)$. We shall evaluate $\det(M)$ by relating to a matrix that has appeared previously.

Let D denote the matrix whose ith row is

$$(\sigma_1(a_i), \ldots, \sigma_r(a_i), \sigma_{r+1}(a_i), \bar{\sigma}_{r+1}(a_i), \\ \ldots, \sigma_{r+s}(a_i), \bar{\sigma}_{r+s}(a_i)).$$

Compute the product DD^t by using Corollary 5.3 to obtain

$$DD^t = |\mathrm{T}(a_i a_j)|,$$

where T is the trace map from K to Q. Thus we know

$$(1) \qquad \Delta(a_1, \ldots, a_n) = (\det D)^2 \neq 0$$

because the discriminant of any basis of a separable extension is nonzero (see Section 5).

Now relate M and D. Column $r+1$ of row i in M has entry $\mathrm{Re}(\sigma_{r+1}(a_i)) =$ real part of $\sigma_{r+1}(a_i)$. Column $r+2$ of row i in M has entry $\mathrm{Im}(\sigma_{r+1}(a_i)) =$ imaginary part of $\sigma_{r+1}(a_i)$.

In matrix D add column $r+2$ to column $r+1$ to obtain in row i the entry $2\,\mathrm{Re}(\sigma_{r+1}(a_i))$. Now subtract one-half of column $r+1$ from column $r+2$ to obtain in row i the entry $-i\,\mathrm{Im}(\sigma_{r+1}(a_i))$. Repeat this procedure in each pair of conjugate columns of D and one transforms D into a matrix "almost" equal to M. The only difference being a factor of 2 in s of the columns and a factor $-i$ in s of the columns. Hence

$$(2) \qquad \det M = (-2i)^{-s} \det D \neq 0.$$

This proves the result.

We can draw a few corollaries from this computation.

11.4 Corollary. The discriminant $\Delta(a_1, \ldots, a_n)$ of the basis $a_1, \ldots, a_n$ for K over Q is positive if and only if s is even.

Proof. $\Delta = \Delta(a_1, \ldots, a_n) = (\det D)^2 = (\det M)^2(-2i)^{2s}$. Since M has only real entries we see $(\det M)^2 > 0$. Thus Δ has the same sign as $(i)^{2s} = (-1)^s$. The result follows at once.

It is also possible to compute the volume of the fundamental parallelopiped of $v(A)$. Since the vectors $v(a_1), \ldots, v(a_n)$ give a basis of the lattice $v(A)$, the fundamental parallelopiped is the set

$$T = \{\textstyle\sum r_i v(a_i) \mid 0 \leqslant r_i < 1\}.$$

Recall how the volume of a parallelopiped in n-dimensional space is computed. We form the matrix whose ith row is the vector $v(a_i)$ expressed in terms of the canonical basis. Then the absolute value of the determinant of this matrix is the volume. Hence we have

$$\operatorname{vol}(T) = \pm\det(M).$$

Of course, T depends upon the choice of basis for $\mathscr{L}$. If $a_1', \ldots, a_n'$ is another basis then there is a matrix with integer entries n_{ij} such that

$$a_i = \sum n_{ij} a_j'.$$

Necessarily this transformation is invertible and maps $\mathscr{L}$ onto $\mathscr{L}$. It follows that $\det|n_{ij}| = \pm 1$. This means $\operatorname{vol}(T)$ is not dependent upon the choice of basis for $\mathscr{L}$.

Now combine Eqs. (1) and (2) of the proof of Theorem 11.3 and obtain

$$\operatorname{vol}(T) = 2^{-s}|\Delta|^{1/2},$$

where Δ is the discriminant of (any) Z-basis for $\mathfrak{A}$.

11.5 Corollary. Let $\mathfrak{A}$ be a nonzero ideal of R and Δ the discriminant of some Z-basis for $\mathfrak{A}$. Then the fundamental parallelopiped of $v(\mathfrak{A})$ has volume $2^{-s}|\Delta|^{1/2}$.

It will be useful to have an expression for $\operatorname{vol}(T)$ that depends only upon the discriminant of R rather than the discriminant of $\mathfrak{A}$. To obtain this we do the following.

11.6 Lemma. Let $\mathfrak{A}$ be a nonzero ideal of R. Then

$$\Delta(\mathfrak{A}/Z) = \mathscr{N}(\mathfrak{A})^2\,\Delta(R/Z).$$

PROOF. It is sufficient to prove this equality holds after localization at every prime in Z. Let p be a prime, $S = Z-(p)$ and let $\mathfrak{A}_S = aR_S$ with a in $\mathfrak{A}$. Since R_S is a PID (by Theorem 3.15) such an element a exists. Then select a basis $x_1, \ldots, x_n$ of R over Z and it follows that $ax_1, \ldots, ax_n$ is a Z_S basis of $\mathfrak{A}_S$. Thus $\Delta(\mathfrak{A}_S/Z_S)$ is the principal ideal generated by

$$\Delta(ax_1, \ldots, ax_n) = \det|\mathrm{T}_{L/Q}(ax_i ax_j)|.$$

Let r_a denote the matrix of the map $y \to ya$ with respect to the basis $x_1, \ldots, x_n$. Then

$$|\mathrm{T}_{L/Q}(ax_i ax_j)| = r_a|\mathrm{T}(x_i x_j)|\,r_a^{\,t}.$$

But now $N(a) = \det(r_a)$ so we have

$$\Delta(ax_1, \ldots, ax_n) = \mathrm{N}(a)^2\,\Delta(x_1, \ldots, x_n).$$

$\mathrm{N}(\mathfrak{A}_S) = Z_S \mathrm{N}(a)$ so we obtain

$$\Delta(\mathfrak{A}_S/Z_S) = \mathcal{N}(\mathfrak{A}_S)^2 \Delta(R_S/Z_S).$$

After applying Lemma 7.1 the result follows.

11.7 Corollary. If $\mathfrak{A}$ is a nonzero ideal of R and Δ_R the discriminant of R over Z, then the volume of the fundamental parallelopiped of $v(\mathfrak{A})$ is $2^{-s}\mathcal{N}(\mathfrak{A})|\Delta_R|^{1/2}$.

We are now in a position to prove the main step along the way to proving the finiteness of the class group.

11.8 Theorem. Let $\mathfrak{A}$ be a nonzero ideal in R. There exists an element $a \neq 0$ in $\mathfrak{A}$ such that

$$|\mathrm{N}(a)| \leqslant \frac{n!}{n^n}\left(\frac{4}{\pi}\right)^s \mathcal{N}(\mathfrak{A})|\Delta_R|^{1/2}.$$

PROOF. Consider the set $X_t = \{(x_1, \ldots, x_{r+s})\}$ in which $x_1, \ldots, x_r$ are real, $x_{r+1}, \ldots, x_{r+s}$ are complex and

$$(3) \qquad |x_1| + \cdots + |x_r| + 2|x_{r+1}| + \cdots + 2|x_{r+s}| < t.$$

This is a subset of V whose volume is given by Eq. 10.8. We want to apply Theorem 10.3 (Minkowski's) to obtain a nonzero point in X_t and in $v(A)$. We require $\mathrm{vol}(X_t) > 2^n \mathrm{vol}(T)$, $T =$ fundamental parallelopiped of $v(A)$. This inequality holds if

$$(4) \qquad t^n = \varepsilon + n!(2^{n-r}/\pi^s)\mathcal{N}(\mathfrak{A})|\Delta_R|^{1/2}$$

and ε is any positive real number. Assume that t is this number and then by Theorem 10.3 there exists a in $\mathfrak{A}$ with $a \neq 0$ such that

$$v(a) = (\sigma_1(a), \ldots, \sigma_{r+s}(a)) = (x_1, \ldots, x_{r+s})$$

is in X_t. We can assume in fact that $\varepsilon = 0$ in (4) because there exist only a finite number of points in $v(\mathfrak{A})$ in any sphere, and hence in any X_t. If we consider the sets X_t with t given by Eq. (4) and with ε decreasing to zero, there must be some point $v(a)$ in all of them.

Now with this value of t we estimate $\mathrm{N}(a)$. By the results in Section 5 we have

$$\begin{aligned}|\mathrm{N}(a)| &= \prod_1^{r+s}|\sigma_i(a)|\prod_1^{s}|\bar{\sigma}_{r+j}(a)| \\ &= |\sigma_1(a)|\cdots|\sigma_r(a)||\sigma_{r+1}(a)|^2\cdots|\sigma_{r+s}(a)|^2 \\ &= |x_1|\cdots|x_r||x_{r+1}|^2\cdots|x_{r+s}|^2.\end{aligned}$$

Now use the arithmetic–geometric mean inequality. It yields

$$n^n |x_1| \cdots |x_r| \cdot |x_{r+1}|^2 \cdots |x_{r+s}|^2 \leqslant \{|x_1| + \cdots + |x_r| + 2|x_{r+1}| + \cdots + 2|x_{r+s}|\}^n.$$

Use this equation along with Eqs. (3) and (4) where $\varepsilon = 0$ to get $|\mathrm{N}(a)| \leqslant n^{-n}t^n$. Since $n = r+2s$ the theorem now follows.

Recall that the class group $\mathbf{C}(R)$ is the collection of equivalence classes $[\mathfrak{B}]$ of fractional ideals $\mathfrak{B}$ of R. Two fractional ideals $\mathfrak{B}_1$ and $\mathfrak{B}_2$ are in the same class if and only if there is some $x \neq 0$ in K for which $\mathfrak{B}_1 = x\mathfrak{B}_2$.

11.9 Theorem (The Minkowski bound). Let $[\mathfrak{B}]$ denote a class in $\mathbf{C}(R)$. There exists an ideal $\mathfrak{B}_1$ in $[\mathfrak{B}]$ with $\mathfrak{B}_1 \subseteq R$ and

$$|\mathcal{N}(\mathfrak{B}_1)| \leqslant \frac{n!}{n^n}\left(\frac{4}{\pi}\right)^s |\Delta_R|^{1/2}.$$

PROOF. Let $[\mathfrak{B}^{-1}] = [\mathfrak{A}]$. If $\mathfrak{A}$ is not in R, select an element y in $\mathfrak{A}^{-1}$, $y \neq 0$, and replace $\mathfrak{A}$ by $\mathfrak{A}y$. Then $[\mathfrak{A}] = [\mathfrak{A}y]$ and $\mathfrak{A}y \subseteq R$. So assume $\mathfrak{A} \subseteq R$ to start with. Let M denote the constant on the right of the inequality. According to Theorem 11.8 there is an element $a \neq 0$ in $\mathfrak{A}$ such that

$$|\mathrm{N}(a)\mathrm{N}(\mathfrak{A})^{-1}| \leqslant M.$$

Set $\mathfrak{B}_1 = a\mathfrak{A}^{-1}$ and observe $\mathfrak{B}_1$ is in $[\mathfrak{B}]$ and $\mathfrak{B}_1 \subseteq R$. Then

$$|\mathcal{N}(\mathfrak{B}_1)| = |\mathcal{N}(a\mathfrak{A}^{-1})| = |\mathrm{N}(a)\mathcal{N}(\mathfrak{A})^{-1}| \leqslant M$$

as required.

The importance of this theorem lies in the fact that the bound depends only upon the field K and not upon the ideal class. This result makes it easy to prove the next result.

11.10 Theorem. The class group $\mathbf{C}(R)$ is finite.

PROOF. Every class $[\mathfrak{B}]$ in $\mathbf{C}(R)$ contains an integral ideal $\mathfrak{B}_1$ with $|\mathcal{N}(\mathfrak{B}_1)| \leqslant M$ with M the bound given above. It is sufficient to prove there exist only a finite number of ideals in R having norms bounded by M.

Let $\mathfrak{B} = \mathfrak{P}_1^{a_1} \cdots \mathfrak{P}_t^{a_t}$ be a nonzero ideal, $\mathfrak{P}_i$ are distinct primes and the a_i are positive integers. Let $(p_i) = \mathfrak{P}_i \cap Z$, $p_i =$ prime integer. Then suppose

$$\mathcal{N}(\mathfrak{B}) = \prod p_i^{a_i f_i} \leqslant M$$

with $f_i =$ relative degree of $\mathfrak{P}_i$ (see Corollary 8.5). First, each prime p_i must be less than or equal to M. Thus only a finite number of p_i can arise this way. Next the exponent $a_i f_i$ is bounded and the number of $\mathfrak{P}_j$ that can appear is finite since there exist only a finite number of $\mathfrak{P}_j$ which contain a given p_i. These facts taken together imply there exist only a finite number of ideals $\mathfrak{B}_1$ in R with $\mathcal{N}(\mathfrak{B}_1) \leqslant M$.

This proof that $\mathbf{C}(R)$ is finite is not the shortest proof possible but it is very useful because the bound is good enough to be practical for computation.

EXAMPLE. We return to the example $K = Q(\theta)$ with $\theta^3 = 2$ which was considered at the end of Section 7. There is one real root of $X^3 - 2$ and two complex roots. Hence in this case $r = 1$ and $s = 1$. The discriminant $\Delta = \Delta(R/Z) = -2^2 \cdot 3^3$ and $n = 3$. If one computes the bound in Theorem 11.9 we see that every ideal class $[\mathfrak{B}]$ in $\mathbf{C}(R)$ contains an ideal $\mathfrak{B}_1$ with $\mathcal{N}(\mathfrak{B}_1) < 4$. Thus $\mathcal{N}(\mathfrak{B}_1) = 2$ or 3 and in either case $\mathfrak{B}_1$ must be a prime ideal dividing $2R$ or $3R$. We have already determined the structure of these last two ideals. Namely $2R$ and $3R$ are each the cube of a prime ideal which is principal; $2R = \theta^3 R$ and $3R = (1+\theta)^3 R$. It follows that any ideal $\mathfrak{B}_1$ with $\mathrm{N}(\mathfrak{B}_1) < 4$ is a principal ideal. Hence $\mathbf{C}(R)$ is the group of order one and R is a PID.

We can obtain more information from Theorem 11.9. For any ideal $\mathfrak{B} \neq 0$, the norm $\mathcal{N}(\mathfrak{B})$ is an integer no smaller than one. It follows that

$$|\Delta_R|^{1/2} \geqslant \frac{n^n}{n!}\left(\frac{\pi}{4}\right)^s \geqslant \frac{n^n}{n!}\left(\frac{\pi}{4}\right)^{n/2}.$$

Let a_n denote the right-most expression. We find

$$a_{n+1}/a_n = (\pi/4)^{1/2}(1+1/n)^n.$$

This number is always >1 so $a_{n+1} > a_n$. Since $a_2 > 1$ we see $|\Delta_R| > 1$. This proves the following statement.

11.11 Corollary (Minkowski). Let R be the ring of algebraic integers in an algebraic number field $\neq Q$. Then $\Delta(R/Z) \neq Z$. In particular some prime in Z must ramify in R.

Next we apply the ideas above to solve another problem. Let $\mathbf{U}$ denote the multiplicative group of units in the ring R of algebraic integers in K. (K is still an algebraic number field.) We shall apply the Minkowski lemma to determine the structure of $\mathbf{U}$.

Let a be a nonzero element of K. Define the function $\ell(a)$ by

$$\ell(a) = (\ln|\sigma_1(a)|, \ldots, \ln|\sigma_r(a)|, 2\ln|\sigma_{r+s}(a)|, \ldots, 2\ln|\sigma_{r+s}(a)|).$$

For convenience let $\ell_i(a) = \ln|\sigma_i(a)|$ if $1 \leqslant i \leqslant r$ and $\ell_i(a) = 2\ln|\sigma_i(a)|$ if $r < i \leqslant r+s$.

Then in view of Theorem 5.3 we find

$$\ln|\mathrm{N}(a)| = \sum \ell_i(a). \tag{11.12}$$

Furthermore $\ell(ab) = \ell(a) + \ell(b)$.

11.13 Proposition. The function ℓ maps the units **U** into an $(r+s-1)$-dimensional lattice in the vector space V_0 of dimension $r+s$ over $\mathscr{R}$.

PROOF. Let V_0 denote the space of $(r+s)$-tuples over the reals so $\ell(\mathbf{U}) \subseteq V_0$ Any element $u \in \mathbf{U}$ must have norm $\mathrm{N}(u) = \pm 1$ since $\mathrm{N}(u)\,\mathrm{N}(u)^{-1} = 1$. Thus (11.12) implies $\sum \ell_i(u) = 0$. This means $\ell(\mathbf{U})$ lies in the hyperplane

$$\{(x_1, \ldots, x_{r+s}) \in V \mid \sum x_i = 0\}.$$

The image $\ell(\mathbf{U})$ is an abelian group (in fact ℓ is a group homomorphism) so $\ell(\mathbf{U})$ will be a lattice if every sphere in V contains only a finite number of points in $\ell(\mathbf{U})$. It is certainly sufficient to show that any "cube" contains only a finite number of points in $\ell(\mathbf{U})$. Let m be a positive constant and consider the set $\mathbf{U}_m$ of all elements $u \in \mathbf{U}$ for which $|\ell_i(u)| \leqslant m$. That is $\ell(u)$ is in the cube with side m and center at the origin. Then $\delta_i \ln|\sigma_i(u)| \leqslant m$ implies

$$|\sigma_i(u)| \leqslant e^{m/\delta_i}, \qquad \delta_i = 1 \quad \text{or} \quad 2.$$

It follows that the set $\mathbf{U}_m$ is mapped by the function v, defined in Eq. 11.1, to a bounded subset of $v(R)$ in m-dimensional space. This means there can be only a finite number of points in $v(\mathbf{U}_m)$. But v is a one-to-one function, so $\mathbf{U}_m$ is a finite set and that proves the proposition.

The main goal is to prove that $\ell(\mathbf{U})$ is actually $r+s-1$ dimensional (rather than just contained in a lattice of this dimension).

We first work with the lattice $v(R)$ in the n-dimensional real space $V = \mathscr{R}^r \times \mathscr{C}^s$. Let $y = (y_1, \ldots, y_{r+s})$ denote a vector in V and Y the linear transformation

$$Y(x_1, \ldots, x_{r+s}) = (x_1 y_1, \ldots, x_{r+s} y_{r+s}).$$

The matrix for Y is almost diagonal, namely,

$$Y = \operatorname{diag}\{y_1, \ldots, y_r, y_{r+1}^*, \ldots, y_{r+s}^*\},$$

where y_{r+j}^* is a 2×2 matrix which represents the linear transformation on the complex field determined by multiplication with y_{r+j}. If $y_{r+j} = a + ib$, with a, b real, then

$$y_{r+j}^* = \begin{vmatrix} a & b \\ -b & a \end{vmatrix}.$$

It follows easily that

$$|\det Y| = |y_1| \cdots |y_r| |y_{r+1}|^2 \cdots |y_{r+s}|^2.$$

Select Y so that

$$\det Y = 1. \tag{11.14}$$

Then Y is a volume preserving linear transformation on V. In particular the lattices $v(R)$ and $Yv(R)$ have fundamental parallelopipeds with the same volume. This volume is

$$\operatorname{vol}(Yv(R)) = 2^{-s}|\Delta|^{1/2},$$

where $\Delta =$ discriminant of R over Z.

Let $c_1, \ldots, c_{r+s}$ be positive constants. Let $X = \{(x_1, \ldots, x_{r+s}) \in V\}$ which satisfy $|x_i| < c_i$ when $1 \leqslant i \leqslant r$ and $|x_{r+j}|^2 < c_{r+j}$ when $1 \leqslant j \leqslant s$.

This is a set such as that described in Eq. 10.5. We have

$$\operatorname{vol}(X) = 2^r \pi^s c_1 \cdots c_{r+s}.$$

The object is to apply Minkowski's Theorem 10.3 to obtain certain points in $Yv(R)$. For this it is necessary that $\operatorname{vol}(X)$ is large enough. We require

$$\operatorname{vol}(X) > 2^n 2^{-s}|\Delta|^{1/2}.$$

Since $Yv(R)$ has only a finite number of points in any such X it follows that $Yv(R)$ and X have a common nonzero point even if

$$\operatorname{vol}(X) = 2^n 2^{-s}|\Delta|^{1/2}.$$

So we assume this holds for the choice of constants $c_1, \ldots, c_{r+s}$. Suppose $0 \neq a \in R$ and $Yv(a) \in X$. Then

$$Yv(a) = (\sigma_1(a)y_1, \ldots, \sigma_{r+s}(a)y_{r+s})$$

and

$$|\sigma_i(a)y_i| < c_i, \qquad i \leqslant r; \tag{11.15}$$

$$|\sigma_{r+j}(a)y_{r+j}|^2 < c_{r+j}.$$

In view of Eq. 11.14 we obtain

$$|\mathrm{N}(a)| = |\sigma_1(a)| \ldots |\sigma_r(a)||\sigma_{r+1}(a)|^2 \cdots |\sigma_{r+s}(a)|^2 < c_1 \cdots c_{r+s}.$$

Notice the element a depends upon Y but the bound for $|\mathrm{N}(a)|$ does not. It has been seen earlier that only a finite number of ideals of R have norms which lie below some bound. Let $Ra_1, \ldots, Ra_N$ be all the distinct principal ideals with norm $< c_1 \cdots c_{r+s}$. Then Ra must coincide with one of these so there exists a unit u in R and some index k such that $a = ua_k$. It will be necessary to estimate the size of $\sigma_i(u)$ in terms of Y. Note that u depends upon Y since a does. From (11.15) one sees

$$|\sigma_i(a)y_i| = |\sigma_i(u)y_i\sigma_i(a_k)| < c_i$$

$$|\sigma_{r+j}(a)y_{r+j}|^2 = |\sigma_{r+j}(u)y_{r+j}\sigma_{r+j}(a_k)|^2 < c_{r+j}.$$

Let $b_i = \min\{|\sigma_i(a_k)|, k = 1,2,\ldots,N\}$. Then

$$(11.16) \qquad |\sigma_i(u)||y_i| < c_i/b_i, \qquad 1 \leqslant i \leqslant r$$
$$|\sigma_{r+j}(u)||y_{r+j}| < (c_{r+j})^{1/2}/b_{r+j}$$

So far Y is arbitrary except for the restriction (11.14). Now let W be a constant and select the y_i so that

$$|y_2| = \cdots = |y_{r+s}| = W, \qquad |y_1| = 1/W^{r+s-1}.$$

The condition (11.14) still holds. Let u_1 denote the unit corresponding to this choice of Y. The condition (11.16) now reads

$$|\sigma_i(u_1)| < c_i/Wb_i, \qquad i \neq 1, \quad i \leqslant r$$
$$|\sigma_{r+j}(u_1)| < (c_{j+r})^{1/2}/Wb_{r+j}$$
$$|\sigma_1(u_1)| < W^{r+s-1}c_1/b_1.$$

We may select W so large that $|\sigma_t(u_1)| < 1$ for all $t \neq 1$. Having done this we observe $\ell_i(u_1) < 0$ for $i \neq 1$. Moreover $N(u_1) = \pm 1$ because u_1 is a unit in $\boldsymbol{R}$ so by Eq. (11.12) we obtain

$$\ell_1(u_1) = -\sum_{i>1} \ell_i(u_1) > 0.$$

In a similar way one produces units $u_2, \ldots, u_{r+s-1}$ in R which satisfy the following conditions:

$$(11.17) \qquad \text{(a)} \quad \ell_i(u_j) < 0 \qquad \text{if} \quad i \neq j,$$
$$\text{(b)} \quad \sum_{i=1}^{r+s-1} \ell_i(u_j) > 0 \qquad \text{for all} \quad i.$$

The crucial point to notice is that

$$\sum_{i=1}^{r+s} \ell_i(u_j) = 0 \qquad \text{by (11.12)},$$

so by dropping the last term $\ell_{r+s}(u_j)$ (which is negative) we obtain (b).

This last statement produces units in R which are very "large" at the ith conjugate, $\sigma_i(u_i)$, but "small" at all the other conjugates, $\sigma_j(u_i)$. They will be the important units needed to show that the lattice $\ell(\mathbf{U})$ has dimension $r+s-1$.

We first cut down the dimension of the space. Let pr denote the projection map from the $(r+s)$-dimensional space V_0 to an $(r+s-1)$-dimensional space V_1 defined by

$$\operatorname{pr}(x_1, \ldots, x_{r+s-1}, x_{r+s}) = (x_1, \ldots, x_{r+s-1}).$$

Then $\ell(\mathbf{U})$ in V_0 is projected onto a lattice $\mathrm{pr}\,\ell(\mathbf{U}) = \mathscr{L}$.

We shall prove $\mathscr{L}$ has dimension $r+s-1$.

11.18 Proposition. Let $u_1, \ldots, u_{r+s-1}$ be units of R which satisfy (11.17). Then the vectors $\mathrm{pr}\,\ell(u_i)$ are linearly independent over $\mathscr{R}$.

PROOF. To prove this we form the matrix M whose ith row is the vector $\mathrm{pr}\,\ell(u_i)$. The proposition will be proved if we show M is nonsingular. We simplify the notation. $M = |m_{ij}|$ is $(r+s-1)\times(r+s-1)$ and

$$\text{(a)}\quad m_{ij} < 0 \qquad \text{if}\quad i \neq j,$$

$$\text{(b)}\quad \sum_j m_{ij} > 0 \qquad \text{for each}\quad i.$$

If M is singular there exist real numbers x_j not all zero such that

$$\sum_j m_{ij} x_j = 0 \qquad \text{for each}\quad i.$$

Select the index k so that $|x_k| \geqslant |x_i|$ all i and assume $x_k > 0$. (Just multiply all the xs by -1 if $x_k < 0$.) Now we have

$$0 = x_k m_{kk} + \sum_{j\neq k} m_{kj} x_j > x_k m_{kk} + \left(\sum m_{kj}\right) x_k > 0$$

because of (b). This contradiction proves Proposition 11.18.

11.19 Theorem (Dirichlet). The group of units in R is the direct product of a finite cyclic group and a free abelian group of rank $r+s-1$. Equivalently there exists units $u_1, \ldots, u_{r+s-1}$ in R such that every unit u in R can be uniquely expressed as

$$u = w u_1^{a_1} \cdots u_{r+s-1}^{a_{r+s-1}}$$

for some root of unity w and integers a_i.

PROOF. By Proposition 11.13 we know $\ell(\mathbf{U})$ has dimension at most $r+s-1$ and by Proposition 11.18 it has dimension at least $r+s-1$ so there must exist units $u_1, \ldots, u_{r+s-1}$ such that $\ell(\mathbf{U})$ has Z-basis $\ell(u_i)$. For any unit u in $\mathbf{U}$ there exist unique integers such that

$$\ell(u) = \sum a_i \ell(u_i)$$

and so $\ell(u \prod u_i^{-a_i}) = 0$.

The proof will be complete if we prove that any unit w for which $\ell(w) = 0$ is a root of unity.

We have $\ell(w) = 0$ if and only if $|\sigma_i(w)| = 1$ for all i. Thus

$$v(w) = (\sigma_1(w), \ldots, \sigma_{r+s}(w))$$

lies in a bounded subset of the lattice $v(R)$. This means only finitely many w

can arise by Theorem 10.2 and the fact that v is one to one. This means that the kernel of ℓ is a finite subgroup of $\mathbf{U}$ and hence is cyclic. (Every finite subgroup of the multiplicative group of a field is cyclic.)

EXAMPLES. Let $D > 0$ be a square free integer and $K = Q(\sqrt{D})$. Then $2 = n$ and $r = 2$, $s = 0$ so the group of units in the ring of integers of K is the direct product of a finite cyclic group and an infinite cyclic group.

In case $D = 2$, the infinite cyclic group is generated by $1+\sqrt{2}$.

If $K = Q(\sqrt{-D})$ then $s = 1$ and $r = 0$ so the group of units in the ring of integers is a finite cyclic group.

EXERCISE 1. Let D be a positive, square free integer and R the ring of integers in $Q(\sqrt{-D})$. Let U denote the group of units in R. Show U has order two except in the two cases $D = 1$, $D = 3$, where U has order four or six, respectively. (Hint: Use information about cyclotomic fields and their dimensions.)

REMARK. If R is the ring of integers in $Q(\sqrt{D})$ with D a positive integer, then the unit group has the form $\langle \pm\varepsilon^k\rangle$ where ε is a generator of the infinite cyclic part of the unit group. Notice that ε, $-\varepsilon$, $1/\varepsilon$, $1/-\varepsilon$ are also generators of this subgroup. We are dealing here with real numbers and exactly one of the four generators is >1. Suppose $\varepsilon > 1$. Then ε is called the *fundamental unit* of $Q(\sqrt{D})$. For any given D the fundamental unit can be calculated by methods using continued fractions. For more information see Chapter II, Section 7.3 of *Number Theory* by Borevich and Shafarevich. Also see the exercise following this section. We shall close this section with an example showing a field with class number two.

EXAMPLE. Let $K = Q(\theta)$ with $\theta^3 = 11$, $R =$ ring of integers in K. Then $Z[\theta] \subseteq R$. In fact equality holds here but we shall not prove this here. One computes directly that $\Delta(1,\theta,\theta^2) = \Delta = -3^3 11^2$. Then $\Delta(R/Z)$ must divide Δ so we use Δ in the Minkowski bound. We find every ideal class $[\mathfrak{A}]$ in $\mathbf{C}(R)$ contains an ideal $\mathfrak{B}$ with $\mathcal{N}(\mathfrak{B}) < 17$. This means we can generate $\mathbf{C}(R)$ by classes $[\mathfrak{P}]$ with $\mathfrak{P}$ a prime ideal having norm $\mathcal{N}(\mathfrak{P}) < 17$. So to find these primes, it is necessary to describe pR when p is an integral prime <17. For the primes $p = 2, 5, 7, 13$ we use Theorem 7.6 and Proposition 7.7.

$p = 2$ $\quad X^3 - 11 = (X-1)(X^2+X+1) \bmod 2$

$\quad 2R = \mathfrak{P}_2\mathfrak{P}_2'$, $\quad \mathcal{N}(\mathfrak{P}_2) = 2$, $\quad \mathcal{N}(\mathfrak{P}_2') = 4$

$p = 5$ $\quad X^2 - 11 = (X-1)(X^2+X+1) \bmod 5$

$\quad 5R = \mathfrak{P}_5\mathfrak{P}_5'$, $\quad \mathcal{N}(\mathfrak{P}_5) = 5$, $\quad \mathcal{N}(\mathfrak{P}_5') = 5^2$

$p = 7$ $\quad X^3 - 11$ is irreducible, mod 7; $\quad 7R = \mathfrak{P}_7$, $\quad \mathcal{N}(\mathfrak{P}_7) = 7^3$

$p = 13$ $\quad X^3 - 11$ is irreducible, mod 13; $\quad 13R = \mathfrak{P}_{13}$,

$\quad \mathcal{N}(\mathfrak{P}_{13}) = 13^3$.

The remaining primes 3, 11 are ramified. Since $N(\theta) = 11$ (look at the minimum polynomial) we have $11R = \mathfrak{P}_{11}^3$, $\mathfrak{P}_{11} = \theta R$, $\mathcal{N}(\mathfrak{P}_{11}) = 11$.

For the last case we make a general observation. For any integer k, $\theta+k$ has minimum polynomial $(X-k)^3 - 11$ and so

$$(*) \qquad \mathrm{N}(\theta+k) = k^3 + 11.$$

In particular $\mathrm{N}(\theta-2) = 3$ so $(\theta-2)R = \mathfrak{P}_3$ is a prime with $\mathcal{N}(\mathfrak{P}_3) = 3$. It is not difficult to show (as in the example on page 37) that $3R = {\mathfrak{P}_3}^3$.

It follows now that $\mathbf{C}(R)$ is generated by the classes $[\mathfrak{P}_2]$, $[\mathfrak{P}_2']$, $[\mathfrak{P}_3]$, $[\mathfrak{P}_5]$, $[\mathfrak{P}_{11}]$, since these represent all the primes with norm <17. The class of a principal ideal is the trivial class in $\mathbf{C}(R)$ so at once we have $[\mathfrak{P}_3] = [\mathfrak{P}_{11}] = 1$, since these are principal ideals. [We write 1 for the identity in $\mathbf{C}(R)$.] Furthermore, $[\mathfrak{P}_2][\mathfrak{P}_2'] = 1$ since $2R$ is principal. Thus, $\mathbf{C}(R)$ is generated by $[\mathfrak{P}_2]$ and $[\mathfrak{P}_5]$.

Now we must look for further relations. By $(*)$ one finds $\mathrm{N}(\theta-1) = 10 = 2\cdot 5$. Thus (by Corollary 8.5) $(\theta-1)R$ is the product of a prime with norm 2 and a prime with norm 5. In each case the prime is unique. Hence

$$(\theta-1)R = \mathfrak{P}_2\mathfrak{P}_5.$$

This means $[\mathfrak{P}_2] = [\mathfrak{P}_5]^{-1}$ in $\mathbf{C}(R)$ so $\mathbf{C}(R)$ is generated by $[\mathfrak{P}_2]$.

To get additional relations it would be helpful to find an element with norm a power of two. The minimum polynomial of θ^2 is $X^3 - 121$ and by the method above one finds

$$\mathrm{N}(\theta^2+k) = k^3 + 121.$$

In particular $\mathrm{N}(\theta^2-5) = -4$ so $(\theta^2-5)R$ is an ideal with norm 4. There are two possible ideals with norm 4; namely, ${\mathfrak{P}_2}^2$ or $\mathfrak{P}_2'$.

First Case. $(\theta^2-5)R = \mathfrak{P}_2'$. Then $2R = \mathfrak{P}_2\mathfrak{P}_2' = (\theta^2-5)\mathfrak{P}_2$. This means $\mathfrak{P}_2$ is the principal ideal generated by $2/(\theta^2-5)$. However, we can prove this element is not in R. We describe a general method for deciding such a question.

Suppose α is any element of R and

$$f(X) = X^3 + a_1X^2 + a_2X + a_3 = 0$$

is the minimum equation for α. Then the minimum equation (over Q) for $1/\alpha$ is

$$X^3f(1/X) = 1 + a_1X + a_2X^2 + a_3X^3.$$

The monic equation for $1/\alpha$ is $(1/a_3)X^3f(1/X)$. The monic equation for $2/\alpha$ is

$$8/a_3 + (4a_1/a_3)X + (2a_2/a_3)X^2 + X^3.$$

So $2/\alpha$ is in R if and only if these coefficients are integers (Proposition 2.4).

In the case $\alpha = \theta^2 - 5$ it turns out that $2/\alpha$ has minimum monic equation with $75/2$ as the coefficient of X^2 so $2/(\theta^2-5)$ is not in R.

SECOND CASE. $\mathfrak{P}_2{}^2 = (\theta^2-5)R$. It now follows that $[\mathfrak{P}_2]^2 = 1$ in $\mathbf{C}(R)$ so $\mathbf{C}(R)$ has order 1 or order 2 depending upon whether $\mathfrak{P}_2$ is a principal ideal or not.

The task of proving $\mathfrak{P}_2$ is not principal is not at all an easy one. We shall use a method suggested by Artin [1] p. 170 (used in a different example.)

This method requires that we know the units in R. By the unit Theorem 11.19, the unit group is the direct product of a finite group and an infinite cyclic group. Since R, and $Q(\theta)$, can be imbedded into the reals by taking θ to be the real cube root of 11, it follows that the finite group of units is $\langle \pm 1 \rangle$ and so all units in R have the form $\pm u^k$ for some fundamental unit u.

We shall not describe how the unit u can be found although there exists an algorithm by which this can be done. It turns out that

$$u = 89 + 40\theta + 18\theta^2.$$

It is straightforward to verify that $\mathrm{N}(u) = 1$ so indeed u is a unit. Moreover, there are techniques for estimating the fundamental unit (Artin [1], p. 169). This enables one to verify that u is indeed a fundamental unit.

Now suppose $\mathfrak{P}_2 = \alpha R$ is a principal ideal. Then

$$\mathfrak{P}_2{}^2 = \alpha^2 R = (\theta^2-5)R.$$

It follows that $\alpha^2 = (\theta^2-5)w$ for some unit w in R. Necessarily $(\theta^2-5)w$ is a square in R and so certainly must be a square modulo any prime ideal in R. After we multiply $(\theta^2-5)w$ by a suitable square of a unit we obtain

$$x = \pm u^d(\theta^2-5) \equiv (\text{square}) \bmod \mathfrak{P}$$

where one of the signs is fixed and $d = 0$ or 1. We are able to make this assertion because $w = \pm u^k$ for some k.

We first take $\mathfrak{P} = \mathfrak{P}_3 = (\theta-2)R$. In $R/\mathfrak{P}_3$ we map θ onto 2 since $\theta - 2 \in \mathfrak{P}_3$. Thus using the form of u yields

$$x \equiv \pm(89+40(2)+18(2)^2)^d(4-5) \equiv \pm(1)^d(-1).$$

We know x is congruent to a square in $R/\mathfrak{P}_3$ but $R/\mathfrak{P}_3$ has order three. Since -1 is not a square mod 3, the negative sign must be the one. That is

$$x = -u^d(\theta^2-5), \qquad d = 0 \text{ or } 1.$$

Next use a calculation above to find

$$\mathrm{N}(\theta+9) = 740 = 2^2\cdot 5\cdot 37.$$

This means $(\theta+9)R$ is divisible by a prime $\mathfrak{P}_{37}$ which has norm 37 and relative degree 1. In $R/\mathfrak{P}_{37}$ we must have x map onto a square. Now compute with θ

mapping onto -9:

$$x \equiv -(89-40\cdot 9+18\cdot 9^2)^d(9^2-5)$$
$$\equiv -(3)^d(2) \bmod 37.$$

We know this is a square mod 37.

However we shall see this is not the case by evaluating the Legendre symbol,

$$((3)^d(-2)/37) = (3/37)^d(-1/37)(2/37) = -1.$$

This follows because

$$(3/37) = (37/3) = (1/3) = +1,$$
$$(-1/37) = +1, \qquad (2/37) = -1.$$

This shows x is not a square modulo $\mathfrak{P}_{37}$ and so is not a square in R. This final contradiction proves $\mathfrak{P}_2$ is not a principal ideal. Thus $\mathbf{C}(R)$ has order two.

EXERCISE 1. Let $d>0$ be a square free integer and m the smallest positive integer such that one of the numbers dm^2-4, dm^2+4 is a square. Let a be the least positive integer for which $a^2 = dm^2+4\varepsilon$, $\varepsilon = \pm 1$. Then the fundamental unit in the field $Q(\sqrt{d})$ is $u = (a+m\sqrt{d})/2$ and $\mathrm{N}(u) = \varepsilon$.

{*Hint.* For any unit u in the ring of integers of $Q(\sqrt{d})$ the minimum polynomial of u over Q has the form $X^2-aX+\varepsilon$, $\varepsilon = \mathrm{N}(u) = \pm 1$. Solve for u and use $(a^2-4\varepsilon)^{1/2} = m\sqrt{d}$ for some m. Now verify the least possible m does give the fundamental unit.}

EXERCISE 2. Verify the following table, which gives the fundamental unit in $Q(\sqrt{d})$.

d	2	3	5	6	7	10
u	$1+\sqrt{2}$	$2+\sqrt{3}$	$(1+\sqrt{5})/2$	$5+2\sqrt{6}$	$8+3\sqrt{7}$	$3+\sqrt{10}$

EXERCISE 3. Imitate the procedure in the example above to show $Q(\sqrt{10})$ has class number 2. In fact the unique prime divisor of (2) is nonprincipal.

Chapter II

COMPLETE FIELDS

1. VALUATIONS

Let K be any field and $x \to |x|$ a function from K to the reals.

Definition. The function $|x|$ is called a valuation if

(i) $|x| > 0$ except that $|0| = 0$,
(ii) $|x||y| = |xy|$,
(iii) $|x+y| \leqslant |x|+|y|$.

If the valuation satisfies the stronger condition

(iii)* $|x+y| \leqslant \max\{|x|, |y|\}$

then it is called *nonarchimedean* valuation. All others are called *archimedean* valuations. The valuation is *nontrivial* if $|x| \neq 1$ for some $x \neq 0$.

EXAMPLE 1. Suppose K is any subfield of the real field. Then the usual absolute value $|x|$ is an archimedean valuation of K.

EXAMPLE 2. Let R be a Dedekind ring with quotient field K and let $\mathfrak{P}$ be a nonzero prime ideal in R. For any nonzero x in R let $v_{\mathfrak{P}}(x)$ denote the power to which $\mathfrak{P}$ appears in the factorization of Rx. We may then write

$$Rx = \prod \mathfrak{P}^{v_{\mathfrak{P}}(x)}, \qquad \mathfrak{P} \text{ runs through the primes.}$$

We abuse the notation slightly and also let $\mathfrak{P}$ denote the maximal ideal in the DVR $R_{\mathfrak{P}}$. We extend the definition of $v_{\mathfrak{P}}$ to all of $R_{\mathfrak{P}}$ using the same defining property. If y is in K but not in $R_{\mathfrak{P}}$, then y^{-1} is in $R_{\mathfrak{P}}$. We define $v_{\mathfrak{P}}(y)$ to be $-v_{\mathfrak{P}}(y^{-1})$. If we interpret $yR_{\mathfrak{P}}$ as a fractional ideal then $v_{\mathfrak{P}}(y)$ is still the power of $\mathfrak{P}$ appearing in the factorization. In fact for all $y \neq 0$ in K,

$$yR_{\mathfrak{P}} = \mathfrak{P}^{v_{\mathfrak{P}}(y)}.$$

Let $\mathfrak{P} = (\pi)$ in $R_{\mathfrak{P}}$ so every nonzero element in K can be expressed as $y = u\pi^n$ for a unit u in $R_{\mathfrak{P}}$ and some integer n. Clearly $v_{\mathfrak{P}}(y) = n$.

Observe that $v_{\mathfrak{P}}$ satisfies the following:

1. $v_{\mathfrak{P}}(y)$ is an integer for each $y \neq 0$ in K,
2. $v_{\mathfrak{P}}(xy) = v_{\mathfrak{P}}(x) + v_{\mathfrak{P}}(y)$,
3. $v_{\mathfrak{P}}(x+y) \geqslant \min\{v_{\mathfrak{P}}(x), v_{\mathfrak{P}}(y)\}$.

The first two follow at once. We shall prove 3. Let $x = u_1\pi^m$, $y = u_2\pi^n$, with u_1, u_2 units in $R_{\mathfrak{P}}$. Suppose $m \geqslant n$. Then

$$x + y = (u_2^{-1}u_1\pi^{m-n} + 1)u_2\pi^n.$$

The element in parenthesis belongs to $R_{\mathfrak{P}}$ so

$$v_{\mathfrak{P}}(x+y) \geqslant n = \min\{v_{\mathfrak{P}}(x), v_{\mathfrak{P}}(y)\}.$$

Definition. A function $v(x)$ which satisfies (1)–(3) is called an *exponential valuation* on K.

Remark. For convenience let us set $v_{\mathfrak{P}}(0) = +\infty$ so that Statement 3 is meaningful even if $x+y = 0$.

One easily obtains a valuation of K from the exponential valuation $v_{\mathfrak{P}}$. Select a real number c such that $0 < c < 1$. Now define

$$|x| = c^{v_{\mathfrak{P}}(x)}.$$

That $|x|$ is a nonarchimedean valuation follows from Statements 1–3. This valuation will be called the $\mathfrak{P}$ *adic valuation* on K.

There is much freedom here due to the choice of the constant c from the interval $(0, 1)$. If a second constant d were used to define a valuation $|x|_1$, we would have $|x|$ and $|x|_1$ equivalent in the following sense:

Definition. Two valuations $|x|$, $|x|_1$ on K are *equivalent* if whenever $|x| < 1$ then also $|x|_1 < 1$ for x in K.

There is a very precise relation that holds between equivalent valuations on any field K.

1.1 Proposition. Let $|x|$ and $|x|_1$ be nontrivial but equivalent valuations on K. Then there is a real number a such that $|x|^a = |x|_1$ for all x in K.

PROOF. Since the valuations are nontrivial, there exists an element y in K with $|y| > 1$. Let a denote the real number

$$a = \log|y|_1/\log|y|.$$

Now let x be any nonzero element in K. There is a real number b such that

$$|x| = |y|^b.$$

Now let m_i/n_i be a sequence of rational numbers (with $n_i > 0$) converging to b from above. Then

$$|x| = |y|^b < |y|^{m_i/n_i}$$

and so

$$|x^{n_i}/y^{m_i}| < 1.$$

It follows that

$$|x^{n_i}/y^{m_i}|_1 < 1$$

and so

$$|x|_1 \leqslant |y|_1^{m_i/n_i}.$$

This means $|x|_1 \leqslant |y|_1{}^b$.

If we repeat this procedure with a sequence of rationals converging to b from below we can obtain the reverse inequality. Thus

$$|x| = |y|^b \quad \text{implies} \quad |x|_1 = |y|_1{}^b$$

for all $x \neq 0$.

This implies

$$\log|x|/\log|x|_1 = 1/a$$

and so $|x|^a = |x|_1$ as required.

When we consider various questions about valuations, it is often necessary to consider archimedean and nonarchimedean valuations separately. We consider next one of the important features of nonarchimedean valuations.

1.2 Proposition. Let $|x|$ be a nonarchimedean valuation on K. Let $R = \{x \in K \,|\, |x| \leqslant 1\}$, $\mathfrak{P} = \{x \,|\, |x| < 1\}$. Then R is a local ring with $\mathfrak{P}$ as its maximal ideal and K as its quotient field.

R is a DVR if and only if the set of nonzero values $|K^*|$ is a multiplicative subgroup of the reals isomorphic to the additive group of integers.

PROOF. We first show that R is a ring. Whenever $x, y \in R$ then (ii) and (iii)* insure that xy and $x+y$ are in R. Furthermore $|-1|^2 = |1| = |1|^2$ by (ii) so $|1| = |-1| = 1$.Thus $|y| = |-y|$ and so $x - y \in R$ which shows R is a ring containing 1. For any $z \in K$, $z \neq 0$ we have $|z||z^{-1}| = 1$ so either z or z^{-1} belongs to R. Thus $K =$ quotient field of R.

Certainly $\mathfrak{P} \subseteq R$ and by (iii)* $\mathfrak{P}$ is an ideal of R. To see that $\mathfrak{P}$ is the unique maximal ideal of R let y be any element of R not in $\mathfrak{P}$. Then $|y| = 1$. Also it follows $|y^{-1}| = 1$ so y^{-1} is in R. Thus every element in R outside $\mathfrak{P}$ is a unit. Hence $\mathfrak{P}$ is a maximal ideal and is the only one.

Now suppose R is a DVR so then $\mathfrak{P} = R\pi$. Every element of K (nonzero) can be expressed as $x = u\pi^n$ for some unit u in R and an integer n. If $c = |\pi|$ then $|x| = c^n$. The function $|x| \to n$ now establishes an isomorphism of the group of nonzero values with the additive group of integers.

Conversely let ϕ denote the isomorphism of $|K^*|$ with Z. From the equation $\phi(|x|^{-1}) = -\phi(|x|)$ we conclude that $\phi(|R|)$ or $-\phi(|R|)$ contains all the positive integers. Replace ϕ by $-\phi$ if necessary to assume 1 is in $\phi(|R|)$. Let π be an element of R such that $\phi(|\pi|) = 1$. For any x in R, $\phi(|x|) = n$ is a positive integer and

$$\phi(|x\pi^{-n}|) = 0.$$

Since ϕ is an isomorphism it must be that

$$|x\pi^{-n}| = 1.$$

Thus $x\pi^{-n} = u$ is a unit in R so that $x = u\pi^n$. Every nonzero element of R is a unit times a power of π. It follows at once that the only ideals in R are powers of $R\pi$ so R is a DVR as required.

Remark. The ring R obtained from a nonarchimedean valuation ring as in Proposition 1.2 is called a *valuation ring*. In the case where the value group is an infinite cyclic subgroup of the reals, the value group is necessarily a *discrete subgroup* and so the valuation ring is called a discrete valuation ring (DVR).

Next we describe a useful test to determine if a valuation is archimedean or not.

1.3 Proposition. A valuation $|x|$ of K is nonarchimedean if and only if the values $|n1|$ are bounded as n runs through the rational integers Z.

PROOF. If $|x|$ satisfies (iii)* then

$$|n1| = |1 + \cdots + 1| \leqslant |1|,$$

so $|1|$ is a bound on the values $|n1|$.

Conversely suppose $|n1| \leqslant N$ for all integers n. For any x, y in K we have for any positive integer n,

$$|x+y|^n = \left|\sum_r \binom{n}{r} x^r y^{n-r}\right| \leqslant \sum_r \left|\binom{n}{r}\right| |x|^r |y|^{n-r}.$$

Now if $|x| \geqslant |y|$ then $|x|^r |y|^{n-r} \leqslant |x|^n$. Since $\binom{n}{r}$ is an integer and $|x| =$

$\max\{|x|, |y|\}$, it follows that

$$|x+y|^n \leqslant N(n+1)\max\{|x|, |y|\}^n.$$

Now then

$$|x+y| \leqslant N^{1/n}(n+1)^{1/n}\max\{|x|, |y|\}.$$

Since this is true for all positive n it must be that the nonarchimedean axiom (iii)* holds.

It follows from this result that a field of characteristic p has only nonarchimedean valuations.

We shall now determine all the (nonequivalent) valuations of Q.

Let $|x|$ be a nontrivial valuation of Q. Let m and n be integers >1. We may write

$$m = a_0 + a_1 n + \cdots + a_r n^r$$

with a_i an integer, $0 \leqslant a_i < n$ and $n^r \leqslant m$. Let $N = \max\{1, |n|\}$. By the triangle inequality,

$$|m| \leqslant \sum |a_i||n|^i \leqslant \sum |a_i| N^r.$$

The number r satisfies $r \leqslant \log m/\log n$ and the numbers a_i are less than n, so

$$|a_i| = |1+\cdots+1| \leqslant a_i|1| < n.$$

Substitute this information into the previous inequality to get

$$* \qquad |m| \leqslant (1+\log m/\log n)\, nN^{\log m/\log n}$$

In this inequality replace m by m^s and take s roots on both sides (s an integer).

$$|m| \leqslant (1+s\log m/\log n)^{1/s} n^{1/s} N^{\log m/\log n}$$

Now let s increase without bound. The terms in the inequality involving s converge to 1. It follows

$$** \qquad |m| \leqslant N^{\log m/\log n}$$

We now consider two cases.

Case 1. $n > 1$ implies $|n| > 1$.

In this case we always have $N = |n|$ and the condition (**) now yields

$$|m|^{1/\log m} \leqslant |n|^{1/\log n}.$$

We may reverse the roles of m and n to obtain the inequality in the reverse direction also. Thus

$$c = |m|^{1/\log m} = |n|^{1/\log n}$$

and we have $|n| = c^{\log n}$ for all integers $n > 1$. Since $|-n| = |n|$ and $|a/b| = |a|/|b|$ we find that

$$|x| = c^{\log x}$$

for all positive rational x. When $c = e$ the valuation is the usual absolute value on Q. Any constant c has the form e^a for some a so the valuation is just the usual absolute value raised to the power a.

In Case 1 only the usual absolute value arises (up to equivalence).

Case 2. For some $n > 1$ we have $|n| \leqslant 1$. In this case $N = 1$ and by (**) we have $|m| \leqslant 1$, for all integers $m > 1$. It follows from Proposition 1.3 that the valuation is nonarchimedean. Let $R = \{x \in Q \mid |x| \leqslant 1\}$ denote the valuation ring and $\mathfrak{P}$ is maximal ideal as in Proposition 1.2. Then $Z \subset R$ and $\mathfrak{P} \neq (0)$ because the valuation is not trivial. In fact $\mathfrak{P} \cap Z$ is nonzero because $|m| = 1$ for all m in Z, $m \neq 0$ would imply the valuation is trivial. Thus $\mathfrak{P} \cap Z = (p)$ is a prime ideal. If m is in Z but not in (p) then m is a unit in R and $|m| = 1$. Thus $|mp^r| = |p|^r$ and this valuation on Q is equivalent to the $p$adic valuation.

We have proved the next statement.

1.4 Proposition. A nonarchimedean valuation of Q is equivalent to a $p$adic valuation for some prime p. An archimedean valuation of Q is equivalent to the usual absolute value.

We shall introduce some new terminology.

Definition. An equivalence class of valuations on a field K will be called a *prime* of K. We shall use letters $\mathfrak{P}$, $\mathfrak{p}$ to denote primes of K. If we want to select a particular valuation from $\mathfrak{P}$ we may denote it by a symbol such as $|x|_{\mathfrak{P}}$.

For the rational field Q the primes are in one-to-one correspondence with the prime integers except that the class of archimedean valuations is not obtained this way. In order to have a consistent terminology we shall call the equivalence class of archimedean valuations on Q the *infinite prime* of Q. All others will be called *finite primes*.

We shall now make some normalizations. From each prime of Q we shall select a particular valuation.

Let p be a prime integer and $\mathfrak{P}$ the prime of Q corresponding to the $p$adic valuation. Let $|x|_p$ be the valuation in $\mathfrak{P}$ which satisfies

$$|p|_p = 1/p.$$

Let $\mathfrak{P}_\infty$ denote the infinite prime of Q and let $|x|_\infty$ denote the usual absolute value.

The collections of valuations defined here will be called the *normalized valuations* of Q.

The reason for making these choices can be seen (in part) from the following.

Product Formula. If $| \ |_{\mathfrak{P}}$ denotes the normalized valuation in the prime of $\mathfrak{P}$ on Q then for each nonzero x in Q we have

$$\prod_{\mathfrak{P}} |x|_{\mathfrak{P}} = 1,$$

where the product is taken over all primes of Q.

PROOF. The function

$$\pi(x) = \prod_{\mathfrak{P}} |x|_{\mathfrak{P}}$$

is well defined because there exist only a finite number of primes $\mathfrak{P}$ such that $|x|_{\mathfrak{P}} \neq 1$ for any given x. To prove the formula, it is first useful to observe that $\pi(xy) = \pi(x)\pi(y)$ because each valuation is multiplicative. Thus it is sufficient to prove $\pi(p)_{\mathfrak{P}} = 1$ for each prime integer p. But this is trivial because $|p|_{\mathfrak{P}} = 1$ unless $\mathfrak{P}$ is the infinite prime or the $p$adic prime. But then

$$|p|_p |p|_\infty = 1$$

by definition of the normalized valuations.

One of our goals will be to prove a product formula theorem for finite extensions of Q. This will require detailed information about extending a valuation from Q to some finite extension field. It will turn out that the extensions of nonarchimedean valuations can be described by using the information in Chapter I. However some new ideas are required to discuss the extensions of archimedean valuations. We shall return to this after some preliminary work.

2. COMPLETIONS

Let K be a field with a valuation $|x|$. A sequence of elements $\{a_n\}$ in K is called a *Cauchy sequence* if

$$\lim_{m,n\to\infty} |a_n - a_m| = 0.$$

Notice that the limit makes sense because the values $|a_n - a_m|$ are real numbers.

A sequence $\{a_n\}$ converges to a if

$$\lim_{n\to\infty} |a - a_n| = 0.$$

It may or may not happen that a Cauchy sequence in K converges to an element of K. For example a Cauchy sequence of rationals with respect to the usual absolute value need not converge to a rational number.

Definition. The field K is *complete* with respect to the valuation if every Cauchy sequence converges to an element of K.

Our object is to start with a field K and valuation and imbed K into a complete field having a valuation which extends the one on K. The process is a generalization of the familiar procedure of obtaining the reals from the rationals.

We begin by defining an equivalence relation on the collection of all Cauchy sequences in K. Say $\{a_n\} \sim \{b_n\}$ if

$$\lim |a_n - b_n| = 0.$$

By using the properties of valuations and the fact that we are dealing with Cauchy sequences, one proves this is indeed an equivalence relation. Let $\{a_n\}^*$ denote the equivalence class of the CS $\{a_n\}$ and let $\hat{K}$ denote the collection of all such equivalence classes.

We leave for the reader the verification of the following statements.

(2.1) Let $\{a_n\}$ and $\{b_n\}$ be Cauchy sequences in K.

(1) $\{a_n + b_n\}$ and $\{a_n b_n\}$ are CS in K.
(2) If $\lim |a_n| \neq 0$ then $a_n \neq 0$ for all $n \geqslant n_0$ for some n_0. In this case $\{a_n^{-1}\}$ $n \geqslant n_0$ is a CS in K.

We use these facts to define addition and multiplication in $\hat{K}$.
Define

$$\{a_n\}^* + \{b_n\}^* = \{a_n + b_n\}^*$$
$$\{a_n\}^* \{b_n\}^* = \{a_n b_n\}^*$$
$$\{a_n\}^{*-1} = \{a_n^{-1}\}^* \quad \text{when } a_n^{-1} \quad \text{defined.}$$

One checks that these operations are well defined and that $\hat{K}$ becomes a field. The original field K is imbedded in $\hat{K}$ by identifying an element x in K with the constant sequence $\{x\}$. The valuation $|x|$ on K is extended to $\hat{K}$ by defining

$$|\{a_n\}^*| = \lim |a_n|.$$

One easily verifies this function is actually a valuation on $\hat{K}$ which agrees with the original valuation on K.

The heart of the matter is to prove $\hat{K}$ is complete with respect to this valuation. Let $[\alpha_n]$ denote a CS in $\hat{K}$. This means each α_n is a CS, say $\alpha_n = \{a_m^{(n)}\}^*$, in K. Let α be the sequence $\{a_m^{(m)}\}^*$.

In order for this to make sense, $\{a_m^{(m)}\}$ must be a CS in K. This follows because

$$|a_r^{(r)} - a_s^{(s)}| \leqslant |a_r^{(r)} - a_r^{(s)}| + |a_r^{(s)} - a_s^{(s)}|.$$

For fixed s, $\{a_n^{(s)}\}$ is CS so the second term on the right converges to zero. The sequence $[\alpha_n]$ is Cauchy so $|a_t^{(n)} - a_t^{(m)}|$ converges to zero as n, m, t get large. Thus the first term on the right also goes to zero. Thus $\alpha = \{a_n^{(n)}\}^*$ is in $\hat{K}$. Moreover we claim $[\alpha_n]$ converges to α. This follows from

$$\lim |\alpha_n - \alpha| = \lim \lim |a_m^{(n)} - a_m^{(m)}| = 0.$$

Thus $\{\alpha_n\}$ has a limit in $\hat{K}$ and $\hat{K}$ is complete.

We shall now establish the uniqueness of this construction. In fact a slightly more general result is useful.

The phrase "$(K_0, |\ \ |_0)$ is a completion of $(K, |\ \ |)$" will mean K is a field with valuation $|\ \ |$ and K_0 is a field which is complete with respect to a valuation $|\ \ |_0$ and K is a subfield of K_0 such that $|\ \ |_0$ agrees with $|\ \ |$ on K. Moreover we require that every element in K_0 be a limit of a sequence of elements in K.

2.2 Theorem. Let K be a field with a valuation $|\ \ |$, and L a field with a valuation $|\ \ |_1$. Suppose there is an imbedding σ of K into L such that $|x| = |\sigma(x)|_1$ for all x in K. Let $(\hat{K}, |\ \ |)$ and $(\hat{L}, |\ \ |_1)$ be completions of $(K, |\ \ |)$ and $(L, |\ \ |_1)$, respectively.

Then there exists a unique imbedding $\hat{\sigma}$ of $\hat{K}$ into $\hat{L}$ such that $|x| = |\hat{\sigma}(x)|_1$ for all x in $\hat{K}$, and $\hat{\sigma}(x) = \sigma(x)$ for all x in K.

PROOF. Take $\{a_n\}^*$ in $\hat{K}$ with each a_n in K. Since $\{a_n\}$ is a CS with respect to $|\ \ |$, it follows that $\{\sigma(a_n)\}$ is CS in L with respect to $|\ \ |_1$. We define $\hat{\sigma}\{a_n\}^* = \{\sigma(a_n)\}^*$. One can verify that $\hat{\sigma}$ is well defined and gives an imbedding of $\hat{K}$ into $\hat{L}$ which preserves the valuations.

If θ is another imbedding of $\hat{K}$ into $\hat{L}$ which preserves the valuation and also agrees with σ on K we can show $\theta = \hat{\sigma}$. Take $\alpha = \{a_n\}^*$ in $\hat{K}$. Then $\alpha = \lim A_n$ where A_n is the constant sequence $\{a_n, a_n, \ldots\}$. Since θ preserves the valuation we may compute $\theta(\alpha)$ by exchanging it with the limit sign. Then since θ agrees with σ on K (constant sequences) we have

$$\theta(\alpha) = \lim \theta(A_n) = \lim \sigma(A_n) = \hat{\sigma}(\alpha).$$

This proves the uniqueness.

2.3 Corollary. The completion $(\hat{K}, |\ \ |)$ of $(K, |\ \ |)$ is unique up to an isomorphism which preserves the valuation on K.

PROOF. Take $L = K$, $\sigma =$ identity in the theorem. If $(\hat{K}, |\ \ |)$ and $(\hat{K}_1, |\ \ |_1)$ are two completions then there exists a valuation-preserving monomorphisms σ_1 of $\hat{K}_1$ into $\hat{K}$ and σ_2 of $\hat{K}$ into $\hat{K}_1$ such that the compositions $\sigma_1 \sigma_2$ and $\sigma_2 \sigma_1$ are identity on K and preserve the valuation. The identity map of $\hat{K}_1$ also has this property so by the uniqueness statement we have $\sigma_2 \sigma_1 = 1$. It follows that σ_2 and σ_1 are both isomorphisms.

When $K = Q$ and $|x|$ denotes the usual absolute value then Q_∞, the completion, is isomorphic to the real field.

If $|\ \ |_p$ is a *p*adic valuation on Q we shall write Q_p to denote the completion with respect to this valuation.

We shall require some additional information about Q_p. It is most convenient to adopt a more general point of view.

Let R be a DVR with maximal ideal $\mathfrak{P} = R\pi$ and quotient field K. Let $|\ \ |$ denote the $\mathfrak{P}$adic valuation on K and $K_\mathfrak{P}$ the $\mathfrak{P}$adic completion of K. The valuation is nonarchimedean on K and so by Proposition 1.3 it is nonarchimedean on $K_\mathfrak{P}$. So we may speak of the valuation ring in $K_\mathfrak{P}$. Let

$$\hat{R} = \{x \in K_\mathfrak{P} \mid |x| \leqslant 1\},$$

$$\hat{\mathfrak{P}} = \{x \in K_\mathfrak{P} \mid |x| < 1\}.$$

We assume at the start that the valuation on R has values $|u\pi^n| = c^n$ with $c = |\pi|$ some positive constant less than one.

Let us show $\hat{R}$ is a DVR. This can be accomplished if we show the set of values of elements from $K_\mathfrak{P}$ is an infinite cyclic group (Proposition 1.2). If α is in $K_\mathfrak{P}$ then $|\alpha|$ is a limit of terms $|a_n|$ with a_n in K. Thus $|\alpha|$ is a limit of sequence of powers c^{m_i}. For $\alpha \neq 0$ this sequence of powers of c must have a nonzero limit. The only finite limit of a sequence containing infinitely many different powers of c is zero. Since this is not allowed for $\alpha \neq 0$ it must be that only finitely many distinct powers of c appear and the sequence $|a_n|$ is eventually constant—say with value c^k. Then $|\alpha| = c^k$ and so every value $|\alpha|$ is already a value $|a_n|$ with a_n in K. Thus it follows from Proposition 1.2 that $\hat{R}$ is a DVR and $\hat{\mathfrak{P}} = x\hat{R}$ for some x.

Suppose $|x| = c^n$ for some integer n. Necessarily $|x| < 1$ so $n \geqslant 1$. Now $|\pi| = c$ so $|\pi^n/x| = 1$ implies π^n/x is in $\hat{R}$ and is a unit in $\hat{R}$. Then $\hat{R}\pi^n = (\pi^n/x)\hat{\mathfrak{P}} = \hat{\mathfrak{P}}$. This implies $n = 1$ and $\hat{\mathfrak{P}} = \pi\hat{R}$. We summarize this.

2.4 Proposition. If $|x|$ is a nonarchimedean valuation on K whose valuation ring R is a DVR, then the valuation ring $\hat{R}$ of the completion $K_\mathfrak{P}$ is also a DVR. Moreover the maximal ideals of R and $\hat{R}$ can be generated by the same element.

One more fact can be gleaned from the discussion above.

2.5 Corollary. In the context above, every element α in $K_\mathfrak{P}$ can be represented by a class $\alpha = \{a_n\}^*$ in which $|a_n|$ is constant.

Proof. We have seen above that $\alpha = \{a_n\}^*$ and the collection of values $|a_n|$ is finite. Thus $|a_n|$ is constant for all $n \geqslant N$ for some sufficiently large N. If we drop the first N terms from $\{a_n\}$ we will not change the equivalence class $\{a_n\}^*$ so the result follows.

2.6 Corollary. The units of $\hat{R}$ are the elements $\{a_n\}^*$ in $K_{\mathfrak{P}}$ for which $|a_n| = 1$ for all n.

PROOF. Every element of $\hat{R}$ (and even $K_{\mathfrak{P}}$) is $\{a_n\}^*$ with $|a_n|$ constant. For this to be in $\hat{R}$ the constant must be $\leqslant 1$ and must $= 1$ if the element is a unit of $\hat{R}$.

2.7 Corollary. $R/\mathfrak{P}^n \cong \hat{R}/(\hat{\mathfrak{P}})^n$ for any positive integer n.

PROOF. Take $\mathfrak{P} = R\pi$ so that $\hat{\mathfrak{P}} = \hat{R}\pi$. Select any α in $\hat{R}$ with α not in $\hat{\mathfrak{P}}$. By Corollary 2.6 we may assume $\alpha = \{a_n\}^*$ with $|a_n| = 1$ for all n. Since this is a Cauchy sequence, there is some N such that

$$|a_{n+1} - a_n| < \tfrac{1}{2}, \qquad n \geqslant N.$$

If we drop the first N terms from the sequence a_n we will not change α so assume $N = 0$. Then $a_{n+1} - a_n$ has value < 1 so is not a unit in R. We have

$$a_{n+1} \equiv a_n \bmod R\pi \qquad \text{for all} \quad n$$

and so

$$a_{n+1} \equiv a_1 \bmod R\pi.$$

Thus

$$\{a_n\}^* \equiv \{a\}^* \bmod \pi\hat{R}$$

where $\{a\}$ is the constant sequence with all terms equal to a_1. Now $\{a\}^*$ is in R so the coset $\alpha + \hat{\mathfrak{P}}$ is in $R + \hat{\mathfrak{P}}$. Since α was an arbitrary element of $\hat{R}$ outside of $\hat{\mathfrak{P}}$ it follows that $\hat{R} = R + \hat{\mathfrak{P}}$. Now multiply this by π and make the appropriate substitutions to get $\hat{R} = R + (\hat{\mathfrak{P}})^2$. By induction one obtains $\hat{R} = R + (\hat{\mathfrak{P}})^n$ for any positive integer n. One more observation is required. By considering the values we find $\mathfrak{P}^n = R \cap (\hat{\mathfrak{P}})^n$. Thus

$$\hat{R}/(\hat{\mathfrak{P}})^n \cong R + (\hat{\mathfrak{P}})^n/(\hat{\mathfrak{P}})^n \cong R/R \cap (\hat{\mathfrak{P}})^n$$

and so we obtain the desired result.

Next we describe an alternate form for the elements in $\hat{K}$. Let S be a set of representatives of the cosets of $\mathfrak{P}$ in R. Assume $0 \in S$. Let $\{s_i\}$ be any sequence of elements in S. For a fixed integer r and any $n \geqslant 0$ let

$$a_n = \pi^r(s_0 + s_1\pi + \cdots + s_n\pi^n).$$

Then $a_n - a_m$ is divisible by π^{r+t+1} with $t = \min\{n, m\}$ so it follows that $\{a_n\}$ is a Cauchy sequence in K. The class $\{a_n\}^*$ in $\hat{K}$ may be considered as a power series

$$\pi^r(s_0 + s_1\pi + \cdots)$$

and the a_n are the partial sums.

2.8 Proposition. Every element $\alpha \neq 0$ in $K_{\mathfrak{P}}$ has a unique representation as a power series,

$$\alpha = \pi^r(s_0 + s_1\pi + \cdots)$$

with the s_i in S and $s_0 \neq 0$.

PROOF. We know $\hat{R}$ is a DVR with quotient field $K_{\mathfrak{P}}$ so every element $\neq 0$ of $K_{\mathfrak{P}}$ has a unique expression as $\pi^r u$ with r some integer and u a unit in $\hat{R}$. It is sufficient to show u can be expressed by a unique power series.

The coset $u + \pi\hat{R}$ contains a unique element s_0 of S and $s_0 \neq 0$ because u is a unit. Thus $u - s_0 = \pi x_0$ for some x_0 in $\hat{R}$. Assume we have found $s_0, s_1 \ldots, , s_k$ in S and x_k in R such that

$$u - s_0 - s_1\pi - \cdots - s_k\pi^k = \pi^{k+1}x_k.$$

Then there is a unique s_{k+1} in S and some x_{k+1} in $\hat{R}$ such that

$$x_k - s_{k+1} = x_{k+1}\pi \in \hat{R}\pi.$$

This gives the next s and x. This method of successive approximation enables us to obtain the sequence of partial sums

$$b_n = \sum_0^n s_i\pi^i$$

which converges to u because the "remainder" $\pi^{n+1}x_{n+1}$ converges to zero.

To show the uniqueness of the expression first observe that the integer r is determined from $|\alpha| = c^r$. The uniqueness for units of $\hat{R}$ follows easily.

It is possible to do calculations with these power series representations by imitating the procedures for the more familiar power series encountered in calculus.

EXAMPLE. Let $R = Z_{(3)}$ and $\hat{R} = Q_3 =$ 3adic completion of Q. The elements in Q_3 can be expressed as series

$$3^r(s_0 + s_1 3 + s_2 3^2 + \cdots),$$

where $s_i \in \{0, 1, 2\}$. The rational number $-\frac{1}{8}$ has the form

$$-\tfrac{1}{8} = 1/(1-3^2) = 1 + 3^2 + 3^4 + \cdots.$$

Similarly

$$-1 = 2/1 - 3 = 2(1+3+3^2+\cdots).$$

EXERCISE 1. In Q_3, $1+2\cdot3+3^2+\cdots+3^{2n}+2\cdot3^{2n+1}+\cdots$ converges to a rational number. Find it.

EXERCISE 2. More generally show that a periodic series $\sum a_i p^i$ with $a_i = a_{i+m}$ for some fixed m, converges in Q_p to a rational number.

EXERCISE 3. Show the usual Taylor series in powers of z for the function $f(z) = (1-4z)^{1/2}$ has integer coefficients. Use this to show $1-4kp$ has a square root in Q_p for any k in $Z_{(p)}$.

EXERCISE 4. Let $\mathbf{U}_p$ denote the multiplicative group of units in the valuation ring of Q_p. Show the group of nonzero elements in Q_p is isomorphic to the direct product $\langle p \rangle \times \mathbf{U}_p$.

EXERCISE 5. Let p be an odd prime and u an element in $\mathbf{U}_p$. Let $\bar{u}$ denote the image of u in GF(p) under the natural map of the valuation ring onto the field of p elements. Show u is the square of an element $\mathbf{U}_p$ if and only if $\bar{u}$ is a square of an element in GF(p).

EXERCISE 6. Let $\bar{u}$ denote the image of u under the mapping of the valuation ring in Q_2 onto $Z/(8)$. The units in $Z/(8)$ have square equal to $\bar{1}$. Show that $u \in \mathbf{U}_2$ is a square if and only if $\bar{u} = \bar{1}$.

EXERCISE 7. Show $Q_p{}^*/(Q_p{}^*)^2$ has order 4 when p is odd and order 8 when $p = 2$. Here $Q_p{}^*$ means the multiplicative group of nonzero elements in Q_p.

3. EXTENSIONS OF NONARCHIMEDEAN VALUATIONS

Let K be a field with a nonarchimedean valuation $|x|_{\mathfrak{p}}$. Let R be the valuation ring and assume R is a DVR with maximal ideal $\mathfrak{p} = \pi R$. Let $v_{\mathfrak{p}}$ denote the exponential valuation defined by $v_{\mathfrak{p}}(u\pi^n) = n$ when u is a unit of R.

Let L denote a finite-dimensional, separable extension of K and R' the integral closure of R in L.

We shall consider the problem of finding all valuations on L which, when restricted to K, give a valuation equivalent to $| \ |_{\mathfrak{p}}$.

It is not difficult to describe some valuations on L which are closely related to $| \ |_{\mathfrak{p}}$. In the ring R' we factor π as

$$\mathfrak{p}R' = \pi R' = \mathfrak{P}_1^{e_1} \cdots \mathfrak{P}_g^{e_g} \tag{1}$$

with the $\mathfrak{P}_i$ distinct primes in R. Let $| \ |_i$ denote the $\mathfrak{P}_i$adic valuation of L. We shall now show that the restriction of $| \ |_i$ to K gives a valuation equivalent to $| \ |_{\mathfrak{p}}$.

The valuation ring of $| \ |_i$ is the localization $R'_{\mathfrak{P}_i}$. Let the generator of the maximal ideal be τ. The factorization of $\pi R'$ above now yields $\pi = u\tau^{e_i}$ for some unit u of $R'_{\mathfrak{P}_i}$. That is

$$\pi R'_{\mathfrak{P}_i} = (\mathfrak{P}_i R'_{\mathfrak{P}_i})^{e_i}.$$

Now for any $w\pi^n$ in K with w a unit in R we have

$$|w\pi^n|_i = |wu^n\tau^{ne_i}|_i = |\tau|_i^{ne_i}.$$

If we evaluate the original valuation at $w\pi^n$ we have

$$|w\pi^n|_{\mathfrak{p}} = |\pi|_{\mathfrak{p}}{}^n.$$

If a is some real number such that $|\tau^{e_i}|_i{}^a = |\pi|_{\mathfrak{p}}$ then it follows $|x|_i{}^a = |x|_{\mathfrak{p}}$ for all x in K. Thus $|\ \ |_i$ and $|\ \ |_{\mathfrak{p}}$ are equivalent on K.

Next we prove the converse of this result. Namely if $|\ \ |$ is a valuation on L which is equivalent to $|\ \ |_{\mathfrak{p}}$ on K, then $|\ \ | = |\ \ |_i$ for one of the $|\ \ |_i$ defined above.

Let $R_0 = \{x \in L \mid |x| \leqslant 1\}$. R_0 is a local ring with maximal ideal $\mathfrak{M}$ and $\mathfrak{M} \cap R = \mathfrak{p}$. We shall first show $R' \subseteq R_0$. Suppose there is an x in R' with x not in R_0. Then $|x| > 1$. Hence $|x^{-1}| < 1$ so x^{-1} is in $\mathfrak{M}$.

The element x is integral over R so there is a relation

$$x^n + a_1 x^{n-1} + \cdots + a_n = 0$$

with a_i in R. From this we obtain

$$1 = \sum_{i=0}^{n-1} a_i (x^{-1})^{n-i}$$

which means 1 is in $\mathfrak{M}$. This is impossible because $|1| = 1$. Thus $R' \subseteq R_0$. Let $\mathfrak{P} = R' \cap \mathfrak{M}$. $\mathfrak{P}$ is a prime ideal of R' which contains $\mathfrak{p}$. The localization $R_{\mathfrak{P}}'$ is also in R_0 because all elements in R' outside $\mathfrak{M}$ are units in R_0. Let the maximal ideal of $R_{\mathfrak{P}}'$ be generated by τ. Every element in L has the form $u\tau^n$ for some unit u in $R_{\mathfrak{P}}'$. This element u is also a unit in R_0 so $|u| = 1$. Thus $|u\tau^n| = |\tau|^n$. It follows that $|\ \ |$ is the $\mathfrak{P}$adic valuation on L.

3.1 Theorem. Let R be a DVR with maximal ideal $\mathfrak{p}$ and quotient field K. Let R' denote the integral closure of R in a finite dimensional separable extension L of K. Let $\mathfrak{p}R'$ have the factorization (1) above. Then the inequivalent valuations of L which give the $\mathfrak{p}$adic valuation on K are the $\mathfrak{P}_i$adic valuations. The exponential valuations are related by $e_i v_{\mathfrak{p}}(x) = v_{\mathfrak{P}_i}(x)$ for x in K where e_i is the ramification index.

All parts of this have been done except that we must show the $\mathfrak{P}_i$adic valuations on L are mutually inequivalent. This is immediate however because equivalent valuations have the same valuation ring. For $\mathfrak{P}_i \neq \mathfrak{P}_j$ we have

$$R'_{\mathfrak{P}_i} \neq R'_{\mathfrak{P}_j}$$

so the $\mathfrak{P}_i$adic valuation is not equivalent to the $\mathfrak{P}_j$adic one if $\mathfrak{P}_i \neq \mathfrak{P}_j$.

The relation between the exponential valuations was proved above when it was observed that $|\tau^{e_i}|_i = |\pi|_{\mathfrak{p}}$.

We have seen that the nonarchimedean valuations of Q are in one-to-one correspondence with the prime integers p. The valuation of Q corresponding to p is the $p$adic valuation.

If K is any algebraic number field, then a nonarchimedean valuation on K must restrict to some $p$adic valuation on Q. Then Theorem 3.1 is used to determine the possible extensions to K. Thus all the nonarchimedean valuations of K are known once the factorization of primes from Q are known.

This theorem can be made more precise if we impose further conditions upon K. We shall prove that when K is complete there is only one extension of the valuation to L. This is equivalent to the assertion that $g = 1$ in the factorization (1) of $\mathfrak{p}R'$. The proof of this will be accomplished by considering some more general properties of complete fields.

Now keep the same notation and assumptions given at the beginning of the section and assume further that K *is complete* with respect to the $\mathfrak{p}$adic valuation.

We shall temporarily adopt a more general point of view.

Let A be an algebra over R. That is A is a ring with identity and R is in the center of A. We shall say A is complete if every Cauchy sequence in A converges to an element of A. Here the terms Cauchy sequence and convergence are interpreted as follows. The sequence $\{a_n\}$ is Cauchy if there is a function $\mathrm{N}(n,m)$ with integer values such that

$$a_n - a_m \in A\pi^{N(n,m)}$$

and $\mathrm{N}(n,m)$ goes to infinity with n and m. The sequence converges to the element a if

$$a_n - a \in A\pi^{N(n)}$$

where $\mathrm{N}(n)$ is an integer valued function that goes to infinity with n.

3.2 Theorem. Let A be an R-algebra which is complete. Suppose e is an idempotent element in $A/A\pi$. Then there exists an idempotent element E in A such that $e = E + A\pi$. If e_1 and e_2 are idempotents in $A/A\pi$ such that $e_1 e_2 = e_2 e_1 = 0$ then we may select E_1, E_2 idempotent in A such that $E_i + A\pi = e_i$ and $E_1 E_2 = E_2 E_1 = 0$.

PROOF. The proof requires some very formal manipulations at the start. For each positive integer n,

$$1 = [X + (1-X)]^{2n} = \sum_{j=0}^{2n} \binom{2n}{j} X^{2n-j}(1-X)^j. \tag{2}$$

Let

$$f_n(X) = \sum_{j=0}^{n} \binom{2n}{j} X^{2n-j}(1-X)^j.$$

Then $f_n(X)$ has integral coefficients and

$$f_n(X) \equiv 0 \bmod X^n, \qquad f_n(X) \equiv 1 \bmod (1-X)^n. \tag{3}$$

The second congruence requires that Eq. (2) be used for $1-f_n(X)$. If we square both sides of the congruences (3) we see they both hold when $f_n(X)$ is replaced by $f_n(X)^2$. It follows then

$$f_n(X) \equiv f_n(X)^2 \bmod X^n(1-X)^n. \tag{4}$$

We may also observe that (3) holds when $f_n(X)$ is replaced by $f_{n+1}(X)$ so by the same reasoning one obtains

$$f_{n+1}(X) \equiv f_n(X) \bmod X^n(1-X)^n. \tag{5}$$

Now we turn our attention to the theorem. Let e be an idempotent element in $A/A\pi$ and let a be any element of A such that $e = a + A\pi$. Since $e = e^2$ we must have $a^2 - a \in A\pi$. Thus

$$a^n(1-a)^n \in A\pi^n \qquad \text{for all} \quad n. \tag{6}$$

It follows from (5) that

$$f_{n+1}(a) - f_n(a) \in A\pi^n.$$

Since A is complete, the sequence $a_n = f_n(a)$ must converge to an element E in A. The congruence (4) shows $E = E^2$ so E is idempotent. Finally we observe $f_1(X) \equiv X \bmod X(1-X)$ so from (5) again

$$f_n(a) \equiv f_{n-1}(a) \equiv \cdots \equiv f_1(a) \equiv a \bmod A\pi$$

and it follows that $E + A\pi = e$ as required.

Now consider the two idempotents e_1, e_2. First note that $e_1 + e_2$ is idempotent in $A/A\pi$ so there is an idempotent E in A with $E + A\pi = e_1 + e_2$. Take any element a in A with $a + A\pi = e_1$. Let $b = EaE$. Then

$$\bar{b} = \bar{E}\bar{a}\bar{E} = e_1,$$

where the bars indicate cosets of $A\pi$. Thus $b^2 - b$ is in $A\pi$ and $\lim f_n(b) = E_1$ is an idempotent in A such that $E_1 + A\pi = e_1$. We look at the definition of b to see that $b = bE = Eb$ so we also obtain $f_n(b) = f_n(b)E = Ef_n(b)$. Passing to the limit shows $E_1 = E_1 E = EE_1$. It follows now $E_2 = E - E_1$ is an idempotent and $E_1 E_2 = E_2 E_1 = 0$, $\bar{E}_2 = e_2$.

We shall obtain a number of consequences from this result.

3.3 Proposition. Let K be a complete field with respect to a nonarchimedean valuation whose valuation ring, R, is a DVR. Let L be a finite-dimensional separable extension field of K and R' the integral closure of R in L. Then R' is a DVR and L is complete in the valuation induced by R'.

PROOF. Let $\mathfrak{p}$ denote the maximal ideal of R and suppose $\mathfrak{p}R'$ has the factorization (1).

We intend to apply Theorem 3.2 with $R' = A$ so let us first show R' is complete.

R' is a finitely generated free R-module so let $x_1, \ldots, x_m$ be an R-basis. Let $\{a_n\}$ be a Cauchy sequence in R' in the sense defined for A above. Write

$$a_n = a_n^{(1)}x_1 + \cdots + a_n^{(m)}x_m$$

with the $a_n^{(i)}$ in R. For each i, $\{a_j^{(i)}\}$ is CS in R so has a limit in R. It follows that $\{a_n\}$ has a limit in R'.

Now we compute $R'/\mathfrak{p}R'$ by CRT to get

$$R'/\mathfrak{p}R' \cong R'/\mathfrak{P}_1^{e_1} \oplus \cdots \oplus R'/\mathfrak{P}_1^{e_g}.$$

If $g > 1$ there is an idempotent element $e = (1,0,\ldots,0)$ and $e \neq 1$. By Theorem 3.2 there is an idempotent E in R' in the coset corresponding to e. But R' is an integral domain so $E = 0$ or $E = 1$, since $E(1-E) = 0$. It follows that $g = 1$. Thus R' has only one maximal ideal $\mathfrak{P}$ and since R' is a Dedekind ring, it must be a DVR.

3.4 Corollary. Let K be complete with respect to a nonarchimedean valuation $|x|_\mathfrak{p}$ and let L be a finite dimensional extension field of K. Then there is a unique extension of the valuation on K to L and it is given by the formula

$$|y| = |\mathrm{N}_{L/K}(y)|_\mathfrak{p}^{1/n}$$

for all y in L where $n = (L:K)$.

Proof. The existence and uniqueness have already been proved. It remains to verify the formula is correct. Let R, R' denote the valuation rings in K and L, respectively with maximal ideals πR, $\tau R'$, respectively. We have $\pi R' = \tau^e R'$. Let $c = |\tau|$ so for any element $y = u\tau^m$ of L we have $|u\tau^m| = c^m$ when u is a unit in R'. In particular

$$|\pi| = |\pi|_\mathfrak{p} = c^e.$$

Now let f denote the relative degree of $\tau R'$ over R. We have

$$\mathrm{N}_{L/K}(\tau R') = \pi^f R$$

by Chapter I, Proposition 8.4. Thus

$$\mathrm{N}(\tau) = w\pi^f$$

with w a unit of R. We now have

$$|\mathrm{N}(y)|_\mathfrak{p} = |\mathrm{N}(u\tau^m)|_\mathfrak{p} = |\mathrm{unit}\cdot\pi^{fm}|_\mathfrak{p} = c^{efm}.$$

From Theorem 6.6 of Chapter I we obtain $ef = (L:K) = n$ so the formula

$$|\mathrm{N}(y)|_\mathfrak{p}^{1/n} = |y|$$

follows at once.

This formula for the extension of valuations will be useful in the last section of this chapter.

As another corollary to Theorem 3.2 we obtain a result about polynomials over R.

3.5 Proposition (Hensel's Lemma). Assume R (and K) as in Proposition 3.3, and let $\bar{R}$ denote the residue field of R. Suppose $F(X)$ is a monic polynomial in $R[X]$ which factors as $\bar{F}(X) = g(x)h(x)$ modulo $\mathfrak{p}$ with $g(X)$ and $h(X)$ relatively prime polynomials in $\bar{R}[X]$. Then $F(X) = G(X)H(X)$ with polynomials $G(X)$, $H(X)$ in $R[X]$ which satisfy $\bar{G}(X) = g(X)$, $\bar{H}(X) = h(X)$ and degree $G(X) =$ degree $g(X)$.

PROOF. Let A denote the R-algebra $R[X]/(F(X))$. Then A is a finitely generated free R-module [because $F(X)$ is monic]. The argument in the proof of Proposition (3.3) shows A is complete so Theorem 3.2 can be applied. We can identify $A/\mathfrak{p}A$ with $\bar{R}[X]/(\bar{F}(X))$ and since $g(X)$ and $h(X)$ are relatively prime it follows from CRT that

$$\bar{A} = A/\mathfrak{p}A \cong \bar{R}[X]/(g(X)) \oplus \bar{R}[X]/(h(X)).$$

Let e_1, e_2 denote the identity elements in the first and second factors, respectively. There are idempotents E_1, E_2 in A whose sum is 1 and whose product is zero and which satisfy $E_i + A\mathfrak{p} = e_i$. It follows that

$$A = AE_1 \oplus AE_2.$$

Let x denote the image of X in A. The characteristic polynomial of x on A is $F(X)$ since $A = R[X]/(F)$. Also by direct computation, the characteristic polynomial of x is a product $G_1(X)H_1(X)$ with $G_1(X)$ the characteristic polynomial of x on AE_1 and $H_1(X)$ the characteristic polynomial of x on AE_2. Here G_1, H_1 may be taken as monic polynomials. After we pass to the residue field $\bar{R}$ and compare the two decompositions of $\bar{A}$, it can be seen that the characteristic polynomial of x on AE_1 is $g(X)$ except that it need not be monic. Thus $\bar{G}_1(X)$ is a scalar multiple of $g(X)$. The scalars in $\bar{R}$ are the images of units in R so we may multiply G_1 by a unit and H_1 by the inverse to get polynomials G, H such that $\bar{G} = g$ and $F = GH = G_1 H_1$.

3.6 Corollary. Same notation as just above. Suppose $\bar{F}(X)$ has a root in $\bar{R}$ with multiplicity one. Then $F(X)$ has a root in R with multiplicity one.

PROOF. If $\bar{F}(X) = (X - \bar{a})h(X)$ and $h(a) \neq 0$ then these factors are relatively prime. Thus $F(X)$ has a monic factor of degree one which is relatively prime to the other factor.

3.7 Corollary. In the $p$adic field Q_p, the polynomial $X^{p-1} - 1$ has $p-1$ distinct roots.

PROOF. The polynomial $X^{p-1}-1$ is in $R[X]$ and has $p-1$ distinct roots in $\bar{R}$ because $\bar{R}$ is the field of p elements. Hence the polynomial splits in R by Corollary 3.6.

We close this section by describing the relation between ramification numbers for primes in a number field and in the completions.

Let K be an algebraic number field which is the quotient field of the DVR R with maximal ideal $\mathfrak{p}$. Let L be a finite dimensional extension of K, R' the integral closure of R and

$$\mathfrak{p}R' = \mathfrak{P}_1^{e_1} \cdots \mathfrak{P}_g^{e_g}, \qquad \mathfrak{P}_i \text{ distinct primes in } R'.$$

Let $|x|_{\mathfrak{p}}$ and $|y|_{\mathfrak{P}}$ denote the padic valuation on K and the $\mathfrak{P}$adic valuation on L, respectively, with $\mathfrak{P} = \mathfrak{P}_1$. Assume $|x|_{\mathfrak{p}} = |x|_{\mathfrak{P}}$ for x in K.

Now complete both fields. Let $K_{\mathfrak{p}}, L_{\mathfrak{P}}$ denote the completions; $\hat{R}, \hat{R}'$ the valuation rings, $\hat{\mathfrak{p}}, \hat{\mathfrak{P}}$ the maximal ideals.

3.8 Proposition. The following equations hold:

(a) $\hat{\mathfrak{p}} = \mathfrak{p}\hat{R}, \qquad \hat{\mathfrak{P}} = \mathfrak{P}\hat{R}'$,

(b) $\hat{\mathfrak{p}}\hat{R}' = (\hat{\mathfrak{P}})^{e_1}$,

(c) $e(\hat{\mathfrak{P}}/\hat{R}) = e(\mathfrak{P}/R) = e$,

(d) $f(\hat{\mathfrak{P}}/\hat{R}) = f(\mathfrak{P}/R) = f$,

(e) $(L_{\mathfrak{P}} : K_{\mathfrak{p}}) = ef$.

PROOF. Proposition 2.4 says $\mathfrak{p}$ and $\hat{\mathfrak{p}}$ can be generated by the same element of R and that $\mathfrak{P}$ and $\hat{\mathfrak{P}}$ can be generated by the same element of R'. Thus Statement (a) holds.

The ideals $\mathfrak{P}_2, \ldots, \mathfrak{P}_g$ each contain elements of R' outside of $\mathfrak{P}$ so $\mathfrak{P}_i\hat{R}' = \hat{R}'$ because $\hat{R}'$ is a valuation ring with $\mathfrak{P}\hat{R}'$ as maximal ideal. Now we have

$$\hat{\mathfrak{p}}\hat{R}' = (\mathfrak{p}R)\hat{R}' = (\mathfrak{P}_1^{e_1} \cdots \mathfrak{P}_g^{e_g})\hat{R}' = (\mathfrak{P}_1\hat{R}')^{e_1} = \hat{\mathfrak{P}}^{e_1}.$$

This proves (b) and (c) is simply a restatement of this. The equality of the relative degrees is a consequence of Corollary 2.7 which implies

$$(\hat{R}/\hat{\mathfrak{P}} : \hat{R}/\hat{\mathfrak{p}}) = (R'/\mathfrak{P} : R/\mathfrak{p}).$$

Finally (e) follows from Theorem 6.6 of Chapter I.

EXERCISE 1. Let θ denote a primitive p^a root of unity, p a prime, and let $L = Q(\theta)$. Use the facts about the ramification of (p) in L to conclude $(Q_p(\theta) : Q_p) = (L : Q)$.

EXERCISE 2. Conclude from Exercise 1 that the only roots of unity in $Q_p(\theta)$ having p power order are those in the cyclic group $\langle\theta\rangle$.

EXERCISE 3. Show the group of roots of unity in $Q_p(\theta)$ has order $p^a(p-1)$.

4. ARCHIMEDEAN VALUATIONS

In this section we determine all the fields which are complete with respect to an archimedean valuation. This information is necessary to determine how archimedean valuations on noncomplete fields extend to finite-dimensional field extensions.

4.1 Theorem (Ostrowski). Let K be a field which is complete with respect to an archimedean valuation $|x|$. Then K is isomorphic to either the real or complex field and the valuation is equivalent to the usual absolute value.

PROOF. Since the valuation is archimedean, the values $|n|$ for n in Z are unbounded. Thus K has characteristic zero and the restriction of $|x|$ to the rationals, Q, must be an archimedean valuation on Q. We know all the valuations on Q so we may replace the original valuation by an equivalent one to obtain that $|x|$ is the usual absolute value of x when x is in Q.

The field K is complete, so the completion of Q must be contained in K. Let $\mathcal{R}$ denote this completion. We know $\mathcal{R}$ is isomorphic to the field of reals and the valuation on $\mathcal{R}$ is the usual absolute value.

It may happen that K contains an element i such that $i^2+1=0$. If so then $\mathcal{C}=\mathcal{R}(i)$ is isomorphic to the complex field.

If K does not contain a root of X^2+1 we adjoin one to K to obtain a field $K(i)$. The valuation on K is extended to $K(i)$ by the rule

$$|a+ib| = (|a|^2+|b|^2)^{1/2}, \qquad a, b \in K.$$

It is straightforward to check that this does give a valuation on $K(i)$ and $K(i)$ is complete under this valuation.

The result of this argument is simply that we may assume $\mathcal{C} \subseteq K$ since a proof of the theorem for $K(i)$ will also prove the theorem for K.

We have arranged a normalization of the valuation so that on $\mathcal{R}$ it is the usual absolute value. Next we show it is the usual absolute value on $\mathcal{C}$ also.

4.2 Lemma. Let $|x|$ be a valuation on $\mathcal{C}$ which coincides with the usual absolute value on $\mathcal{R}$. Then $|a+ib|=(a^2+b^2)^{1/2}$ for a, $b \in \mathcal{R}$.

PROOF. Write $\|a+ib\|$ for $(a^2+b^2)^{1/2}$.

Now first notice that $i^4=1$ implies $|i|=1$. So for $\alpha=a+ib \in \mathcal{C}$ we have

$$|\alpha| = |a+ib| \leqslant |a|+|b| \leqslant \sqrt{2}(a^2+b^2)^{1/2} = \sqrt{2}\,\|\alpha\|.$$

The function $f(\alpha)=|\alpha|/\|\alpha\|$ for $\alpha \neq 0$ is bounded by $\sqrt{2}$. Since $f(\alpha^n)=f(\alpha)^n$ it follows that $f(\alpha) \leqslant 1$. However $f(\alpha^{-1})=f(\alpha)^{-1}$ implies $f(\alpha) \geqslant 1$ for all $\alpha \neq 0$. Thus $f(\alpha)=1$ as required.

Now we come to the main part of the proof. We have $\mathscr{C} \subseteq K$ and it is necessary to show equality. Suppose there is some $z \in K$ but $z \notin \mathscr{C}$. Let

$$m = \operatorname*{glb}_{\alpha \in \mathscr{C}} |z-\alpha|.$$

Let us prove first there is some α_0 in $\mathscr{C}$ for which $m = |z-\alpha_0|$.

For any positive number ε, the set of complex numbers α for which $|z-\alpha| \leqslant m+\varepsilon$ is contained in the set

$$\{\beta \in \mathscr{C} \mid |\beta| \leqslant m+\varepsilon+|z|\}.$$

This is a disc and the function $f(\beta) = |z-\beta|$ is a continuous function from the disc into the reals. The minimum of this function is attained at some α_0 in $\mathscr{C}$. The assumption that z was not in $\mathscr{C}$ implies $z-\alpha_0$ is not in $\mathscr{C}$. We may now replace our original z with $z-\alpha_0$ to obtain

$$\text{(a)} \quad z \notin \mathscr{C} \qquad \text{and} \qquad \text{(b)} \quad m = |z| \leqslant |z-\alpha|, \qquad \alpha \in \mathscr{C}.$$

Notice that $m = |z| \neq 0$ because z is not in $\mathscr{C}$. The next step shows $|z-\alpha| = m$ whenever α is in $\mathscr{C}$ and $|\alpha| < m$.

Let n denote any positive integer and ω a primitive nth root of unity in $\mathscr{C}$. The factorization

$$z^n - \alpha^n = (z-\alpha)(z-\omega\alpha)\cdots(z-\omega^{n-1}\alpha)$$

implies

$$|z-\alpha||z-\omega\alpha|\cdots|z-\omega^{n-1}\alpha| = |z^n-\alpha^n| \leqslant |z|^n + |\alpha|^n.$$

Each term $|z-\omega^i\alpha| \geqslant m$ so

$$|z-\alpha|\, m^{n-1} \leqslant |z|^n(1+|\alpha|^n/|z|^n) = m^n(1+|\alpha|^n/m^n).$$

This implies

$$|z-\alpha| \leqslant m(1+|\alpha|^n/m^n).$$

This holds for any integer n so if $|\alpha| < m$ we let n increase without bound. Then it follows $|z-\alpha| \leqslant m$.

The minimal choice of m forces $|z-\alpha| = m$. If we now replace z by $z-\alpha$, for any α in $\mathscr{C}$ with $|\alpha| < m$, then conditions (a) and (b) above are satisfied. We may repeat the above procedure and obtain $|z-\alpha-\beta| = m$ whenever $\beta \in \mathscr{C}$ and $|\beta| < m$. In particular $|z-2\alpha| = m$. Repeat this and by induction we obtain for any positive integer n,

$$|z-n\alpha| = m, \qquad \alpha \in \mathscr{C}, \qquad |\alpha| < m.$$

But any complex number β can be written as

$$\beta = n\alpha, \qquad |\alpha| < m, \qquad n = \text{positive integer}$$

because $m \neq 0$. Thus $|z-\alpha| = m$ for all $\alpha \in \mathscr{C}$. This implies for any $\alpha, \beta \in \mathscr{C}$,

$$|\alpha-\beta| \leqslant |z-\alpha|+|z-\beta| = 2m$$

which is clearly not the case if $\alpha = 3m$, $\beta = 0$. This contradiction is a result of the assumption that $K \neq \mathscr{C}$. Hence $K = \mathscr{C}$ and by the Lemma 4.2, the valuation is uniquely determined.

This theorem allows one to describe all the archimedean valuations of an algebraic number field.

Let K be an algebraic number field and $|x|_1$ an archimedean valuation on K. The completion, $\hat{K}$, of K with respect to the valuation must be a copy of the reals or complexes with the usual valuation. This means there is an isomorphism, ϕ, of $\hat{K}$ with $\mathscr{R}$ or $\mathscr{C}$ such that

$$|x|_1 = |\phi(x)|$$

for all x in $\hat{K}$ and where $|\phi(x)|$ denotes the usual absolute value on $\mathscr{R}$ or $\mathscr{C}$. We compose ϕ with the natural imbedding of K into $\hat{K}$ to see that $|x|_1$ is determined by the imbedding of K into $\mathscr{R}$ or $\mathscr{C}$. So far we have proved the following:

4.3 Lemma. Every archimedean valuation of K is equivalent to one obtained by the formula $|x|_1 = |\phi(x)|$ for all x in K where ϕ is an imbedding of K into $\mathscr{R}$ or $\mathscr{C}$ and $|\phi(x)|$ is the usual absolute value.

We shall next determine which of these are inequivalent. The notation of Section 11, Chapter I will be used. Namely let $\sigma_1, \ldots, \sigma_r$ be the distinct imbeddings of K into $\mathscr{R}$ and let $\sigma_{r+1}, \ldots, \sigma_{r+s}, \bar{\sigma}_{r+1}, \ldots, \bar{\sigma}_{r+s}$ be the $2s$ distinct imbeddings of K into $\mathscr{C}$. Here $\bar{\sigma}$ means the map defined by taking the complex conjugate of $\sigma(x)$ for the value of $\bar{\sigma}(x)$.

Since a complex number and its conjugate have the same absolute value we find

$$|\sigma(x)| = |\bar{\sigma}(x)|$$

for all x. So the imbedding $\bar{\sigma}_{r+j}$ gives rise to the same valuation as the imbedding σ_{r+j}. It will be seen that these are the only relations between the valuations.

Let $|x|_i = |\sigma_i(x)|$ for $1 \leqslant i \leqslant r+s$. Each archimedean valuation of K is equivalent to one of the $|x|_i$ and these are inequivalent. This has been proved already in the proof of the Dirichlet Unit Theorem (11.19). It was seen there that for each i there exists u_i in K such that

$$|u_i|_i > 1 \qquad \text{and} \qquad |u_i|_j < 1 \; i \neq j.$$

This means $|\ \ |_i$ is not equivalent to any other $|\ \ |_j$. We summarize these calculations.

4.4 Theorem. Let K be an algebraic number field; $\sigma_1, \ldots, \sigma_r$ the real imbeddings of K; $\sigma_{r+1}, \ldots, \sigma_{r+s}$ one member of each conjugate pair of complex imbeddings. Each archimedean valuation is equivalent to one and only one of the valuations defined by $|x|_i = |\sigma_i(x)|$.

The terminology introduced for the rationals will also be used in the case of an algebraic number field, K. A *prime* of K is an equivalence class of valuations. A prime is called an *infinite prime* if it contains an archimedean valuation. The other primes are called *finite*. An infinite prime is called a *real prime* of K if the completion at that prime is the real field. The other infinite primes are *complex*.

If we now consider a finite-dimensional extension L of K and an archimedean valuation on K, then the valuations of L which restrict to the given one on K are easily described.

Let σ be an imbedding of K into $\mathscr{R}$ or $\mathscr{C}$ such that the valuation on K is $|x|_1 = |\sigma(x)|$. We may regard σ as an imbedding into a field containing an algebraic closure of K. By Galois theory there exist $(L:K) = n$ extensions of σ to imbeddings of L into $\mathscr{C}$. (Even though $\sigma(K) \subset \mathscr{R}$ we cannot assert the extended map will carry L into $\mathscr{R}$.)

We shall not try to describe any more precisely what the extensions are but at least we see extensions to L can be described by Galois theory. In the next section a method will be described.

5. LOCAL NORMS AND TRACES AND THE PRODUCT FORMULA

Let K denote any field and $L = K(\theta)$ a finite-dimensional, separable extension. Let $f(X)$ be the minimum polynomial of θ over K. Take $\mathfrak{p}$ either an archimedean prime of K or a prime whose valuation ring R is a DVR. Let $\mathfrak{P}_1, \ldots, \mathfrak{P}_g$ denote the primes of L which extend $\mathfrak{p}$ and let L_i denote the completion of L at $\mathfrak{P}_i$.

5.1 Theorem. Let $f(X) = f_1(X) \cdots f_t(X)$ be the factorization of $f(X)$ as a product of irreducible polynomials over $K_\mathfrak{p}$. If the $f_i(X)$ are suitably numbered then $g = t$ and

$$L_i \cong K_\mathfrak{p}[X]/(f_i(X)).$$

We also have $L \otimes_K K_\mathfrak{p} \cong L_1 \oplus \cdots \oplus L_g$.

PROOF. Since $f(X)$ is a separable polynomial we obtain (by CRT)

$$L \otimes K_\mathfrak{p} = K_\mathfrak{p}[X]/(f(X)) \cong \sum \oplus K_\mathfrak{p}[X]/(f_i(X)).$$

We show the L_i are the direct summands of $L \otimes K_\mathfrak{p}$.

Let $x \to x_i$ denote the imbedding of L into L_i. Then the map on $L \otimes K_\mathfrak{p}$ determined by

$$\varphi_i : x \otimes k \to x_i k$$

is a homomorphism into L_i. This is in fact onto L_i as one easily verifies by examining the Cauchy sequences in L_i. The only fields which are homomorphic images of $L \otimes K_\mathfrak{p}$ are the direct summands. So each L_i is a direct summand. No two of the φ_i have the same image because the composite map sending x to $\varphi_i(x \otimes 1)$ is the natural imbedding of L into L_i and the L_i are distinct completions of L. Thus the direct sum $L_1 \oplus \cdots \oplus L_g$ is a direct summand of $L \otimes K_\mathfrak{p}$. But now

$$\sum (L_i : K_\mathfrak{p}) = \sum e_i f_i = (L : K) = (L \otimes K_\mathfrak{p} : K_\mathfrak{p})$$

by Proposition 3.8 and Corollary 6.7 of Chapter I. Thus $L \otimes K_\mathfrak{p} = L_1 \oplus \cdots \oplus L_g$.

Next we establish a further connection between the extension L/K and the various completions $L_i/K_\mathfrak{p}$.

5.2 Theorem. For each element y in L we have

(i) $\operatorname{char\,poly}_{L/K}(y) = \prod_i \operatorname{char\,poly}_{L_i/K_\mathfrak{p}}(y)$

(ii) $\mathrm{N}_{L/K}(y) = \prod_i \mathrm{N}_{L_i/K_\mathfrak{p}}(y)$

(iii) $\mathrm{T}_{L/K}(y) = \sum_i \mathrm{T}_{L_i/K_\mathfrak{p}}(y)$

PROOF. Let $x_1, \ldots, x_n$ be a K-basis for L. For y in L let r_y denote the matrix $|a_{ij}|$ defined by $x_i y = \sum a_{ij} x_j$.

Now $x_1 \otimes 1, \ldots, x_n \otimes 1$ is a $K_\mathfrak{p}$ basis for $L \otimes K_\mathfrak{p}$ and multiplication by $y \otimes 1$ induces a linear transformation with matrix $|a_{ij}|$ also.

Use a basis for $L \otimes K_\mathfrak{p}$ which is compatible with the decomposition $L_1 \oplus \cdots \oplus L_g$. It follows that a matrix for $r_y \otimes 1$ has the form

$$\operatorname{diag}\{r_1, \ldots, r_g\},$$

where r_i is a matrix for the regular representation of y on L_i over $K_\mathfrak{p}$. Now it follows that

$$\begin{aligned} \operatorname{char\,poly}_{L/K}(y) &= \det(XI - r_y) \\ &= \prod \det(XI - r_i) \\ &= \prod \operatorname{char\,poly}_{L_i/K_\mathfrak{p}}(y). \end{aligned}$$

The statements (ii) and (iii) follow by examining the coefficients of the characteristic polynomials.

Now assume $\mathfrak{p}$ is a nonarchimedean prime on K so it may be considered as an ideal in the valuation ring R. Let R' be the integral closure of R in L and $\mathfrak{p}R'$ has the factorization

$$\mathfrak{p}R' = \mathfrak{P}_1^{e_1} \cdots \mathfrak{P}_g^{e_g}.$$

If L_i is the $\mathfrak{P}_i$adic completion of L then $(L_i : K_\mathfrak{p}) = e_i f_i$. We have seen that the valuation on L_i can be given in the form

$$|y|_i = |\mathrm{N}_{L_i/K_\mathfrak{p}}(y)|_\mathfrak{p}^{1/e_i f_i}.$$

We consider this for y in L and observe that

$$\prod_i |y|^{e_i f_i} = \prod |\mathrm{N}_{L_i/K_\mathfrak{p}}(y)|_\mathfrak{p} = |\mathrm{N}_{L/K}(y)|_\mathfrak{p}.$$

If we replace the valuation $|\ \ |_i$ on L_i by its $e_i f_i$ power we then obtain the following.

5.3 Lemma. If the valuations $|\ \ |_i$ on L are suitably normalized, the following formula holds:

$$\prod_i |y|_i = |\mathrm{N}_{L/K}(y)|_\mathfrak{p}.$$

A similar normalization can be obtained for archimedean valuations. Assume $|x|_\mathfrak{p}$ is the valuation on K which is obtained by a particular imbedding σ of K into $\mathcal{R}$ or $\mathscr{C}$ by the formula $|x|_\mathfrak{p} = |\sigma(x)|$, where the last denotes the usual absolute value. All the extensions of $|x|_\mathfrak{p}$ to L are determined by imbeddings σ_i of L into $\mathscr{C}$ which agree with σ on K. The valuations on L are normalized as follows:

I. $\sigma(K) \subset \mathcal{R}$ so $\mathfrak{p}$ is a real infinite prime. Then

$$\begin{aligned} |y|_i &= |\sigma_i(y)| && \text{if} \quad \sigma_i(L) \subseteq \mathcal{R} \\ &= |\sigma_i(y)|^2 && \text{if} \quad \sigma_i(L) \not\subset \mathcal{R}. \end{aligned}$$

In this case the term $|\sigma_i(y)|^2$ can be written as $|\sigma_i(y)\overline{\sigma_i}(y)|$. So

$$\prod |y|_i = \prod |\sigma_k(y)|$$

where σ_k runs through all extensions of σ to imbeddings of L into $\mathscr{C}$. These images may be regarded as landing in a normal extension of K so the product is $|\mathrm{N}_{L/K}(y)|_\mathfrak{p}$.

II. $\sigma(K) \not\subset \mathcal{R}$ so $\mathfrak{p}$ is a complex infinite prime.
Then $|y|_i = |\sigma_i(y)|$ where σ_i runs through all extensions of σ. As above,

$$\prod |y|_i = |\mathrm{N}_{L/K}(y)|_\mathfrak{p}.$$

These normalizations enable us to prove a product formula for algebraic number fields.

Product Formula. Let K be an algebraic number field. For each prime $\mathfrak{P}$ of K (finite or infinite) there is a valuation $|x|_\mathfrak{P}$ in $\mathfrak{P}$ such that for each $x \neq 0$ in K the formula holds

$$\prod_\mathfrak{P} |x|_\mathfrak{P} = 1.$$

PROOF. Let $\mathfrak{p}$ be any prime of the rational field Q and $\mathfrak{P}_1, \ldots, \mathfrak{P}_g$ the distinct primes of K which extend $\mathfrak{p}$. We may select valuations $|x|_{\mathfrak{P}_i}$ in $\mathfrak{P}_i$ such that

$$\prod_i |x|_{\mathfrak{P}_i} = |\mathrm{N}_{K/Q}(x)|_{\mathfrak{p}}$$

where $|y|_{\mathfrak{p}}$ is the normalized valuation on Q defined in Section 1. Let us write $\mathfrak{P}_i|\mathfrak{p}$ to denote that $\mathfrak{P}_i$ is a prime extending $\mathfrak{p}$. Then

$$\prod_{\mathfrak{P}} |x|_{\mathfrak{P}} = \prod_{\mathfrak{p}} \prod_{\mathfrak{P}_i|\mathfrak{p}} |x|_{\mathfrak{P}_i} = \prod_{\mathfrak{p}} |\mathrm{N}_{K/Q}(x)|_{\mathfrak{p}} = 1$$

by the product formula for Q.

EXAMPLES. Let $K = Q(\theta)$ with $\theta^3 = 2$.

There is one real imbedding of K obtained by taking $\sigma_1(\theta)$ = real cube root of $2 = 2^{1/3}$. The pair of complex imbeddings are found by $\sigma_2(\theta) = \omega 2^{1/3}$ and $\bar{\sigma}_2(\theta) = \bar{\omega} 2^{1/3}$ where ω is a primitive cube root of unity. Thus K has two archimedean primes, one real and one complex.

To discuss the nonarchimedean primes of K it is necessary to know either how $X^3 - 2$ factors over Q_p, the $p$adic completion of Q or how the prime p factors in the ring of algebraic integers in K. This latter information has been given for some primes p in Chapter I, Section 7.

For example the prime $p = 7$ remains prime. That is there is a unique prime of K, say $\mathfrak{P}$, containing 7. Hence the 7adic valuation on Q has a unique extension to K. After taking completions we find $(K_{\mathfrak{P}} : Q_7) = 3$ by Proposition 3.7(d) and (e).

The prime $p = 29$ is contained in two primes $\mathfrak{P}_1$, $\mathfrak{P}_2$ of K having relative degrees $f_1 = 1$ and $f_2 = 2$. Thus the completions satisfy

$$(K_{P_1} : Q_{29}) = 1, \qquad (K_{P_2} : Q_{29}) = 2.$$

This gives an example where $K \neq Q$ but $K_{\mathfrak{P}_1} = Q_{29}$. This will happen, of course, whenever $ef = 1$.

The ramified primes $p = 2$ and $p = 3$ are each contained in unique primes $\mathfrak{P}_2$, $\mathfrak{P}_3$ of K with relative degrees equal to one and ramification numbers equal to three. The completions $K_{\mathfrak{P}_2}, K_{\mathfrak{P}_3}$ have dimension three over Q_2, Q_3, respectively.

We can compute the extended valuations by the formulas obtained in this section. For example $\mathrm{N}_{K_{\mathfrak{P}_2}/Q_2} = \mathrm{N}_{K/Q}$ on K since 2 has a unique prime of K over it. Thus the valuation on $K_{\mathfrak{P}_2}$ is given by

$$|x| = |\mathrm{N}_{K_{\mathfrak{P}_2}/Q_2}(x)|_2^{1/3}, \qquad x \text{ in } K_{\mathfrak{P}_2}.$$

When we restrict x to lie in K we obtain

$$|x| = |\mathrm{N}_{K/Q}|_2^{1/3}.$$

For example, if we normalize $|\ \ |_2$ so that $|2|_2 = \frac{1}{2}$ then

$$|\theta| = |\mathrm{N}_{K/Q}(\theta)|_2^{1/3} = |2|_2^{1/3} = (\tfrac{1}{2})^{1/3}.$$

EXERCISE 1. Let K be an algebraic number field, $L = K(\theta)$ an extension generated by an element θ with minimum polynomial $f(X)$. Let $\mathfrak{p}$ be a prime of K and suppose $f(X)$ factors as $f_1(X) \cdots f_g(X)$ over the completion $K_{\mathfrak{p}}$. Then the prime $\mathfrak{p}$ of K has exactly g extensions to primes $\mathfrak{P}_1, \ldots, \mathfrak{P}_g$ of L and with suitable numbering $e(\mathfrak{P}_i/\mathfrak{p}) f(\mathfrak{P}_i/\mathfrak{p}) = \text{degree } f_i(X)$.

EXERCISE 2. Let $L = Q(\theta)$ where θ is a root of $X^4 - 14 = 0$.

(i) Show the prime 5 of Q has two factors $\mathfrak{P}_1, \mathfrak{P}_2$ in L and $(L_{\mathfrak{P}_i} : Q_5) = 2$ for $i = 1, 2$.

Procedure. Use Exercise 5 following Chapter II, Section 2 to see that 14 is a square but not a fourth power in Q_5.

(ii) Show that the prime 11 has three factors $\mathfrak{P}_1, \mathfrak{P}_2, \mathfrak{P}_3$ in L and $L_{\mathfrak{P}_1} = L_{\mathfrak{P}_2} = Q_{11}$ while $(L_{\mathfrak{P}_3} : Q_{11}) = 2$.

(iii) The prime 13 has four prime factors in L.

Let $k = \mathrm{GF}(q)$, q a prime power, and $K = k(x)$ the field of rational functions in one indeterminant.

EXERCISE 3. For each monic irreducible polynomial $p(x)$ in $k[x]$ there is a prime of K containing the valuation defined by

$$|y|_p = q^{-v(y)}$$

where v is the exponential valuation

$$v(p(x)^t a(x)/b(x)) = t \deg p(x)$$

if $a(x), b(x)$ are polynomials not divisible by $p(x)$.

There is a prime p on K containing the valuation

$$|a(x)/b(x)|_p = q^{\deg a(x) - \deg b(x)}.$$

Show that these primes are mutually inequivalent and every prime of K is equivalent to one of these.

EXERCISE 4. With the valuations of K normalized as above show the product formula holds; that is $y \neq 0$ in K implies

$$\prod_{\text{all } p} |y|_p = 1.$$

EXERCISE 5. Let p be the prime of K corresponding to the irreducible polynomial $p(x) = x$. The completion K_p is naturally isomorphic to the

Laurent series field $k\langle x\rangle$ consisting of all elements

$$x^n(a_0+a_1x+\cdots)$$

with n any integer and the $a_i \in k$.

EXERCISE 6. If p is the prime corresponding to an irreducible polynomial of degree f then K_p is isomorphic to the Laurent series field $F\langle x\rangle$ with $F = \mathrm{GF}(q^f)$.

EXERCISE 7. The completion of K with respect to the remaining prime is isomorphic to $k\langle y\rangle$ with $y = 1/x$.

Chapter III

DECOMPOSITION GROUPS AND THE ARTIN MAP

1. DECOMPOSITION AND INERTIA GROUPS

We want to study the decomposition of primes of an algebraic number field K in a finite-dimensional Galois extension L with Galois group G. There are connections between ramification numbers, relative degrees of a prime and certain subgroups of G. Parts of the discussion work equally well for finite and infinite primes so we shall make the appropriate definitions.

Let $\mathfrak{p}$ be an infinite prime of K and $\mathfrak{P}_1, \ldots, \mathfrak{P}_g$ the distinct primes of L which extend $\mathfrak{p}$. We say $\mathfrak{P}_i$ is *unramified* if the completions $L_{\mathfrak{P}_i}$ and $K_\mathfrak{p}$ are equal; that is if $\mathfrak{P}_i$ and $\mathfrak{p}$ are both real or both complex. In this case we set $e(\mathfrak{P}_i/\mathfrak{p}) = f(\mathfrak{P}_i/\mathfrak{p}) = 1$. In the remaining case $\mathfrak{p}$ is real and $\mathfrak{P}_i$ complex. We set $e(\mathfrak{P}_i/\mathfrak{p}) = 2, f(\mathfrak{P}_i/\mathfrak{p}) = 1$. We then have

$$(L_{\mathfrak{P}_i} : K_\mathfrak{p}) = e_i f_i$$

$$\sum (L_{\mathfrak{P}_i} : K_\mathfrak{p}) = (L : K),$$

just as in the case of finite primes. These definitions are the same whether L is Galois over K or not. Just as in the finite case the Galois group is transitive on the $\mathfrak{P}_i$ and all the e_i are equal.

Now let $\mathfrak{p}$ denote any prime of K, finite or infinite, and let its decomposition in L be

$$\mathfrak{p} = (\mathfrak{P}_1 \cdots \mathfrak{P}_g)^e.$$

Set $\mathfrak{P} = \mathfrak{P}_1$ and

$$G(\mathfrak{P}) = \{\sigma \in G | \sigma(\mathfrak{P}) = \mathfrak{P}\}.$$

We call $G(\mathfrak{P})$ the *decomposition group of* $\mathfrak{P}$.

1.1 Lemma. $|G(\mathfrak{P})| = ef$ with $f =$ relative degree of $\mathfrak{P}$ over $\mathfrak{p}$.

PROOF. The Galois group G is transitive on the $\mathfrak{P}_i$ so the subgroup fixing one of them has index $[G : G(\mathfrak{P})] = g$. The result now follows from the relation $|G| = (L : K) = efg$.

Let $K_\mathfrak{p}$ and $L_\mathfrak{P}$ denote the completions at the indicated primes. Each element σ in $G(\mathfrak{P})$ leaves fixed the $\mathfrak{P}$adic valuation on L and so there is a unique element σ^* in $G(L_\mathfrak{P}/K_\mathfrak{p})$ with the property $\sigma^* = \sigma$ on L (by Chapter II, Theorem 2.2 applied to $\sigma : L \to L$).

This shows that the correspondence $\sigma \to \sigma^*$ is one to one. It is a group homomorphism because of the uniqueness property. That is $\sigma^*\tau^*$ and $(\sigma\tau)^*$ are both extensions of $\sigma\tau$ and hence are equal. Finally we recall $(L_\mathfrak{P} : K_\mathfrak{p}) = ef = |G(\mathfrak{P})|$ so $G(\mathfrak{P})$ maps onto the full Galois group of $L_\mathfrak{P}$ over $K_\mathfrak{p}$.

From now on we shall identify $G(\mathfrak{P})$ with $G(L_\mathfrak{P}/K_\mathfrak{p})$.

Now assume $\mathfrak{p}$ is a *finite* prime of K.

Let R, R' denote the valuation rings in $K_\mathfrak{p}$ and $L_\mathfrak{P}$, respectively; the maximal ideals are $\mathfrak{p}$ and $\mathfrak{P}$; the residue fields are $\bar{R}$ and $\bar{R}'$. These are finite fields with $(\bar{R}' : \bar{R}) = f$. For $\sigma \in G(\mathfrak{P})$ let $\bar{\sigma}$ denote the automorphism of $\bar{R}'$ defined by

$$\bar{\sigma}(x + \mathfrak{P}) = \sigma(x) + \mathfrak{P}.$$

The mapping $\sigma \to \bar{\sigma}$ is evidently a homomorphism of $G(\mathfrak{P})$ into $G(\bar{R}'/\bar{R})$. The kernel of this map is denoted by $T(\mathfrak{P})$ and is called the *inertia group* of $\mathfrak{P}$. Clearly $T(\mathfrak{P})$ reflects the ramification of $\mathfrak{P}$. We shall describe $T(\mathfrak{P})$ more precisely and make this connection clear. First the following definition is needed.

Definition. A finite prime $\mathfrak{p}$ of K is *totally ramified* in L if $\mathfrak{p}$ has only one prime divisor $\mathfrak{P}$ in L and the relative degree $f(\mathfrak{P}/\mathfrak{p})$ equals one.

When this situation holds it necessarily follows that the ramification number of $\mathfrak{P}$ over $\mathfrak{p}$ equals the degree of the field extension.

EXERCISE. Let $K \subset E \subset L$ be a tower of algebraic number fields and let $\mathfrak{p}$ be a prime of K which is totally ramified in L. Then $\mathfrak{p}$ is totally ramified in E.

Now we describe some properties of $T(\mathfrak{P})$ and $G(\mathfrak{P})$.

1.2 Theorem

(a) The map $\sigma \to \bar{\sigma}$ carries $G(\mathfrak{P})$ onto $G(\bar{R}'/\bar{R})$;
(b) $|T(\mathfrak{P})| = e(\mathfrak{P}/\mathfrak{p}) = e$;

(c) the subfield E of $L_{\mathfrak{P}}$ left fixed by $T(\mathfrak{P})$ is an unramified extension of $K_{\mathfrak{p}}$ and

$$f(E/K_{\mathfrak{p}}) = (E : K_{\mathfrak{p}}) = f(L_{\mathfrak{P}}/K_{\mathfrak{p}}) = f;$$

(d) the extension $L_{\mathfrak{P}}/E$ is totally ramified having $e(L_{\mathfrak{P}}/E) = (L_{\mathfrak{P}} : E) = e$.

PROOF. Let $\bar{R} \cong GF(q)$, $\bar{R}' \cong GF(q^f)$ and let $d = q^f - 1$. Consider the polynomial

$$A(X) = X^d - 1 \qquad \in R[X].$$

When we pass to $\bar{R}$, $\bar{A}(X)$ splits in $\bar{R}'$ and in fact the roots of $\bar{A}(X)$ are precisely the nonzero elements in $\bar{R}'$. One of the elements in $\bar{R}'$ has degree f over $\bar{R}$ so $\bar{A}(X)$ must factor as

$$\bar{A}(X) = b(X)c(X)$$

with $b(X)$ irreducible of degree f over $\bar{R}$. The polynomials $b(X)$ and $c(X)$ are relatively prime because $\bar{A}(X)$ has no repeated roots. Hensel's lemma (Chapter II, Proposition 3.5) may be applied to obtain the factorization

$$A(X) = B(X)C(X)$$

with $B(X)$ in $R[X]$ irreducible of degree f and $\bar{B}(X) = b(X)$. By Chapter II, Corollary 3.6 a simple root of $b(X)$ in $\bar{R}'$ is the image of a root θ in R' of $B(X)$.

Now let $E = K_{\mathfrak{p}}(\theta)$ and S = valuation ring in E. Since θ is a dth root of unity all roots of $B(X)$ are powers of θ. Thus E is the splitting field of $B(X)$ so E is normal over $K_{\mathfrak{p}}$. Moreover $(E : K_{\mathfrak{p}}) = f =$ degree $B(X)$. Any element σ in $G(\mathfrak{P})$ permutes the roots of $B(X)$ in exactly the same way as $\bar{\sigma}$ permutes the roots of $\bar{B}(X)$ since the roots correspond one to one. Thus σ fixes θ if and only if $\bar{\sigma}$ fixes $\bar{\theta}$. It follows that $T(\mathfrak{P})$ is exactly the subgroup fixing E because any $\bar{\sigma}$ fixing $\bar{\theta}$ is the identity on $\bar{R}'$. Now

$$[G(\mathfrak{P}) : T(\mathfrak{P})] = (E : K_{\mathfrak{p}}) = f$$

and so $|T(\mathfrak{P})| = e$ because $|G(\mathfrak{P})| = ef$.

Since $|G(\bar{R}'/\bar{R})| = f$ it follows that $G(\mathfrak{P})$ maps onto $G(\bar{R}'/\bar{R})$. This proves Statements (a), (b), and (c).

To prove Statement (d) we observe that

$$\bar{R}' \supseteq \bar{S} \supseteq \bar{R}(\theta) = \bar{R}'$$

so in fact $\bar{S} = \bar{R}'$. Thus $f(L_{\mathfrak{P}}/E) = 1$ and $(L_{\mathfrak{P}} : E) = e(L_{\mathfrak{P}}/E)$ as required.

We now translate this local data into global data.

The chain of subgroups

$$1 < T(\mathfrak{P}) < G(\mathfrak{P}) < G = G(L/K)$$

corresponds by Galois theory to the chain of subfields

$$L > K^{T(\mathfrak{P})} > K^{G(\mathfrak{P})} > K.$$

We call $K^{T(\mathfrak{P})}$ and $K^{G(\mathfrak{P})}$ the *inertia field* and *decomposition field* of $\mathfrak{P}$ over K. Let the primes in these fields be denoted by

$$\mathfrak{P}_T = \mathfrak{P} \cap K^{T(\mathfrak{P})}, \qquad \mathfrak{P}_Z = \mathfrak{P} \cap K^{G(\mathfrak{P})}.$$

1.3 Theorem

(a) $\mathfrak{P}_Z$ is unramified in $K^{T(\mathfrak{P})}$. The only divisor of $\mathfrak{P}_Z$ in $K^{T(\mathfrak{P})}$ is $\mathfrak{P}_T$ and $f(\mathfrak{P}_T/\mathfrak{P}_Z) = f(\mathfrak{P}/\mathfrak{p})$.

(b) $\mathfrak{P}_T$ is totally ramified in L. $\mathfrak{P}$ is the only divisor in L of $\mathfrak{P}_T$ and $e(\mathfrak{P}/\mathfrak{P}_T) = e(\mathfrak{P}/\mathfrak{p})$.

PROOF. The group $G(\mathfrak{P})$ is the Galois group of L over $K^{G(\mathfrak{P})}$ and so $G(\mathfrak{P})$ is transitive on the divisors in L of $\mathfrak{P}_Z$. However $\mathfrak{P}$ is one of those divisors and $G(\mathfrak{P})$ leaves $\mathfrak{P}$ fixed. Hence $\mathfrak{P}_Z$ has only one divisor in L. Now we complete all the fields at the primes above $\mathfrak{P}_Z$. The statements about relative degree and ramification hold in the complete case by Theorem 1.2 and so they hold also in the global case.

The group $G(\mathfrak{P})$ need not be normal in G so this complicates the discussion of the factorization of $\mathfrak{p}$ in $K^{G(\mathfrak{P})}$. We shall be able to describe the situation shortly but for now we give the easier case. We shall refer to this in the case where G is abelian.

1.4 Proposition. Suppose $G(\mathfrak{P})$ is normal in G. Then $\mathfrak{p}$ has the factorization $\mathfrak{p} = \mathfrak{p}_1 \cdots \mathfrak{p}_g$ in $K^{G(\mathfrak{P})}$ where $e(\mathfrak{p}_i/\mathfrak{p}) = f(\mathfrak{p}_i/\mathfrak{p}) = 1$.

PROOF. From above we see $e(\mathfrak{P}/\mathfrak{P}_Z) = e(\mathfrak{P}/\mathfrak{p})$ and $f(\mathfrak{P}/\mathfrak{P}_Z) = f(\mathfrak{P}/\mathfrak{p})$ so the multiplicative property of the es and fs implies

$$e(\mathfrak{P}_Z/\mathfrak{p}) = f(\mathfrak{P}_Z/\mathfrak{p}) = 1.$$

Our additional hypothesis that $G(\mathfrak{P})$ is normal in G means $K^{G(\mathfrak{P})}$ is Galois over K so all prime divisors of $\mathfrak{p}$ in $K^{G(\mathfrak{P})}$ have the same ramification number and same relative degree as $\mathfrak{P}_Z$ over $\mathfrak{p}$.

2. THE FROBENIUS AUTOMORPHISM

We continue in the context of the previous section except that we suppose $\mathfrak{p}$ is unramified in L.

In this case the inertia group $T(\mathfrak{P})$ has order $e = 1$ so in particular $G(\mathfrak{P}) \cong G(\bar{R}'/\bar{R})$. This last group is the Galois group of a finite field and so it is cyclic of order f. Then the decomposition group $G(\mathfrak{P})$ is cyclic of order f.

Even more can be said. The Galois group of $\bar{R}' = GF(q^f)$ over $\bar{R} = GF(q)$ is generated by a distinguished automorphism

$$\bar{x} \to \bar{x}^q.$$

This means there is a *unique* automorphism $\sigma \in G(\mathfrak{P})$ which satisfies

$$\sigma(x) \equiv x^q \bmod \mathfrak{P}, \qquad x \in R' \tag{2.1}$$

This automorphism is called the *Frobenius automorphism* of $\mathfrak{P}$. To indicate the dependence upon $\mathfrak{P}$, L, and K we denote it by

$$\sigma = \left[\frac{L/K}{\mathfrak{P}}\right].$$

This notion plays an important part in the sequel.

EXERCISE. Show there is a primitive q^f-1 root of unity, β, in $L_{\mathfrak{P}}$ and the Frobenius automorphism σ is uniquely determined by the condition $\sigma(\beta) = \beta^q$.

We shall make a number of calculations to determine the behavior of the Frobenius automorphism as a function of $\mathfrak{P}, K, L$.

Suppose $\mathfrak{P}_j$ is another prime of L dividing $\mathfrak{p}$. There exists $\tau \in G(L/K)$ with $\tau(\mathfrak{P}) = \mathfrak{P}_j$. If σ is an element of $G(\mathfrak{P})$ then $\tau\sigma\tau^{-1}$ is in $G(\mathfrak{P}_j)$ and conversely. So we have

$$G(\mathfrak{P}_j) = G(\tau(\mathfrak{P})) = \tau G(\mathfrak{P})\tau^{-1}.$$

2.2 Property

$$\left[\frac{L/K}{\tau(\mathfrak{P})}\right] = \tau\left[\frac{L/K}{\mathfrak{P}}\right]\tau^{-1}.$$

PROOF. Any element in the ring of algebraic integers of L can be written as $\tau^{-1}(x)$ with x an algebraic integer. By (2.1) we obtain

$$\left[\frac{L/K}{\mathfrak{P}}\right]\tau^{-1}(x) \equiv \tau^{-1}(x)^q \bmod \mathfrak{P}.$$

Apply τ to this and obtain

$$\tau\left[\frac{L/K}{\mathfrak{P}}\right]\tau^{-1}(x) \equiv x^q \bmod \tau(\mathfrak{P}).$$

The uniqueness of the Frobenius automorphism gives the desired result.

Suppose $L \supset E \supset K$ and $\mathfrak{P} \cap E = \mathfrak{p}_0$. Since $\mathfrak{p}$ is unramified in L, $\mathfrak{p}_0$ is also unramified in L. There is defined then a Frobenius automorphism of $\mathfrak{P}$ for the extension L/E. It is related to that for L/K as follows.

2.3 Property

$$\left[\frac{L/K}{\mathfrak{P}}\right]^{f(\mathfrak{p}_0/\mathfrak{p})} = \left[\frac{L/E}{\mathfrak{P}}\right].$$

PROOF. The residue fields of the rings of integers in L, E, K modulo $\mathfrak{P}$, $\mathfrak{p}_0$, $\mathfrak{p}$ are related by

$$GF(q^f) \supset GF(q^{f_0}) \supset GF(q)$$

with $f_0 = f(\mathfrak{p}_0/\mathfrak{p})$. The generating automorphism for the Galois group of $GF(q^f)$ over $GF(q^{f_0})$ is

$$\bar{x} \to \bar{x}^{q^{f_0}}$$

and this is the f_0 power of the generating automorphism when $GF(q)$ is the ground field. The property now follows from the definitions of the Frobenius automorphism of $\mathfrak{P}$ over E and over K.

Suppose we also know E is normal over K. Thus the expression

$$\left[\frac{E/K}{\mathfrak{p}_0}\right]$$

is defined.

2.4 Property

$$\left[\frac{E/K}{\mathfrak{p}_0}\right] = \left[\frac{L/K}{\mathfrak{P}}\right]\Bigg|E \qquad \text{(restriction to } E\text{)}.$$

PROOF. For an algebraic integer x in E, a congruence of the type

$$\sigma(x) \equiv x^q \bmod \mathfrak{P}$$

is equivalent to a congruence

$$\sigma(x) \equiv x^q \bmod \mathfrak{p}_0$$

because $\mathfrak{P} \cap E = \mathfrak{p}_0$ is sent to itself by $G(\mathfrak{P})$ when E is normal over K. Thus with

$$\sigma = \left[\frac{L/K}{\mathfrak{P}}\right]$$

we have

$$\sigma|E = \left[\frac{E/K}{\mathfrak{p}_0}\right].$$

Next suppose E_1 and E_2 are normal over K and $L = E_1 E_2$. Let $\mathfrak{P} \cap E_i = \mathfrak{p}_i$ for $i = 1, 2$. The expressions

$$\left[\frac{E_1 E_2/K}{\mathfrak{P}}\right], \quad \left[\frac{E_1/K}{\mathfrak{p}_1}\right], \quad \left[\frac{E_2/K}{\mathfrak{p}_2}\right]$$

are defined but they lie in different groups. Consider the mapping

$$G(L/K) \to G(E_1/K) \times G(E_2/K)$$

defined by

$$\sigma \to (\sigma|E_1, \sigma|E_2).$$

This is a one-to-one mapping because any automorphism which is the identity on both E_1 and E_2 is the identity on $L = E_1 E_2$. Identify $G(L/K)$ with its image in the direct product of $G(E_1/K)$ and $G(E_2/K)$. Property 2.4 implies the next statement.

2.5 Property

$$\left[\frac{E_1 E_2/K}{\mathfrak{P}}\right] = \left[\frac{E_1/K}{\mathfrak{p}_1}\right] \times \left[\frac{E_2/K}{\mathfrak{p}_2}\right].$$

Definition. A prime $\mathfrak{p}$ of K is said to *split completely* in L if $\mathfrak{p}$ has $(L:K)$ distinct prime divisors in L.

An equivalent statement is that $\mathfrak{p}$ splits completely in L if for each prime $\mathfrak{P}$ of L dividing $\mathfrak{p}$ we have $e(\mathfrak{P}/\mathfrak{p}) = f(\mathfrak{P}/\mathfrak{p}) = 1$.

2.6 Property. The prime $\mathfrak{p}$ splits completely in L if and only if

$$\left[\frac{L/K}{\mathfrak{P}}\right] = 1.$$

PROOF. The definition of

$$\left[\frac{L/K}{\mathfrak{P}}\right]$$

implies that it generates the decomposition group $G(\mathfrak{P})$. However $\mathfrak{p}$ splits completely if and only if $|G(\mathfrak{P})| = ef = 1$.

2.7 Corollary. Let E_1 and E_2 be normal extensions of K and $L = E_1 E_2$. The prime $\mathfrak{p}$ of K splits completely in L if and only if $\mathfrak{p}$ splits completely in both E_1 and E_2.

PROOF. By Property 2.5,

$$\left[\frac{L/K}{\mathfrak{P}}\right] = 1 \quad \text{if and only if} \quad \left[\frac{E_i/K}{\mathfrak{p}_i}\right] = 1$$

for $i = 1$ and 2.

Factorization in Nonnormal Extensions

In the case L/K is a normal extension, the Frobenius automorphism carries all the information about the factorization of unramified primes. That is

$$\left[\frac{L/K}{\mathfrak{P}}\right]$$

has order f [because it generates $G(\mathfrak{P})$] and the number of prime factors in L of $\mathfrak{p}$ is $g = (L : K)/f$.

We shall now consider how the Frobenius automorphism can be used to describe the factorization of $\mathfrak{p}$ in some nonnormal extension.

Consider a tower $K \subset E \subset L$ with E/K not necessarily normal. Let H be the subgroup of $G(L/K)$ fixing E elementwise. Consider a coset decomposition of $G = G(L/K)$ given by

$$G = H\sigma_1 \cup \cdots \cup H\sigma_k.$$

Any element σ in G permutes these cosets by right multiplication: $H\sigma_i \to H\sigma_i\sigma$. By a *cycle of length t* for σ we mean a sequence

$$H\sigma_i, H\sigma_i\sigma, H\sigma_i\sigma^2, \ldots, H\sigma_i\sigma^{t-1}$$

in which the t cosets are distinct and $H\sigma_i = H\sigma_i\sigma^t$. This coincides with the usual notion of a cycle for permutations. The collection of all cosets is partitioned into disjoint *cycles of* σ.

We want to describe how the prime $\mathfrak{p}$ of K factors into a product of primes of E. We still assume $\mathfrak{p}$ is unramified in L and $\mathfrak{P}$ is a prime factor of $\mathfrak{p}$ in L.

2.8 Proposition. Let σ be the Frobenius automorphism of $\mathfrak{P}$ in L/K. Suppose σ has cycles of length $t_1, \ldots, t_s$ when acting upon the cosets of H in G. Then $\mathfrak{p}$ is the product of s distinct primes in E having relative degrees $t_1, \ldots, t_s$.

PROOF. Let $H\tau$ be a representative of a cycle of length t for σ. Set $\mathfrak{p}_0 = \tau(\mathfrak{P}) \cap E$. Then $\mathfrak{p}_0$ is a prime of E dividing $\tau(\mathfrak{P}) \cap K = \mathfrak{p}$. The relative degree $f(\mathfrak{p}_0/\mathfrak{p}) = f$ can be computed in the following way:

The relative degree of $\tau(\mathfrak{P})$ over $\mathfrak{p}_0$ is the order of the decomposition group $H(\tau(\mathfrak{P}))$ – the subgroup of $H = G(L/E)$ fixing $\tau(\mathfrak{P})$. This subgroup is clearly given by

$$H(\tau(\mathfrak{P})) = H \cap G(\tau(\mathfrak{P})).$$

Now $G(\tau(\mathfrak{P})) = \tau G(\mathfrak{P})\tau^{-1} = \langle \tau\sigma\tau^{-1} \rangle$ because the Frobenius automorphism σ generates $G(\mathfrak{P})$. It takes just one step to show

$$H \cap \langle \tau\sigma\tau^{-1} \rangle = \langle \tau\sigma^t\tau^{-1} \rangle$$

when t is the least positive integer for which $H\tau = H\tau\sigma^t$. So we have $H(\tau(\mathfrak{P})) = \langle \tau\sigma^t\tau^{-1} \rangle$. Now finally

$$\begin{aligned} f(\mathfrak{p}_0/\mathfrak{p}) &= f(\mathfrak{P}/\mathfrak{p})/f(\mathfrak{P}/\mathfrak{p}_0) \\ &= |G(\mathfrak{P})|/|H(\tau(\mathfrak{P}))| \\ &= |\langle \sigma \rangle|/|\langle \tau\sigma^t\tau^{-1} \rangle| = t. \end{aligned}$$

Thus a cycle for σ of length t corresponds to a prime of E dividing $\mathfrak{p}$ and having relative degree t. Next we prove this correspondence is one to one.

Suppose $H\tau$ and $H\lambda$ are cosets of H for which

$$\mathfrak{p}_0 = \lambda(\mathfrak{P}) \cap E = \tau(\mathfrak{P}) \cap E.$$

Then $\lambda(\mathfrak{P})$ and $\tau(\mathfrak{P})$ are primes of L dividing $\mathfrak{p}_0$ and by transitivity properties of the Galois group H, there is some γ in H with $\gamma\lambda(\mathfrak{P}) = \tau(\mathfrak{P})$. It follows $\tau^{-1}\gamma\lambda$ is in $G(\mathfrak{P}) = \langle \sigma \rangle$ so $\gamma\lambda = \tau\sigma^i$ for some i. Hence

$$H\tau\sigma^i = H\gamma\lambda = H\lambda$$

and so $H\tau$, $H\lambda$ belong to the same cycle.

The last step requires that we show every prime divisor in E of $\mathfrak{p}$ has been obtained by this procedure. Each of the s cycles corresponds to a prime $\mathfrak{p}_i$ of E having relative degree $t_i = f(\mathfrak{p}_i/\mathfrak{p})$. But now

$$\sum t_i = [G : H] = (E : K)$$

along with Corollary 6.7, Chapter I, implies that all the prime divisors of $\mathfrak{p}$ in E have been counted.

2.9 Corollary. The number of primes in E dividing $\mathfrak{p}$ which have relative degree one over $\mathfrak{p}$ is equal to the number of coset representatives σ_i which satisfy $\sigma_i G(\mathfrak{P}) \sigma_i^{-1} \subseteq H$.

PROOF. The stated condition is equivalent to $H\sigma_i\sigma = H\sigma_i$ when $\langle \sigma \rangle = G(\mathfrak{P})$. This gives a cycle of length one so the result follows from the last proposition.

This corollary is important in the next chapter.

3. THE ARTIN MAP FOR ABELIAN EXTENSIONS

In this section it is assumed that L is normal over K and $G = G(L/K)$ is *abelian*. For an unramified prime $\mathfrak{p}$ of K and two prime divisors $\mathfrak{P}$ and $\tau(\mathfrak{P})$ in L, τ in G, we have by Property 2.2

$$\left[\frac{L/K}{\mathfrak{P}}\right] = \left[\frac{L/K}{\tau(\mathfrak{P})}\right].$$

This shows the Frobenius automorphism does not really depend on $\mathfrak{P}$ but only upon $\mathfrak{p}$. Accordingly we change notation and write

$$\left[\frac{L/K}{\mathfrak{p}}\right] \quad \text{for} \quad \left[\frac{L/K}{\mathfrak{P}}\right],$$

and call this the *Frobenius automorphism of* $\mathfrak{p}$.

In this way the Frobenius automorphism may be considered a correspondence between unramified primes of K and elements in the abelian Galois group. This can be extended further.

The *ideal group* $\mathbf{I}_K$ of K is the group of fractional R-ideals, $R =$ algebraic integers in K (Chapter I, Section 4). Let S denote a finite set of primes of K including all the primes which ramify in L. Then $\mathbf{I}_K{}^S$ or simply $\mathbf{I}^S$ denotes the subgroup of $\mathbf{I}_K$ generated by all the primes *outside* S. For each element $\mathfrak{A}$ in $\mathbf{I}^S$ we shall define an element $\varphi_{L/K}(\mathfrak{A})$ in G. First factor $\mathfrak{A}$ as

$$\mathfrak{A} = \prod_{\mathfrak{p}} \mathfrak{p}^{a(\mathfrak{p})}$$

and then set

$$\varphi_{L/K}(\mathfrak{A}) = \prod_{\mathfrak{p}} \left[\frac{L/K}{\mathfrak{p}}\right]^{a(\mathfrak{p})}.$$

The product is well defined because G is abelian. The function $\varphi_{L/K}$ is a homomorphism from $\mathbf{I}^S$ into G and is called the *Artin map* for the extension L/K. We emphasize that $\varphi_{L/K}$ is defined only for ideals whose factorization involves only unramified primes.

Of course when $\mathfrak{p}$ is a prime unramified in L, $\varphi_{L/K}(\mathfrak{p})$ is the Frobenius automorphism of $\mathfrak{p}$.

Suppose E is any finite-dimensional extension of K. We may translate the abelian extension L/K by E to obtain an abelian extension EL/E with Galois group H. The restriction of H to L naturally identifies H as a subgroup of G. The next lemma relates the Artin maps for EL/E and L/K. Let $\mathbf{I}_E{}^S$ denote the part of the ideal group of E, generated by primes of E which do not divide any prime in S. We could equally well say $\mathbf{I}_E{}^S$ is generated by primes of E having norms in $\mathbf{I}_K{}^S$.

3.1 Proposition. When $G(EL/E)$ is identified (by restriction) with a subgroup of $G(L/K)$ we have

$$\varphi_{EL|E} = \varphi_{L|K} \cdot \mathrm{N}_{E|K} \qquad \text{on} \quad \mathbf{I}_E{}^S.$$

PROOF. Let $\mathfrak{P}$ denote a prime of EL and let $\mathfrak{P}_L = \mathfrak{P} \cap L$, $\mathfrak{P}_E = \mathfrak{P} \cap E$, $\mathfrak{P}_K = \mathfrak{P} \cap K$. Let $\mathrm{N}_{K/Q}(\mathfrak{P}_K) = q =$ prime power and $\mathrm{N}_{E/K}(\mathfrak{P}_E) = \mathfrak{P}_K{}^f$. Set $\sigma = \varphi_{EL|E}(\mathfrak{P}_E)$. For each algebraic integer x in EL we have

$$\sigma(x) \equiv x^{q^f} \bmod \mathfrak{P}.$$

When x also lies in L then

$$\sigma(x) \equiv x^{q^f} \bmod \mathfrak{P}_L.$$

Here we use the fact that $\sigma(\mathfrak{P}) = \mathfrak{P}$ and $\sigma(\mathfrak{P}_L) = \mathfrak{P}_L$. Now let $\tau = \varphi_{L|K}(\mathfrak{P}_K)$. For x as just above

$$\tau(x) \equiv x^q \bmod \mathfrak{P}_L \qquad \text{and} \qquad \tau^f(x) \equiv x^{q^f} \bmod \mathfrak{P}_L.$$

By the uniqueness property, $\tau^f = \sigma$ on L. Thus

$$\varphi_{EL|K}(\mathfrak{P}_E) = \varphi_{L|K}(\mathfrak{P}_K)^f = \varphi_{L|K}\,\mathrm{N}_{E|K}(\mathfrak{P}_E).$$

This proves the equation for primes in $\mathbf{I}_E{}^S$ and the equation must hold on all of $\mathbf{I}_E{}^S$ because all maps are multiplicative.

3.2 Corollary.

$$\mathrm{N}_{L/K}(\mathbf{I}_L{}^S) \subseteq \ker \varphi_{L/K}.$$

PROOF. In Proposition 3.1 take $E = L$ to get $\varphi_{L/K}\,\mathrm{N}_{L/K} = \varphi_{L/L} = 1$.

This result describes a part of the kernel of the Artin map. One important goal is the description of the kernel and image of the Artin map. In the next chapter it is proved that the Artin map is always onto $G(L/K)$. Later the kernel will be described explicitly. For now we shall work an example which will illustrate the ideas and which will also be important later on.

Let m be a positive integer, θ a primitive mth root of unity, $K = Q =$ rationals, $L = Q(\theta)$. The Galois group G of L over K consists of automorphisms σ_t uniquely determined by the condition

$$\sigma_t(\theta) = \theta^t.$$

Here t must be a positive integer relatively prime to m. Let p denote a prime integer not dividing m. Then (p) is unramified in L and the automorphism σ_p satisfies the requirement (2.1) placed upon the Frobenius automorphism of (p). Thus

$$\varphi_{L/Q}(p) = \sigma_p.$$

For a positive integer $a = \prod p_i^{c_i}$ relatively prime to m we have

$$\varphi_{L/Q}(a) = \prod \varphi_{L/Q}(p_i)^{c_i} = \prod (\sigma_{p_i})^{c_i} = \sigma_a.$$

For a positive integer b relatively prime to m there is a positive integer b^* such that

$$bb^* \equiv 1 \bmod m.$$

We then find

$$\varphi_{L/Q}(1/b) = \varphi_{L/Q}(b^*) = \sigma_{b^*},$$

and more generally

$$\varphi_{L/Q}(a/b) = \sigma_{ab^*}.$$

It is now an easy matter to describe the kernel and image of $\varphi_{L/Q}$.

3.3 Proposition. Let S denote the set of prime ideals containing (m). The Artin map $\varphi_{L/Q}$ carries $\mathbf{I}_Q{}^S$ onto $G(L/Q)$ and the kernel is the set of fractional ideals (a/b) with a, b positive integers satisfying $a \equiv b \bmod m$.

PROOF. We see from the discussion above that $\varphi_{L/K}$ maps onto $G(L/Q)$. The ideal (a/b) is in the kernel precisely when $\sigma_{ab^*} = \sigma_1$; that is when $a \equiv b \bmod m$.

An important part of the theory to be developed later will give a generalization of this theorem to the case of an abelian extension of an arbitrary algebraic number field rather than an abelian extension of Q as given here.

EXERCISE 1. Use the theory of the Frobenius automorphism to describe the factorization of primes of Q in $Q(\theta)$, $\theta^m = 1$. (This gives an alternate approach to the results in Section 9, Chapter I about factorization.)

EXERCISE 2. (a) The Galois group of $Q(\theta_5)/Q$ is generated by $\sigma_2(\theta_5) = (\theta_5)^2$ when θ_5 is a primitive fifth root of unity. Show σ_2 has order four and conclude the ideal (2) remains prime in $Q(\theta_5)$ with relative degree four. Show the same assertion holds for any prime ideal (p) with $p \equiv \pm 2 \bmod 5$.

(b) The prime 19 and any prime $p \equiv -1 \bmod 5$ has Frobenius automorphism of order 2 and so (p) has two prime factors in $Q(\theta_5)$ each with relative degree two.

(c) The primes $p \equiv 1 \bmod 5$ have trivial Frobenius automorphism and (p) splits completely in $Q(\theta_5)$.

(d) Let E be the subfield of $Q(\theta_5)$ fixed by σ_4. Use the formula $\varphi_{E/Q} = \operatorname{res} E\varphi_{Q(\theta_5)/Q}$ to compute the factorization of rational primes in E (res E means restriction to E).

EXERCISE 3. (a) Let $\theta = \theta_{25}$ and $\sigma_n(\theta) = \theta^n$. Compute the order of σ_n for each integer n not divisible by 5. For a prime $p \neq 5$, describing the factorization of (p) in $Q(\theta)$ in terms of the congruence class of p mod 25.

(b) Let E be the subfield of $Q(\theta)$ left fixed by σ_7; $(E:Q) = 5$. Use the restriction of $\varphi_{Q(\theta)/Q}(p)$ to E to show that p splits completely in E if and only if $p \equiv \pm 1, \pm 7 \bmod 25$ where as p splits completely in $Q(\theta)$ only when $p \equiv 1 \bmod 25$.

(c) There is a unique subfield of $Q(\theta)$ properly between E and $Q(\theta)$. Find its Galois group and determine which rational primes split completely in it.

Chapter IV

ANALYTIC METHODS

Nil sapientiae odiosius acumine nimio.

Seneca

In this chapter we begin to study rather delicate properties of primes in number fields. Many results in earlier chapters hold in much more general fields. However we shall use in several ways the assumption that our field is a finite extension of the rationals. The main idea involves the use of infinite series to prove results about the distribution of prime ideals. The Frobenius density theorem shows, roughly speaking, that infinitely many primes have the same Frobenius automorphism. This is the result needed to prove that the Artin map is onto. Other results proved by analytic techniques include Dirichlet's famous theorem about primes in an arithmetic progression.

Much has been written about procedures to avoid the analysis in this algebraic subject. Chevalley and others did accomplish this by introducing ideles and the machinery of cohomology of groups. It is this approach that is called the modern treatment of class field theory. The use of Dirichlet series to prove algebraic theorems goes back more than a century and so it can hardly be called modern. However the approach presented here benefits from Artin's ideas involving the systematic use of the Artin map. Historically the existence theorem was proved by Tagaki before the reciprocity theorem of Artin. The reversal of the order here simplifies the development.

It seems that the analysis required by the student in this treatment is far less than the corresponding amount of material he would have to know before attacking the development of the subject by cohomology of groups and the method of ideles. For this reason I prefer to think of this as the direct approach to class field theory.

1. MODULI AND RAY CLASSES

We consider an algebraic number field K and its ring of algebraic integers R. For a prime $\mathfrak{p}$ of K, $K_{\mathfrak{p}}$ denotes the completion of K with respect to a valuation in $\mathfrak{p}$.

The ideal group $\mathbf{I}_K$ of K is the group of fractional R-ideals of K. It is the free abelian group with the finite primes (integral ideals) as generators.

The multiplicative group of nonzero elements in K is denoted by K^*. There is a natural map i which sends K^* into $\mathbf{I}_K$ by mapping an element α in K onto the principal ideal $(\alpha) = \alpha R = i(\alpha)$. The kernel of i is the group $\mathbf{U}_K$ of units in R. The structure of $\mathbf{U}_K$ is given in Chapter I, Theorem 11.19. The cokernel of i is by definition the class group of K, denoted by $\mathbf{C}_K$. We saw in Chapter I, Section 11 that $\mathbf{C}_K$ is a finite group. We summarize these facts with an exact sequence

$$1 \to \mathbf{U}_K \to K^* \to \mathbf{I}_K \to \mathbf{C}_K \to 1.$$

It is fair to say that our main interest for most of what follows in this book is in the study of certain subgroups of the groups in this sequence and how they relate to the problem of describing, in terms of K, all the abelian extensions of K.

Definition. A *modulus* for K is a formal product

$$\mathfrak{m} = \prod_{\mathfrak{p}} \mathfrak{p}^{n(\mathfrak{p})}$$

taken over all primes $\mathfrak{p}$ of K in which $n(\mathfrak{p})$ is a nonnegative integer and $n(\mathfrak{p}) > 0$ for only a finite number of $\mathfrak{p}$. Furthermore $n(\mathfrak{p}) = 0$ or 1 when $\mathfrak{p}$ is a real infinite prime and $n(\mathfrak{p}) = 0$ when $\mathfrak{p}$ is a complex infinite prime.

A modulus $\mathfrak{m}$ may be considered a product $\mathfrak{m}_0 \mathfrak{m}_\infty$ with $\mathfrak{m}_0$ the product of the finite primes appearing with positive exponent in $\mathfrak{m}$ and $\mathfrak{m}_\infty$ the product of the real primes in $\mathfrak{m}$. Then $\mathfrak{m}_0$ is identified with an integral ideal; that is an ideal in R.

Our intention is to extend the notion of congruence between two elements of R modulo an ideal to a notion of congruence between elements of K^* modulo a modulus.

Let $\mathfrak{p}$ denote a real prime of K so $K_{\mathfrak{p}}$ is isomorphic to the real field. Let $x \to x_{\mathfrak{p}}$ denote the imbedding of K into $K_{\mathfrak{p}}$. For elements α, β in K^* we write

$$\alpha \equiv \beta \bmod \mathfrak{p}$$

to mean $\alpha_{\mathfrak{p}}$ and $\beta_{\mathfrak{p}}$ have the same sign; equivalently we could say $(\alpha/\beta)_{\mathfrak{p}} > 0$.

Now let $\mathfrak{p}$ be a finite prime, α, β elements in K^* and suppose that

$$\alpha = a/c, \qquad \beta = b/d, \qquad a, b, c, d \in R.$$

Then we write

$$\alpha \equiv \beta \bmod \mathfrak{p}^n$$

if $\alpha/\beta = ad/bc$ is in the valuation ring $R_\mathfrak{p}$ of $\mathfrak{p}$ and this element is congruent to 1 modulo $\mathfrak{p}^n$; that is, $(ad-bc)/bc \in \mathfrak{p}^n$.

Some care must be taken with congruences defined for elements of K^* because they can be multiplied but not added. By this we mean

$$\alpha_1 \equiv \beta_1 \qquad \text{and} \quad \alpha_2 \equiv \beta_2 \bmod \mathfrak{p}^n$$

implies

$$\alpha_1 \alpha_2 \equiv \beta_1 \beta_2 \bmod \mathfrak{p}^n$$

but it need *not* follow that

$$\alpha_1 + \alpha_2 \equiv \beta_1 + \beta_2 \bmod \mathfrak{p}^n.$$

For example, with $K = Q$ and p a prime integer we take $\alpha = 1/p$, $\beta = (p+1)/p$. Then

$$\beta/\alpha = p + 1 \equiv 1 \bmod p \qquad \text{and so} \quad \alpha \equiv \beta \bmod p.$$

However we do *not* have

$$\alpha - \alpha \equiv \beta - \alpha \bmod p$$

because $\beta - \alpha = 1 \not\equiv 0 \bmod p$.

Now we extend in the expected way to congruences for a modulus $\mathfrak{m}$ as given in the definition. For α, β in K^* we write

$$\alpha \equiv \beta \bmod \mathfrak{m}$$

if

$$\alpha \equiv \beta \bmod \mathfrak{p}^{n(\mathfrak{p})}$$

for all primes $\mathfrak{p}$ with $n(\mathfrak{p}) > 0$.

We now define two subgroups of K^* associated with a modulus $\mathfrak{m} = \mathfrak{m}_0 \mathfrak{m}_\infty$.

Definition

$$K_\mathfrak{m} = \{a/b \mid a, b \in R, aR, bR \text{ relatively prime to } \mathfrak{m}_0\},$$

$$K_{\mathfrak{m},1} = \{\alpha \in K_\mathfrak{m} \mid \alpha \equiv 1 \bmod \mathfrak{m}\}.$$

Notice that $K_\mathfrak{m}$ depends only upon the finite primes dividing $\mathfrak{m}_0$ and not upon their exponents.

The group $K_{\mathfrak{m},1}$ is sometimes called the "ray mod $\mathfrak{m}$."

Recall that for a set S of primes, $\mathbf{I}^S$ (or $\mathbf{I}_K{}^S$) denotes the part of the ideal group $\mathbf{I}_K$ generated by primes outside S. We shall also use the symbol $\mathbf{I}^\mathfrak{m}$ (or $\mathbf{I}_K{}^\mathfrak{m}$) to denote $\mathbf{I}^S$ where S is the set of primes dividing $\mathfrak{m}_0$. Thus $\mathbf{I}^\mathfrak{m}$ does not depend upon the exponents of the primes dividing $\mathfrak{m}$.

Clearly the image under i of $K_{\mathfrak{m}}$ or $K_{\mathfrak{m},1}$ lands in $\mathbf{I}^{\mathfrak{m}}$. The quotient

$$\mathbf{I}^{\mathfrak{m}}/i(K_{\mathfrak{m},1})$$

is called the *ray class group* mod $\mathfrak{m}$ and the cosets of $i(K_{\mathfrak{m},1})$ in this quotient are the *ray classes* mod $\mathfrak{m}$. The study of the ray class group requires approximation techniques. We shall see momentarily the ray class group is finite.

1.1 Theorem (Approximation Theorem). Let $|\ |_1, \ldots, |\ |_n$ be nontrivial pairwise inequivalent valuations on K and let $\beta_1, \ldots, \beta_n$ be nonzero elements in K. For any positive real ε, there is an α in K such that $|\alpha - \beta_i|_i < \varepsilon$ for all $i = 1, \ldots, n$.

PROOF. The first step is to show there exists elements $y_1, \ldots, y_n$ in K such that

$$|y_i|_i > 1, \qquad |y_i|_j < 1, \qquad i \neq j.$$

Use induction on n. For $n = 2$ the definition of equivalence implies the existence of elements w, z such that

$$|w|_1 > 1, \qquad |w|_2 \leqslant 1,$$
$$|z|_1 \leqslant 1, \qquad |z|_2 > 1.$$

Now set $y = w/z$ to get

$$|y|_1 > 1, \qquad |y|_2 < 1.$$

Suppose we have an element y which satisfies

$$|y|_1 > 1, \qquad |y|_j < 1, \qquad j = 2, \ldots, n-1.$$

By the case $n = 2$ there is an element t which satisfies $|t|_1 > 1$ and $|t|_n < 1$. Now select y_1 in the following way:

$$y_1 = y \qquad \text{if} \quad |y|_n < 1,$$
$$y_1 = y^r t \qquad \text{if} \quad |y|_n = 1,$$
$$y_1 = \frac{y^r t}{1 + y^r} \qquad \text{if} \quad |y|_n > 1.$$

In the last two cases r is an integer yet to be determined. If the second case holds,

$$|y_1|_j = |y|_j^r |t|_j, \qquad 2 \leqslant j \leqslant n$$

and this can be made < 1 with sufficiently large r for all $j \neq 1$.

In the third case we reach the same conclusion because

$$\frac{|y^r|_j}{|1+y^r|_j} < \frac{1}{|y^{-r}|_j - 1}$$

and this has limit $=0$ as $r \to \infty$.

In all cases we have the element y_1 which is "large" at $|\ \ |_1$ and "small" at the other valuations. By symmetry we can obtain $y_1, y_2, \ldots, y_n$ to satisfy our requirement.

Now to finish the proof let

$$\alpha = \sum_i \frac{y_i^r}{1+y_i^r} \beta_i$$

with r an integer to be determined. The triangle inequality implies

$$|\alpha - \beta_i|_i \leqslant \left|\frac{\beta_i}{1+y_i^r}\right|_i + \sum_{j \neq i} \left|\frac{y_j^r}{1+y_j^r}\beta_j\right|_i.$$

For r sufficiently large this expression will be less than the given ε and the theorem is proved.

Before recording some consequences of this approximation theorem it is helpful to see how we can pass from statements about valuations to congruences.

When $\mathfrak{p}$ is a real prime of K, the statement $\alpha\beta \neq 0$ and $|\alpha-\beta|_\mathfrak{p} < \varepsilon$ for small ε means $\alpha_\mathfrak{p}/\beta_\mathfrak{p}$ is positive. That is $\alpha \equiv \beta \bmod \mathfrak{p}$.

Suppose $\mathfrak{p}$ is a finite prime and $\alpha\beta \neq 0$. The $\mathfrak{p}$-adic valuation satisfies $|\alpha|_\mathfrak{p} = c^{v(a)}$ where $v(\alpha)$ is the power of $\mathfrak{p}$ appearing in $i(\alpha)$ and c is some real number $0 < c < 1$.

The condition $|\alpha-\beta|_\mathfrak{p} < \varepsilon$ is equivalent to $|\alpha/\beta - 1|_\mathfrak{p} < \varepsilon/|\beta|_\mathfrak{p} = \varepsilon'$. When ε' is sufficiently small, say $\varepsilon' < c^n$ with n a positive integer, then $v(\alpha/\beta - 1) > n$. In particular $\alpha/\beta - 1$ is in the valuation ring and moreover

$$\alpha/\beta \equiv 1 \bmod \mathfrak{p}^n.$$

In the extended sense of congruences defined above we also have

$$\alpha \equiv \beta \bmod \mathfrak{p}^n.$$

To summarize then for $\alpha\beta \neq 0$ and sufficiently small ε, the inequality $|\alpha-\beta|_\mathfrak{p} < \varepsilon$ implies $\alpha \equiv \beta \bmod \mathfrak{p}^n$.

This idea is extended in the proof of the next result.

When $\mathfrak{m}_1$ and $\mathfrak{m}_2$ are moduli for K such that no prime appears with positive exponent in both $\mathfrak{m}_1$ and $\mathfrak{m}_2$ we say $\mathfrak{m}_1$ and $\mathfrak{m}_2$ are *relatively prime*.

1.2 Proposition. Let $\mathfrak{m}_1, \ldots, \mathfrak{m}_n$ be relatively prime moduli (in pairs) and let $\mathfrak{m}$ denote the product $\mathfrak{m} = \mathfrak{m}_1 \cdots \mathfrak{m}_n$. The natural map from $K_\mathfrak{m}$ into the

Cartesian product $\prod K_{\mathfrak{m}_i}$ induces an isomorphism

$$\frac{K_{\mathfrak{m}}}{K_{\mathfrak{m},1}} \cong \prod \frac{K_{\mathfrak{m}_i}}{K_{\mathfrak{m}_i,1}}.$$

PROOF. By the natural map we mean the one sending α in $K_{\mathfrak{m}}$ to $(\alpha, \alpha, \ldots, \alpha)$. The induced map is the one

$$\alpha K_{\mathfrak{m},1} \to (\ldots, \alpha K_{\mathfrak{m}_i,1}, \ldots).$$

The kernel of this map is the collection of cosets $\alpha K_{\mathfrak{m},1}$ with α in all of the $K_{\mathfrak{m}_i,1}$. However the intersection of these is just $K_{\mathfrak{m},1}$ because the $\mathfrak{m}_i$ are relatively prime. So the induced map is one-to-one. To show it is onto, we select β_i in $K_{\mathfrak{m}_i}$ and find an α in K to satisfy

$$|\alpha - \beta_i|_{\mathfrak{p}} < \varepsilon$$

where $\mathfrak{p}$ runs through the divisors of $\mathfrak{m}_i$ and $i = 1, 2, \ldots, n$. For ε sufficiently small, we saw just above that this implies

$$\alpha/\beta_i \equiv 1 \bmod \mathfrak{m}_i.$$

Thus $\alpha/\beta_i \in K_{\mathfrak{m}_i,1}$ and $\alpha K_{\mathfrak{m}_i,1} = \beta_i K_{\mathfrak{m}_i,1}$. So $\alpha K_{\mathfrak{m},1}$ maps onto $(\ldots, \beta_i K_{\mathfrak{m}_i,1}, \ldots)$ and the proof is complete.

1.3 Corollary. For any modulus $\mathfrak{m}$, the group $K_{\mathfrak{m}}/K_{\mathfrak{m},1}$ is finite.

PROOF. The Proposition 1.2 shows the result is true provided we can prove it true in the special case with $\mathfrak{m}$ the power of a single prime.

Suppose $\mathfrak{m}$ is a real prime of K. Then $K_{\mathfrak{m}}/K_{\mathfrak{m},1}$ is the full group K^* modulo the subgroup of positive elements at $\mathfrak{m}$. This quotient has order two.

Now suppose $\mathfrak{m} = \mathfrak{p}^n$ with $\mathfrak{p}$ a finite prime. Then $K_{\mathfrak{m}}$ is the group of units in the valuation ring $R_{\mathfrak{p}}$ and $K_{\mathfrak{m},1}$ is the subgroup of units congruent to 1 modulo $\mathfrak{p}^n$. It follows that $K_{\mathfrak{m}}/K_{\mathfrak{m},1}$ is the group of units in the ring $R_{\mathfrak{p}}/\mathfrak{p}^n$. Since this is a finite ring with $\mathrm{N}_{K/Q}(\mathfrak{p}^n)$ elements, the unit group is also finite.

EXERCISE. If $\mathfrak{m} = \mathfrak{m}_0 \mathfrak{m}_\infty$ then $K_{\mathfrak{m}}/K_{\mathfrak{m},1}$ has order

$$2^r \mathrm{N}(\mathfrak{m}_0) \prod_{\mathfrak{p} \mid \mathfrak{m}_0} \left(1 - \frac{1}{\mathrm{N}(\mathfrak{p})}\right)$$

where r is the number of real primes dividing $\mathfrak{m}_\infty$ and N means $\mathrm{N}_{K/Q}$.

Proposition 1.2 will be frequently used in a slightly different way which is nothing but a restatement. Namely, given the relatively prime moduli $\mathfrak{m}_1, \ldots, \mathfrak{m}_n$ and $\beta_i \in K_{\mathfrak{m}_i}$ we can find α in $K_{\mathfrak{m}}$ to solve the congruences $\alpha \equiv \beta_i \bmod \mathfrak{m}_i$.

1.4 Corollary. Each coset of $K_{\mathfrak{m},1}$ in $K_{\mathfrak{m}}$ contains an element relatively prime to any given ideal.

PROOF. Let the given ideal be $\prod \mathfrak{q}_j^{a_j} = \mathfrak{A}$ and let $\beta K_{\mathfrak{m},1}$ be some coset in $K_{\mathfrak{m}}$. Select γ such that

$$\gamma \equiv \beta \bmod \mathfrak{m}, \qquad \gamma \equiv 1 \bmod \mathfrak{q}_j,$$

where $\mathfrak{q}_j$ runs through those $\mathfrak{q}$ not dividing $\mathfrak{m}$. Then γ and β lie in the same coset of $K_{\mathfrak{m},1}$ and γ is prime to $\mathfrak{A}$.

1.5 Corollary. For any finite set of primes S, there is a natural isomorphism

$$\mathbf{C}_K \cong \mathbf{I}^S/\mathbf{I}^S \cap i(K^*).$$

PROOF. The inclusion $\mathbf{I}^S \to \mathbf{I}_K$ gives an inclusion

$$\mathbf{I}^S/\mathbf{I}^S \cap i(K^*) \to \mathbf{I}_K/i(K^*) = \mathbf{C}_K.$$

To show this map is onto, it is necessary to prove each ideal class contains a representative not divisible by any prime in S. Let $\mathfrak{B}$ be any ideal in $\mathbf{I}_K$ and $\mathfrak{B} = \mathfrak{B}_1 \mathfrak{B}_2$ with $\mathfrak{B}_1$ prime to S and

$$\mathfrak{B}_2 = \prod_{\mathfrak{p} \in S} \mathfrak{p}^{n(\mathfrak{p})}.$$

Let $\pi_{\mathfrak{p}}$ be an element which generates $\mathfrak{p}$ in $R_{\mathfrak{p}}$ (the localization at $\mathfrak{p}$) and which satisfies

$$\pi_{\mathfrak{p}} \equiv 1 \bmod \mathfrak{q} \qquad \text{for all} \quad \mathfrak{q} \neq \mathfrak{p} \quad \text{in} \quad S.$$

Such elements exist by CRT. Let

$$\alpha = \prod_{\mathfrak{p} \in S} (\pi_{\mathfrak{p}})^{n(\mathfrak{p})}.$$

Then by localization one sees the power of $\mathfrak{p}$ dividing (α) is precisely $n(\mathfrak{p})$. It may be that (α) is divisible by primes outside S but this is no matter. Now $\mathfrak{B}\alpha^{-1}$ is not divisible by any primes in S. The ideal class containing $\mathfrak{B}$ is the same as that containing $\mathfrak{B}\alpha^{-1}$ so each class has a representative in $\mathbf{I}^S$ as required.

1.6 Corollary. Let $\mathfrak{m}$ be any modulus. Then the ray class group $\mathbf{I}^{\mathfrak{m}}/i(K_{\mathfrak{m},1})$ is a finite group.

PROOF. One sees at once that $\mathbf{I}^{\mathfrak{m}} \cap i(K^*)$ is the collection of principal ideals relatively prime to $\mathfrak{m}$ so

$$\mathbf{I}^{\mathfrak{m}} \cap i(K^*) = i(K_{\mathfrak{m}}).$$

So then

$$[\mathbf{I}^{\mathfrak{m}} : i(K_{\mathfrak{m},1})] = [\mathbf{I}^{\mathfrak{m}} : i(K_{\mathfrak{m}})][i(K_{\mathfrak{m}}) : i(K_{\mathfrak{m},1})].$$

The first factor is the class number, $|\mathbf{C}_K|$ by Corollary 1.5 and the last remark and the second is a divisor of $[K_{\mathfrak{m}} : K_{\mathfrak{m},1}]$ which is finite by Corollary 1.3. Since the class number is finite, the result follows.

We shall use $h_{\mathfrak{m}}$ to denote the order of the ray class group mod $\mathfrak{m}$. In case $\mathfrak{m}$ is the trivial modulus, that is, the empty product, then the ray class group is just the class group whose order is denoted by h_K. We extract one fact from the above proof.

1.7 Proposition. h_K divides $h_{\mathfrak{m}}$ for any modulus $\mathfrak{m}$.

2. DIRICHLET SERIES

A *Dirichlet series* is a function of the type

$$f(s) = \sum_{n=1}^{\infty} \frac{a(n)}{n^s} \tag{1}$$

with $a(n)$ complex and $s = \sigma + it$ a complex variable. A special case is the *Riemann ζ-function* defined by

$$\zeta(s) = \sum_{n=1}^{\infty} \frac{1}{n^s}. \tag{2}$$

Series of this form will be used to study properties of primes in number fields. We begin by studying questions about convergence.

Denote by $D\,(b, \delta, \varepsilon)$ the region of the complex plane

$$\{s \mid \operatorname{Re}(s) \geqslant b+\delta, |\arg(s-b)| \leqslant \pi/2-\varepsilon\}.$$

2.1 Proposition. Let $f(s)$ have the form (1) and let $s(x) = \sum a(n)$ taken over $n \leqslant x$. Suppose there exist positive constants a, b such that $|s(x)| \leqslant ax^b$ for all $x \geqslant 1$. Then the following hold:

(a) The series $f(s)$ is uniformly convergent for s in $D(b,\delta,\varepsilon)$, with any positive δ, ε;

(b) $f(s)$ is analytic in the half-plane $\operatorname{Re}(s) > b$;

(c) if

$$\lim_{x\to\infty} \frac{s(x)}{x} = a_0$$

then

$$\lim_{s\to 1} (s-1)f(s) = a_0, \qquad s \in D(1,0,\varepsilon).$$

PROOF. Notice that $a(n) = s(n) - s(n-1)$, so for $v \geqslant u+1$

$$\left|\sum_u^v \frac{a(n)}{n^s}\right| = \left|\sum_u^v \frac{s(n)}{n^s} - \sum_{u-1}^{v-1} \frac{s(n)}{(n+1)^s}\right|$$

$$= \left|\frac{s(v)}{v^s} - \frac{s(u-1)}{u^s} + \sum_u^{v-1} s(n)\left[\frac{1}{n^s} - \frac{1}{(n+1)^s}\right]\right|$$

$$\leqslant \left|\frac{s(v)}{v^s}\right| + \left|\frac{s(u-1)}{u^s}\right| + \sum_u^{v-1} |s(n)| \left|\frac{1}{n^s} - \frac{1}{(n+1)^s}\right|.$$

Recall that $|n^s| = n^\sigma$ when $s = \sigma + it$.

The estimate of $|s(x)|$ allows us to rewrite the first two terms. The expression behind the summation can be rewritten with the use of the identity

$$\frac{1}{n^s} - \frac{1}{(n+1)^s} = s\int_n^{n+1} \frac{dt}{t^{s+1}}.$$

Thus far we see

$$\left|\sum_u^v \frac{a(n)}{n^s}\right| \leqslant \frac{a}{v^{\sigma-b}} + \frac{a}{u^{\sigma-b}} + \sum_u^{v-1} |s|\, an^b \left|\int_n^{n+1} \frac{dt}{t^{s+1}}\right|.$$

We can further change this expression by noting the term involving the summation is less than

$$|s|\, a\int_u^\infty \frac{t^b\, dt}{|t^{s+1}|} \leqslant |s|\, a\int_u^\infty \frac{dt}{t^{\sigma+1-b}} = \frac{a|s|}{(\sigma-b)\, u^{\sigma-b}}.$$

Also $v > u$ so finally

$$\left|\sum_u^v \frac{a(n)}{n^s}\right| \leqslant \frac{2a}{u^{\sigma-b}} + \frac{|s|\, a}{(\sigma-b)\, u^{\sigma-b}}$$

Now $|s|/\sigma - b \leqslant (|s-b| + b)(\sigma-b)^{-1} \leqslant 1/\cos\theta + b/\delta$, with $\theta = \arg(s-b)$. The number b/δ is constant and the restriction $|\theta| \leqslant \pi/2 - \varepsilon$ means $1/\cos\theta \leqslant M$ for some constant M. Thus given any number ε_0 we can find a sufficiently large integer u to insure

$$\frac{2a}{u^{\sigma-b}} + \frac{|s|\, a}{(\sigma-b)\, u^{\sigma-b}} \leqslant \frac{2a + M + b/\delta}{u^{\sigma-b}} < \varepsilon_0.$$

This implies the uniform convergence of $f(s)$ in $D(b, \delta, \varepsilon)$.

To prove (b) first note that any point s in the half-plane $\mathrm{Re}(s) > b$ lies in some $D(b, \delta, \varepsilon)$. Thus at this s the series is a uniformly convergent series of analytic functions and hence is analytic.

Now assume (c) holds. This means $s(x) = a_0 x + e(x)x$ with $\lim e(x) = 0$ as $x \to \infty$. Necessarily $e(x)$ is bounded and so there is some constant a_2 with $|s(x)| \leqslant a_2 x$. By Part (a) $f(s)$ is uniformly convergent in $D(1, \delta, \varepsilon)$ for positive δ, ε.

The proof of (c) is carried out by showing $(s-1)f(s)$ has the same limit at $s = 1$ as $a_0(s-1)\zeta(s)$ with $\zeta(s)$ defined in Eq. (2), and then evaluating the limit for $\zeta(s)$. Notice by Part (a), the ζ-function $\zeta(s)$ is uniformly convergent in $D(1, \delta, \varepsilon)$ just as $f(s)$ is.

Begin with the same sort of computation as used in the proof of (a). Namely,

$$\begin{aligned} |f(s) - a_0 \zeta(s)| &= \left| \sum \frac{a(n) - a_0}{n^s} \right| \\ &= \left| \sum [s(n) - na_0] \left[\frac{1}{n^s} - \frac{1}{(n+1)^s} \right] \right| \\ &\leqslant \left| \sum ne(n) s \int_n^{n+1} \frac{dt}{t^{s+1}} \right| \\ &\leqslant \sum n |e(n)| |s| \int_n^{n+1} \frac{dt}{t^{\sigma+1}} . \end{aligned}$$

Select any $\varepsilon_0 > 0$ and N so large that $|e(n)| < \varepsilon_0$ if $n \geqslant N$. Take M as a bound on $|e(n)|$ for all n. Also notice

$$n \int_n^{n+1} \frac{dt}{t^{\sigma+1}} \leqslant \int_n^{n+1} \frac{t\, dt}{t^{\sigma+1}} .$$

After all these estimates are combined, one sees

$$\begin{aligned} &|s-1| |f(s) - a_0 \zeta(s)| \\ &\quad \leqslant |s(s-1)| M \int_1^N \frac{dt}{t^\sigma} + |s(s-1)| \varepsilon_0 \int_N^\infty \frac{dt}{t^\sigma} . \end{aligned}$$

The first term involving an integral can be evaluated and as $s \to 1$ with $s \in D(1, 0, \varepsilon)$ the limit is zero.

The second term involving the integral is

$$\frac{|s(s-1)| \varepsilon_0}{\sigma - 1} \frac{1}{N^{\sigma-1}} .$$

The restriction that $s \in D(1, 0, \varepsilon)$ means

$$\left| \frac{s-1}{\sigma-1} \right| \leqslant \sec(\pi/2 - \varepsilon) = T_0 .$$

Thus for s near 1 in $D(1,0,\varepsilon)$ we have

$$|s-1||f(s)-a_0\zeta(s)| < \varepsilon_0 T_0.$$

Since ε_0 was arbitrary we find

$$\lim_{s\to 1}\{(s-1)f(s)-a_0(s-1)\zeta(s)\} = 0 \qquad \text{if} \quad s \in D(1,0,\varepsilon).$$

The proof will be complete if we show

$$\lim_{s\to 1}(s-1)\zeta(s) = 1. \tag{3}$$

This will be done by first showing that $(s-1)\zeta(s)$ can be continued to a function analytic in the disk $|s-1|<\frac{1}{2}$ and so the limit can be taken along any convenient path to $s=1$. The convenient path is the real axis to the right of 1. Consider the function

$$\zeta_2(s) = \sum_1^\infty (-1)^{n+1}\frac{1}{n^s}.$$

The sum of the first n coefficients is 0 or 1 so by Part (a) this is uniformly convergent in $D(0,\delta,\varepsilon)$. In particular it is analytic in the disk $|s-1|<\frac{1}{2}$. Next observe

$$\zeta_2(s)+\frac{2}{2^s}\zeta(s) = \zeta(s)$$

in the common region of convergence. It follows

$$\zeta(s) = \left(1-\frac{1}{2^{s-1}}\right)^{-1}\zeta_2(s).$$

The function on the right is the quotient of analytic functions and the denominator is zero only at points where $2^{s-1}=1$; namely $s=1+2k\pi i/\ln 2$ with k an integer. Clearly the only real pole of $\zeta(s)$ can occur at $s=1$. We shall verify the other points are not poles.

Consider the function

$$\zeta_3(s) = \frac{1}{1^s}+\frac{1}{2^s}-\frac{2}{3^s}+\cdots$$
$$+\frac{1}{(3n+1)^s}+\frac{1}{(3n+2)^s}-\frac{2}{(3n+3)^s}+\cdots.$$

For this the sum function $s(x)$ has values 0, 1, or 2 and so $\zeta_3(s)$ is uniformly convergent in $D(0,\delta,\varepsilon)$. As above we find

$$\zeta(s) = \left(1-\frac{1}{3^{s-1}}\right)^{-1}\zeta_3(s)$$

and so the only possible poles of $\zeta(s)$ are the points where $s = 1+2m\pi i/\ln 3$. If this is in fact a pole then $1+2\pi mi/\ln 3 = 1+2\pi ki/\ln 2$ and so $3^k = 2^m$. Since k, m are integers we must have $k = m = 0$ and the pole of $\zeta(s)$ must be $s = 1$. One easily checks that it is a pole of order 1 because $2^{s-1}-1$ has only zeros of order 1. Finally we obtain $(s-1)\zeta(s)$ is analytic at $s = 1$ and in the half-plane $\mathrm{Re}(s) > 0$. The limit at 1 can be evaluated by approaching 1 on the real axis from the right.

Approximate the area under the curve $y = x^{-s}$ (s real) by rectangles with base $[n, n+1]$ and height $1/(n+1)^s$. The area of the rectangles summed from $n = 0$ to ∞ is $\zeta(s)$ so

$$\zeta(s) \leqslant 1 + \int_1^\infty \frac{dx}{x^s} = 1 + \frac{1}{s-1}.$$

In a similar way use rectangles with base $[n, n+1]$ and height n^{-s} to get

$$\frac{1}{s-1} = \int_1^\infty \frac{dx}{x^s} \leqslant \zeta(s).$$

From these inequalities one obtains $1 \leqslant (s-1)\zeta(s) \leqslant s$ for real s and Eq. (3) holds. This completes the proof of Proposition 2.1.

Let K be any algebraic number field and for each integral ideal $\mathfrak{A}$ let $\mathcal{N}(\mathfrak{A})$ denote a positive generator of the ideal $(\mathrm{N}_{K/Q}(\mathfrak{A}))$. Equivalently $\mathcal{N}(\mathfrak{A}) =$ number of elements in $R/\mathfrak{A}$. We shall write N for $\mathrm{N}_{K/Q}$.

Definition. The function $\zeta_K(s) = \sum_{\mathfrak{A}} 1/\mathcal{N}(\mathfrak{A})^s$ is called the *ζ-function of K*. In this (and all similar expressions) the sum is taken over all *integral ideals* of K.

This can also be written as

$$\zeta_K(s) = \sum_1^\infty \frac{a_K(n)}{n^s},$$

where $a_K(n)$ is the number of integral ideals of K with norm exactly n. Notice that the ζ-function of Q is the Riemann ζ-function.

It is useful to also consider more general types of ζ-functions. We describe one such now.

Let $\mathfrak{m}$ be a modulus for K and let k be a coset of $i(K_{\mathfrak{m},1})$ in $\mathbf{I}^{\mathfrak{m}}$. The *$\zeta$-function of the class k* is

$$\zeta(s, k) = \sum_{\mathfrak{A} \in k} \frac{1}{\mathcal{N}(\mathfrak{A})^s}.$$

Notice when $\mathfrak{m}$ is the trivial modulus (empty product) then $\mathbf{I}^{\mathfrak{m}} = I_K$ and $\zeta_K(s) = \sum_k \zeta(s, k)$.

It is necessary to compute the limit of $(s-1)\zeta(s,k)$ as $s \to 1$. This is done by using Proposition 2.1, Part (c). First write

$$\zeta(s,k) = \sum \frac{a(n)}{n^s}.$$

Then $s(n,k) = a(1)+\cdots+a(n)$ is the number of integral ideals in k with norm $\leqslant n$. The above limit is the same as the limit of $s(n,k)/n$ as $n \to \infty$. Evaluation of this limit ultimately depends upon the following idea:

Let V denote d-dimensional Euclidean space and Γ a solid in V, that is, a bounded region. The points in V are d-tuples $(x_1, \ldots, x_d)$ with real coordinates. Let $\mathscr{L}$ be the lattice of points with only integral coordinates. Fix a vector v and let $\mathscr{L}_v$ denote the set of translates $v+\mathscr{L}$. The volume of Γ can be computed in the following way. For a real number $\gamma > 0$ consider the points of $\gamma\mathscr{L}_v$ which lie inside Γ. With each such point as center construct a d-dimensional cube having side of length γ and volume γ^d. Let $T_1(\gamma)$ denote the number of such cubes which lie entirely inside Γ. Then Γ is approximated from the inside by a polyhedron of volume $\gamma^d T_1(\gamma)$. If Γ is a sufficiently nice solid—say one described by analytic conditions on the coordinates—then the volume of Γ is the limit of $\gamma^d T_1(\gamma)$ as $\gamma \to 0$.

In the same way approximate Γ from the outside. Let $T_2(\gamma)$ denote the number of cubes with center at some point of $\gamma\mathscr{L}_v$ and having some point in common with Γ. Again for sufficiently nice Γ one obtains $\operatorname{vol}(\Gamma) = \lim \gamma^d T_2(\gamma)$.

Finally let $T(\gamma)$ denote the number of points of $\gamma\mathscr{L}_v$ in Γ. Then $T_1(\gamma) \leqslant T(\gamma) \leqslant T_2(\gamma)$ so $\operatorname{vol}(\Gamma) = \lim \gamma^d T(\gamma)$.

Now change the point of view. The number of points of $\mathscr{L}_v$ which are in $\gamma^{-1}\Gamma$ is $T(\gamma)$. Change notation so $M(t) = T(t^{-1})$.

2.2 Proposition. If Γ, $\mathscr{L}_v$, $M(t)$ are as above then

$$\operatorname{vol}(\Gamma) = \lim_{t \to \infty} M(t)/t^d.$$

The plan for evaluating $\lim s(n,k)/n$ is to identify $s(n,k)/n$ with $M(t)/t^d$ for a suitable $M(t)$ as above and compute $\operatorname{vol}(\Gamma)$ in place of this limit. The computation is rather long.

2.3 Lemma. Each class k contains an integral ideal.

PROOF. Since $\mathbf{I}^{\mathfrak{m}}/i(K_{\mathfrak{m},1})$ is finite, each prime not dividing $\mathfrak{m}$ has some power in $i(K_{\mathfrak{m},1})$. If $\mathfrak{A} = \mathfrak{A}_1\mathfrak{A}_2^{-1}$ is an ideal in k with $\mathfrak{A}_1, \mathfrak{A}_2$ integral, then $\mathfrak{A}_2{}^t$ is in $i(K_{\mathfrak{m},1})$ for some $t > 1$ and so $\mathfrak{A}\mathfrak{A}_2{}^t$ is an integral ideal in $k\mathfrak{A}_2{}^t = k$.

Let $\mathfrak{C}$ be an integral ideal in the inverse class k^{-1} and $\mathfrak{A}$ an integral ideal in k. Our object is to compute the number of $\mathfrak{A}$ with $\mathscr{N}(\mathfrak{A}) \leqslant n$. If this holds then $\mathfrak{A}\mathfrak{C} \in i(K_{\mathfrak{m},1})$ so

$$(4) \qquad \mathfrak{A}\mathfrak{C} = (\alpha), \qquad \alpha \in \mathfrak{C} \cap K_{\mathfrak{m},1}, \qquad \mathscr{N}(\alpha) \leqslant n\mathscr{N}(\mathfrak{C}).$$

Conversely if α satisfies the last two conditions in Eq. (4) then the ideal $\mathfrak{A} = (\alpha)\mathfrak{C}^{-1}$ belongs to k, is integral and has norm $\leqslant n$.

2.4 Lemma. $s(n,k)$ is the number of principal ideals (α) such that $\alpha \in \mathfrak{C} \cap K_{\mathfrak{m},1}$ and $\mathcal{N}(\alpha) \leqslant n\mathcal{N}(\mathfrak{C})$.

Now write $\mathfrak{m} = \mathfrak{m}_0 \mathfrak{m}_\infty$ with $\mathfrak{m}_0$ an integral ideal and $\mathfrak{m}_\infty$ the product of the infinite primes dividing $\mathfrak{m}$. Let α_0 be one solution of

$$\alpha_0 \equiv 1 \bmod \mathfrak{m}_0, \qquad \alpha_0 \equiv 0 \bmod \mathfrak{C}.$$

Such an α_0 exists because $\mathfrak{C} \in \mathbf{I}^{\mathfrak{m}}$ implies $\mathfrak{C}$ is relatively prime to $\mathfrak{m}_0$.

2.5 Lemma. $s(n,k)$ is the number of ideals (α) such that
(a) $\alpha \equiv \alpha_0 \bmod \mathfrak{m}_0\mathfrak{C}$,
(b) $\alpha \equiv 1 \bmod \mathfrak{m}_\infty$,
(c) $0 < \mathcal{N}(\alpha) \leqslant n\mathcal{N}(\mathfrak{C})$.

This is clear in view of Lemma 2.4 because Conditions (a) and (b) are equivalent to $\alpha \in \mathfrak{C} \cap K_{\mathfrak{m},1}$.

Let $\alpha_1, \ldots, \alpha_d$ be a (free) Z-basis for the ideal $\mathfrak{m}_0\mathfrak{C}$. Here d is the dimension $(K:Q)$. The elements α which satisfy Condition (a) of Lemma 2.5 are those which have the form

$$\alpha = \alpha_0 + \sum n_i \alpha_i.$$

There exist rational numbers h_i such that $\alpha_0 = \sum h_i \alpha_i$.

In the d-dimensional euclidean space V let $\mathcal{L}$ be the lattice of d-tuples having integer coordinates. Let $v = (h_1, \ldots, h_d)$ and $\mathcal{L}_v = v + \mathcal{L}$. The correspondence

$$(x_1, \ldots, x_d) \to \sum x_i \alpha_i$$

gives a one to one correspondence between the points in $\mathcal{L}_v$ and the elements of K^* which satisfy (a) of Lemma 2.5.

Next we consider how to select generators for the principal ideals we are trying to count. If (α) and (β) are equal and also α, β satisfy (a), (b), (c) of Lemma 2.5 then $\alpha = \beta u$ with u a unit of R in $K_{\mathfrak{m},1}$.

It is necessary to digress and discuss this group of units. The full group of units, $\mathbf{U}_K$ is a subgroup of $K_{\mathfrak{m}}$ and so the finiteness of $K_{\mathfrak{m}}/K_{\mathfrak{m},1}$ implies $\mathbf{U}_K/\mathbf{U}_K \cap K_{\mathfrak{m},1}$ is finite.

Use the notation of Section 11, Chapter I. There we saw the function $\ell(a)$ mapped $\mathbf{U}_K$ onto an $(r+s-1)$-dimensional lattice in V_{r+s}, the space of $(r+s)$-tuples over the reals. It follows that $\ell(\mathbf{U}_K \cap K_{\mathfrak{m},1})$ has finite index in $\ell(\mathbf{U}_K)$ and so it too is an $(r+s-1)$-dimensional lattice.

Let $w_1, \ldots, w_{r+s-1}$ be elements in $\mathbf{U}_K \cap K_{\mathfrak{m},1}$ such that the images give a Z-basis for $\ell(\mathbf{U}_K \cap K_{\mathfrak{m},1})$. The vectors $W_i = \ell(w_i)$ have the sum of the coordinates equal to zero. (Chapter I, proof of Proposition 11.13) and so the vector $W = (1, \ldots, 1, 2, \ldots, 2)$ is independent of the space $\ell(\mathbf{U}_K \cap K_{\mathfrak{m},1})$. We take W to have r ones and s twos. Summarize this discussion.

2.6 Lemma. The group $\mathbf{U}_K \cap K_{\mathfrak{m},1}$ is the direct product of a finite cyclic group $\langle w \rangle$ with a free abelian group $\langle w_1, \ldots, w_{r+s-1} \rangle$ of rank $r+s-1$. The images $W_i = \ell(w_i)$ along with the vector W give a basis for the $(r+s)$-dimensional real space V_{r+s}.

The map ℓ sends all of K^* into V_{r+s} so for any α there exist real numbers c, c_i such that

$$\ell(\alpha) = cW + \sum c_i W_i.$$

Let $u = w^a \prod w_i^{a_i}$ be a unit in $\mathbf{U}_K \cap K_{\mathfrak{m},1}$. Then $\ell(w^a) = 0$ and

$$\ell(\alpha u) = cW + \sum (c_i + a_i) W_i.$$

Clearly there is a unique selection of the a_i such that $0 \leqslant c_i + a_i < 1$ for each i. This proves part of the next step.

2.7 Lemma. Let $w_{\mathfrak{m}}$ denote the number of roots of unity in $\mathbf{U}_K \cap K_{\mathfrak{m},1}$. Then $w_{\mathfrak{m}} s(n,k)$ is the number points $(x_1, \ldots, x_d) \in \mathscr{L}_v$ which satisfy the conditions:

(a) $\alpha = \sum x_i \alpha_i$,
(b) $\alpha \equiv 1 \bmod \mathfrak{m}_\infty$,
(c) $0 < \mathscr{N}(\alpha) \leqslant n\mathscr{N}(\mathfrak{C})$,
(d) $\ell(\alpha) = cW + \sum c_i W_i$ with $0 \leqslant c_i < 1$.

Just to be explicit here, we know there exist $s(n,k)$ ideals (α) satisfying the conditions in Lemma 2.5. Each such ideal can be generated by any of the $\alpha' = \alpha u$ with u in $\mathbf{U}_K \cap K_{\mathfrak{m},1}$ and α' will still satisfy (a), (b), (c) of Lemma 2.5. From all these elements exactly $w_{\mathfrak{m}}$ of them satisfy Condition (d) of Lemma 2.7. Finally the correspondence with points in $\mathscr{L}_v$ has already been described above.

For the next step we shall extend the map ℓ to an analogous one on part of V_d. We first adopt the convention of Section 11, Chapter I in which the d-dimensional real vectors are written with r real coordinates and s complex coordinates to give $r + 2s = d$ real dimensions. If $(y_1, \ldots, y_r, y_{r+1}, \ldots, y_{r+s}) = Y$ is such a vector with each $y_i \neq 0$, then we set

$$\ell(Y) = (\ln|y_1|, \ldots, \ln|y_r|, 2\ln|y_{r+1}|, \ldots, 2\ln|y_{r+s}|).$$

Thus ℓ maps certain vectors in V_d into V_{r+s}. Recall also the map $v(\alpha)$ mapping K into V_d by

$$v(\alpha) = (\sigma_1(\alpha), \ldots, \sigma_{r+s}(\alpha)).$$

These maps are consistent in that $\ell[v(\alpha)] = \ell(\alpha)$ for $\alpha \in K^*$.

We also extend the norm map to V_d by

$$\mathrm{N}(y_1, \dots, y_{r+s}) = |y_1| \cdots |y_r| |y_{r+1}|^2 \cdots |y_{r+s}|^2.$$

2.8 Lemma. Let Γ_0 denote the set of all points $(x_1, \dots, x_d)$ with x_i real which satisfy the following conditions:

(a) $0 < \mathrm{N}(\sum x_i v(\alpha_i)) \leqslant 1$,
(b) $\ell(\sum x_i v(\alpha_i)) = cW + \sum c_i W_i$ with $0 \leqslant c_i < 1$,
(c) the first r_0 coordinates of $\sum x_i v(\alpha_i)$ are positive where r_0 is the number of real primes dividing $\mathfrak{m}_\infty$.

Then

$$\text{volume}(\Gamma_0) = \frac{w_{\mathfrak{m}}}{\mathcal{N}(\mathfrak{C})} \lim_{n \to \infty} \frac{s(n, k)}{n}.$$

Proof. For each positive real number t, let $M(t)$ denote the number of points in $t\Gamma_0$ which lie in $\mathscr{L}_v$ (defined above). The volume of Γ_0 is given in terms of $M(t)$ by Proposition 2.2 so it is necessary to identify $M(t)$ in terms of $s(n, k)$. Suppose $(x_1, \dots, x_d) \in \Gamma_0$ and $x_i' = tx_i$. Then

(a′) $0 < N(\sum x_i' v(\alpha_i)) \leqslant t^d$,
(b′) $\ell(\sum x_i' v(\alpha_i)) = (c + \ln t)W + \sum c_i W_i$, $0 \leqslant c_i < 1$,
(c′) the first r_0 coordinates of $\sum x_i' v(\alpha_i)$ are positive.

Now let us assume the first r_0 real imbeddings $\sigma_1, \dots, \sigma_{r_0}$ of K are those corresponding to the real infinite primes dividing $\mathfrak{m}$. If we set $\alpha = \sum x_i' v(\alpha_i)$ and assume $(x_1', \dots, x_d')$ is in $\mathscr{L}_v$ then Conditions (a)–(d) of Lemma 2.7 are satisfied so long as $t^d = n\mathcal{N}(\mathfrak{C})$. With this restriction on t one finds $w_{\mathfrak{m}} s(n, k) = M(t)$. Thus

$$\frac{M(t)}{t^d} = \frac{w_{\mathfrak{m}} s(n, k)}{\mathcal{N}(\mathfrak{C}) n} \qquad \text{for} \quad t^d = \mathcal{N}(\mathfrak{C}) n.$$

The result follows.

The rest of the problem is to compute $\mathrm{vol}(\Gamma_0)$. This will require several changes of the variables. Let

$$y_i = \sum_j x_j \sigma_i(\alpha_j) \qquad \text{for} \quad 1 \leqslant i \leqslant r,$$

$$y_j + iy_{j+s} = \sum_k x_k \sigma_j(\alpha_k) \qquad \text{for} \quad r < j \leqslant r + s.$$

2.9 Statement. The conditions of Lemma 2.8 are equivalent to

(a) $0 < |y_1| \cdots |y_r| (y_{r+1}^2 + y_{r+1+s}^2) \cdots (y_{r+s}^2 + y_{r+2s}^2) \leqslant 1$,
(b) $\ell(y_1, \dots, y_{r+2s}) = cW + \sum c_i W_i$, $0 \leqslant c_i < 1$,
(c) $y_1, \dots, y_{r_0}$ are positive.

Let Γ denote the space of all points $(y_1, \ldots, y_d)$ which satisfy Conditions (a) and (b) and also

(c′) $y_1, \ldots, y_r$ are positive.

Let $J = \partial(y)/\partial(x)$ be the Jacobian of the transformation just defined. Then the volumes of the regions are related by

$$\operatorname{vol}(\Gamma_0) = \int_{\Gamma_0} dx = 2^{r-r_0} \int_{\Gamma} J^{-1}\, dy.$$

The coefficient 2^{r-r_0} appears because Condition (c) allows positive and negative coordinates in $r - r_0$ positions where (c′) requires positive terms.

The Jacobian J is the absolute value of the determinant with i, j entry given by

$$\begin{aligned} \partial y_i/\partial x_j &= \operatorname{Re}(\sigma_i(\alpha_j)), & 1 \leqslant i \leqslant r+s \\ &= \operatorname{Im}(\sigma_{i-s}(\alpha_j)), & r+s < i \leqslant r+2s. \end{aligned}$$

By Corollary 11.7, Chapter I we find

$$J = 2^{-s} \mathcal{N}(\mathfrak{m}_0 \mathfrak{C}) |\Delta|^{1/2},$$

where Δ is the discriminant of R over Z.

2.10 Lemma

$$\operatorname{vol}(\Gamma_0) = \frac{2^{r+r_0+s}}{\mathcal{N}(\mathfrak{m}_0 \mathfrak{C}) |\Delta|^{1/2}} \operatorname{vol}(\Gamma).$$

For the next change we use polar coordinates. Let

$$\rho_i = y_i, \qquad 1 \leqslant i \leqslant r$$

$$\rho_{r+j}(\cos\theta_j + i \sin\theta_j) = y_{r+j} + i y_{r+j+s}.$$

Conditions 2.9 (a), (b), (c′) are equivalent to

(2.11) (a) $0 < P = \rho_1, \ldots, \rho_r \rho_{r+1}^2, \ldots, \rho_{r+s}^2 \leqslant 1,$

(b) $\ln \rho_i = \dfrac{\ln P}{d} + \delta_i \sum c_j \ln |\sigma_i(w_j)|$

with

$$0 \leqslant c_j < 1, \qquad \delta_i = 1 \quad \text{if} \quad 1 \leqslant i \leqslant r$$

and

$$\delta_i = 2 \quad \text{if} \quad r < i \leqslant r+s.$$

(c) $0 \leqslant \theta_j \leqslant 2\pi.$

Here Condition (b) requires comment. In 2.9(b) there is an unrestricted constant c which can be computed. Recall that $\ell(w_i) = W_i$ has the sum of its coordinates equal to zero so the sum of the coordinates of $\ell(y_1, \ldots, y_{r+2s})$ is cd. The sum of the coordinates however is also

$$\ln |y_1| \cdots |y_r| (y_{r+1}^2 + y_{r+1+s}^2) \cdots = \ln P.$$

This accounts for the $\ln P$ term in (b) which is just a coordinate wise statement of 2.9(b).

Let $J =$ Jacobian of the transformation. The terms $\partial y_i/\partial \rho_j$ contribute an $r \times r$ identity block in J. The remaining terms are grouped to give 2×2 blocks of the form

$$B_j = \begin{vmatrix} \cos\theta_j & \sin\theta_j \\ -\rho_{r+j}\sin\theta_j & \rho_{r+j}\cos\theta_j \end{vmatrix}$$

so

$$J = \det \begin{vmatrix} I_r & & & \\ & B_1 & & \\ & & \ddots & \\ & & & B_s \end{vmatrix} = \rho_{r+1} \cdots \rho_{r+s}.$$

We now have

$$\begin{aligned} \operatorname{vol}(\Gamma) &= \int_{\Gamma(\rho,\theta)} \rho_{r+1} \cdots \rho_{r+s}\, d\rho_1 \cdots d\rho_{r+s}\, d\theta_1 \cdots d\theta_s \\ &= (2\pi)^s \int_{\Gamma(\rho)} \rho_{r+1} \cdots \rho_{r+s}\, d\rho_1 \cdots d\rho_{r+s}. \end{aligned}$$

Here $\Gamma(\rho)$ is the region in $r+s$ variables described by Eqs. (2.11) (a) and (b).

Finally the last change of variables is given by Eqs. (2.11) (a) and (b) with $P, c_1, \ldots, c_{r+s-1}$ as the new variables. The restrictions on these variables are $0 < P \leqslant 1, 0 \leqslant c_i < 1$.

To compute the Jacobian of the transformation we differentiate Eq. 2.11(b):

$$\begin{aligned} \partial\rho_i/\partial P &= \rho_i 1/Pd \\ \partial\rho_i/\partial c_j &= \rho_i \delta_i \ln|\sigma_i(w_j)| \end{aligned}$$

Each column of $J = |\partial(\rho)/\partial(P, c)|$ has a factor ρ and row one has $1/Pd$ so

$$J = \frac{\rho_1 \cdots \rho_{r+s}}{Pd} \begin{vmatrix} 1 \cdots 1 \quad 2 \cdots 2 \\ \ell(w_1) \\ \vdots \\ \ell(w_{r+s-1}) \end{vmatrix}.$$

The determinant which is left is exactly the determinant of the transformation changing to the basis $W, W_1, \ldots, W_{r+s-1}$ of V_{r+s} from the basis of unit vectors

$(0, \ldots, 1, 0, \ldots)$. In particular, the determinant is nonzero. We set

$$\begin{vmatrix} 1 \cdots 1 \quad 2 \cdots 2 \\ \ell(w_1) \\ \vdots \\ \ell(w_{r+s-1}) \end{vmatrix} = 2^{-s} d \operatorname{reg}(\mathfrak{m})$$

and call the term $\operatorname{reg}(\mathfrak{m})$ the *regulator* of $\mathfrak{m}$. Thus

$$\begin{aligned} \operatorname{vol}(\Gamma) &= (2\pi)^s \int \rho_{r+1} \cdots \rho_{r+s} J \, dP \, dc_1 \cdots dc_{r+s-1} \\ &= (2\pi)^s \operatorname{reg}(\mathfrak{m}) 2^{-s} \int dP \, dc_1 \cdots dc_{r+s-1} \\ &= (2\pi)^s \operatorname{reg}(\mathfrak{m}) 2^{-s}. \end{aligned}$$

Here we used the form of P to see that all the ρs drop out of the integral.

Finally we can reassemble all these equations to get the desired result.

2.12 Lemma

$$\operatorname{vol}(\Gamma_0) = \frac{2^{r-r_0} \operatorname{reg}(\mathfrak{m})(2\pi)^s}{\mathcal{N}(\mathfrak{m}_0 C) |\Delta|^{1/2}}.$$

Before writing this in its final form one more conventional change will be made. For convenience we define $\mathcal{N}(\mathfrak{m}) = 2^{r_0} \mathcal{N}(\mathfrak{m}_0)$ when $\mathfrak{m}$ is a modulus equal to the product of the integral ideal $\mathfrak{m}_0$ and r_0 real infinite primes.

2.13 Theorem. If $\zeta(s, k)$ is the ζ-function of the class k for the modulus $\mathfrak{m}$ then

$$\lim_{s \to 1} (s-1) \zeta(s, k) = g_{\mathfrak{m}},$$

where $g_{\mathfrak{m}}$ is a positive constant depending only upon K and $\mathfrak{m}$ and not upon k. The exact value is

$$g_{\mathfrak{m}} = \frac{2^r (2\pi)^s \operatorname{reg}(\mathfrak{m})}{\mathcal{N}(\mathfrak{m}) \, w_{\mathfrak{m}} |\Delta|^{1/2}},$$

where

$$\begin{aligned} r &= \text{number of real primes of } K, \\ s &= \text{number of complex primes of } K, \\ \Delta &= \text{discriminant of } K \text{ over } Q, \\ w_{\mathfrak{m}} &= \text{number of roots of unity in } \mathbf{U}_K \cap K_{\mathfrak{m},1}, \\ \operatorname{reg}(\mathfrak{m}) &= \text{regulator of } \mathfrak{m}. \end{aligned}$$

In case $\mathfrak{m} = 1$ this result is still interesting. In this case the ray class group mod $\mathfrak{m}$ is the usual class group $\mathbf{I}_K/i(K^*)$. If we take the union of all the cosets k we get all the K-ideals. So

$$\sum_k \zeta(s,k) = \zeta_K(s).$$

When this is multiplied by $(s-1)$ and the limit taken, each term in the sum has the same limit.

2.14 Theorem

$$\lim_{s\to 1}(s-1)\,\zeta_K(s) = \frac{2^r(2\pi)^s \operatorname{reg}(K)}{w_K|\Delta|^{1/2}}\,h_K,$$

where

w_K = number of roots of unity in K,
h_K = class number of K, and the other symbols as above.

Here we have written $\operatorname{reg}(K)$ for the regulator of the units $\mathbf{U}_K$. In the case $\mathfrak{m} = 1$ as in general the regulator has the following interpretation. We have seen that $\ell(\mathbf{U}_K \cap K_{\mathfrak{m},1})$ is an $(r+s-1)$-dimensional lattice contained in $(r+s)$-dimensional space. It turns out that $\operatorname{reg}(\mathfrak{m})$ is the $(r+s-1)$-dimensional volume of the fundamental parallelopiped for this lattice. Thus $\operatorname{reg}(\mathfrak{m})$ depends only on $\mathfrak{m}$ and not upon the choice of generators w_i.

The result given by the last theorem is sometimes useful for computation of the class number h_K. For some explicit examples see Chapter 5 of Borevich and Shafarevich [3].

We can obtain slightly better estimates for $s(n,k)$ without any further difficulty. Refer back to the proof of Lemma 2.8. There we had

$$M(t) = w_{\mathfrak{m}}\, s(n,k) \qquad \text{if} \quad t^d = n\mathcal{N}(\mathfrak{C}).$$

The d-dimensional cubes of side equal to one and having center at points of $\mathcal{L}_v \cap t\Gamma_0$ may not all lie entirely within Γ_0. For some fixed ε all these cubes will lie inside $(t+\varepsilon)\Gamma_0$ and $(t-\varepsilon)\Gamma_0$ will be contained in the union of these cubes. It follows then that

$$(t-\varepsilon)^d \operatorname{vol}(\Gamma_0) \leqslant M(t) \leqslant (t+\varepsilon)^d \operatorname{vol}(\Gamma_0).$$

Also

$$|\operatorname{vol}(t\Gamma_0) - M(t)| \leqslant [(t+\varepsilon)^d - (t-\varepsilon)^d] \operatorname{vol}(\Gamma_0) < a_1 t^{d-1}$$

for some constant a_1. After making the substitutions indicated above, one finds

2.15 Statement. $s(n,k) \leqslant a_2 n + a_3 n^{1-(1/d)}$ for some positive constants a_2, a_3.

Now consider the function

$$\zeta(s,k) - a_2 \zeta_Q(s) = \sum \frac{b(n)}{n^s}.$$

We have at once from (2.15)

$$\left|\sum_{n \leqslant x} b(n)\right| \leqslant |s(x,k) - a_2 x| \leqslant a_3 x^{1-(1/d)}.$$

This means the series converges for all s with $\mathrm{Re}(s) > 1 - 1/d$ by Proposition 2.1. We have already seen that $\zeta_Q(s)$ can be analytically continued into this region except for the simple pole at $s = 1$. It follows the same must hold for $\zeta(s,k)$.

2.16 Theorem. The function $\zeta(s,k)$ can be analytically continued to the region $\mathrm{Re}(s) > 1 - 1/d$ except for the simple pole at $s = 1$.

This is not the whole story. The functions $\zeta_K(s)$ can be continued to the whole plane except for the pole at $s = 1$. However we shall not discuss this matter any further here.

3. CHARACTERS OF ABELIAN GROUPS

In this section, A denotes a finite abelian group.

Definition. A *character of A* is a homomorphism of A into the multiplicative group of complex numbers of absolute value one. The collection of all characters of A is denoted by $\hat{A}$.

If χ_1, χ_2 are characters of A, then their product is the function sending a to $\chi_1(a)\chi_2(a)$. Clearly $\chi_1 \chi_2$ is also a character. This operation makes $\hat{A}$ into an abelian group. The identity is called the *principal character* and is usually denoted by χ_0. One sees $\chi_0(a) = 1$ for all $a \in A$.

3.1 Proposition. $A \cong \hat{A}$.

Proof. Use induction on $|A|$. Suppose A is cyclic of order m with generator y. Then $y^m = 1$ implies $\chi(y)$ is an mth root of unity. If ω is a fixed primitive mth root of unity then the characters of A are determined by the equations $\chi_r(y) = \omega^r$. There are m choices for r and so m characters. They all are powers of χ_1 so $\hat{A}$ is also cyclic of order m.

Now suppose $A = A_1 \times A_2$ is noncyclic. We shall prove $\hat{A} \cong \hat{A}_1 \times \hat{A}_2$ and the result will follow because $\hat{A}_i \cong A_i$ by induction.

Map $\hat{A}$ into $\hat{A}_1 \times \hat{A}_2$ by sending χ to $(\chi|A_1, \chi|A_2)$. Then map $\hat{A}_1 \times \hat{A}_2$ into $\hat{A}$ by identifying (χ_1, χ_2) with the character sending (a_1, a_2) to $\chi_1(a_1)\chi_2(a_2)$.

One checks easily these two maps are inverses of one another so both are isomorphisms.

3.2 Corollary. A is naturally isomorphic to $\hat{\hat{A}}$ by letting the element $a \in A$ correspond to the character $\chi \to \chi(a)$ on $\hat{A}$.

PROOF. The indicated correspondence is a homomorphism of A into $\hat{\hat{A}}$. If it is one to one, then the groups are isomorphic because they have the same order by Proposition 3.1. Suppose the kernel of this homomorphism is the subgroup B. Then for any character χ of A we have $\chi(b) = 1$ for $b \in B$. This means χ can be viewed as a character on A/B. Thus $|A| = |\hat{A}| \leqslant |A|/|B|$. It follows $B = 1$ as required.

3.3 Proposition (Orthogonality relations)

$$(1)\qquad \sum_{a \in A} \chi_1(a)\chi_2(a) = \begin{cases} 0 & \text{if } \chi_1 \neq \chi_2^{-1}, \\ |A| & \text{if } \chi_1 = \chi_2^{-1}. \end{cases}$$

$$(2)\qquad \sum_{\chi \in \hat{A}} \chi(a)\chi(b) = \begin{cases} 0 & \text{if } ab \neq 1, \\ |A| & \text{if } ab = 1. \end{cases}$$

PROOF. Let χ denote a nonprincipal character and b some element of A with $\chi(b) \neq 1$. Then

$$\sum_{a \in A} \chi(a) = \sum_{a \in A} \chi(ab) = \chi(b) \sum_{a \in A} \chi(a).$$

Since $\chi(b) \neq 1$ it must happen that $\sum \chi(a) = 0$. Now if $\chi = \chi_1 \chi_2$ the first alternative of Eq. (1) is proved. The second alternative is obvious since $\chi_1(a)\chi_1^{-1}(a) = 1$ for each $a \in A$. To prove Eq. (2) just use the identification of A with $\hat{\hat{A}}$ and Eq. (1) with $\hat{A}$ in place of A.

4. *L*-SERIES AND PRODUCT REPRESENTATIONS

We shall extend slightly the ζ-functions considered above. Let $\mathfrak{m}$ be a modulus for K and χ a character of the finite group $\mathbf{I}^{\mathfrak{m}}/i(K_{\mathfrak{m},1})$. We view χ as a function on all of $\mathbf{I}^{\mathfrak{m}}$ by defining $\chi(\mathfrak{C})$ for an ideal $\mathfrak{C}$ to be the value of χ at the coset $i(K_{\mathfrak{m},1})\mathfrak{C}$.

The *L-series* for χ is

$$L(s, \chi) = \sum_{(\mathfrak{m}, \mathfrak{A}) = 1} \chi(\mathfrak{A})/\mathcal{N}(\mathfrak{A})^s,$$

where the sum is taken over all integral ideals prime to $\mathfrak{m}$. Since $\chi(\mathfrak{A})$ depends only upon the class, k, of $\mathfrak{A}$ we may express $L(s, \chi)$ in terms of ζ-functions already introduced. Namely

$$L(s,\chi) = \sum_k \chi(k)\cdot \sum_{\mathfrak{A}\in k} \mathcal{N}(\mathfrak{A})^{-s} \tag{1}$$
$$= \sum_k \chi(k)\zeta(s,k).$$

4.1 Proposition

$$\lim_{s\to 1}(s-1)L(s,\chi) = \begin{cases} 0 & \text{if } \chi \neq \chi_0, \\ h_{\mathfrak{m}} g_{\mathfrak{m}} & \text{if } \chi = \chi_0 \end{cases}$$

where $h_{\mathfrak{m}}$ is the order of the ray class group mod $\mathfrak{m}$ and $g_{\mathfrak{m}}$ is the constant in Theorem 2.13.

PROOF. By Eq. (1) the limit in the proposition is $\sum_k \chi(k) g_{\mathfrak{m}}$ since the functions $(s-1)\zeta(s,k)$ all have the same limit $g_{\mathfrak{m}}$ at $s=1$. Now apply Proposition 3.3.

The next result expresses the L-functions in turns of primes rather than all the ideals. This is the stepping off point to the investigation of primes and their part in the groups mentioned so far.

4.2 Theorem. For all s with $\operatorname{Re}(s) > 1$ the function $L(s,\chi)$ can be represented as a uniformly convergent product

$$L(s,\chi) = \prod_{\mathfrak{p}\nmid\mathfrak{m}} (1-\chi(\mathfrak{p})/\mathcal{N}(\mathfrak{p})^s)^{-1}$$

taken over all primes of K not dividing $\mathfrak{m}$.

PROOF. Let $\mathfrak{p}$ denote any prime ideal. There is an absolutely convergent series

$$\left(1 - \frac{\chi(\mathfrak{p})}{\mathcal{N}(\mathfrak{p})^s}\right)^{-1} = 1 + \frac{\chi(\mathfrak{p})}{\mathcal{N}(\mathfrak{p})^s} + \frac{\chi(\mathfrak{p}^2)}{\mathcal{N}(\mathfrak{p}^2)^s} + \cdots.$$

Suppose $\mathfrak{p}_1, \ldots, \mathfrak{p}_r$ are all the primes in $\mathbf{I}^m$ having norm $\leqslant t$. Then

$$\prod\left(1 - \frac{\chi(\mathfrak{p}_i)}{\mathcal{N}(\mathfrak{p}_i)^s}\right)^{-1} = \sum \frac{\chi(\mathfrak{p}_1^{a_1}\cdots\mathfrak{p}_r^{a_r})}{\mathcal{N}(\mathfrak{p}_1^{a_1}\cdots\mathfrak{p}_r^{a_r})^s}$$
$$= \sum_{\mathfrak{A}}{}^{*} \frac{\chi(\mathfrak{A})}{\mathcal{N}(\mathfrak{A})^s}$$

where the * means the sum is taken over all integral ideals in $\mathbf{I}^{\mathfrak{m}}$ divisible only by primes with norm $\leqslant t$. Now one sees that

$$\left| L(s,\chi) - \prod_{\mathcal{N}(\mathfrak{p})\leqslant t}\left(1 - \frac{\chi(\mathfrak{p})}{\mathcal{N}(\mathfrak{p})^s}\right)^{-1}\right| \leqslant \left|\sum_{\mathcal{N}(\mathfrak{A})>t} \frac{\chi(\mathfrak{A})}{\mathcal{N}(\mathfrak{A})^s}\right|.$$

The rightmost term is the remainder term for $L(s,\chi)$. The convergence of

$L(s, \chi)$ is implied by that for $\zeta(s, k)$ when $\mathrm{Re}(s) > 1$ so the remainder term must have a zero limit as t increases. The result follows.

Some special cases are worth recording. Take $\mathfrak{m} = 1$ and $\chi = \chi_0$; then $L(s, \chi) = \zeta_K(s)$.

4.3 Corollary

(a) $\zeta_K(s) = \prod_{\mathfrak{p}} (1 - 1/\mathcal{N}(\mathfrak{p})^s)^{-1}$,

(b) $\zeta_Q(s) = \prod_{p} (1 - 1/p^s)^{-1}$.

The uniform convergence of the infinite series that arise can be used to obtain uniform convergence of the infinite product. In particular the product may be taken in any order without changing the value.

Let $\log z$ denote the branch of the logarithm function having imaginary part on $(-\pi/2, \pi/2)$ when $\mathrm{Re}(z) > 0$. Then $\log z$ is real for real z and the function has an absolutely convergent series representation

$$-\log(1-z) = z + \frac{z^2}{2} + \frac{z^3}{3} + \cdots$$

for $|z| < 1$. We shall apply this to the infinite products but first we give a lemma that allows us to change a product into a sum.

Lemma. Let $\{u_j\}$ be a sequence of real numbers all $\geqslant 2$ and suppose the function

$$f(s) = \prod_j (1 - u_j^{-s})^{-1}$$

is uniformly convergent in each region $D(1, \delta, \varepsilon)$. Then

$$\log f(s) = \sum_j u_j^{-s} + g(s),$$

where $g(s)$ is bounded in neighborhood of $s = 1$.

PROOF. The uniform convergence allows the following manipulations:

$$\begin{aligned} \log f(s) &= -\sum_j \log(1 - u_j^{-s}) \\ &= \sum_j \sum_{m=1}^{\infty} \frac{1}{m u_j^{sm}} \\ &= \sum_j \frac{1}{u_j^s} + \sum_j \sum_{m=2}^{\infty} \frac{1}{m u_j^{sm}} \\ &= \sum_j \frac{1}{u_j^s} + g(s). \end{aligned}$$

Let $\sigma = \operatorname{Re}(s)$ so we obtain

$$|g(s)| \leqslant \sum_j \sum_{m=2}^{\infty} \frac{1}{m u_j^{m\sigma}}.$$

Estimate the inner sum by using

$$\sum_2^{\infty} \frac{1}{m u^{m\sigma}} \leqslant \sum_2^{\infty} \frac{1}{2}\left(\frac{1}{u^{\sigma}}\right)^m = \frac{1}{2}\left\{\frac{1}{1-u^{-\sigma}} - u^{-\sigma} - 1\right\} < \frac{1}{u^{2\sigma}}.$$

Thus

$$|g(s)| \leqslant \sum_j \frac{1}{u_j^{2\sigma}}.$$

The convergence of $f(2\sigma)$ for $2\sigma \geqslant 1+\delta$ implies the finiteness of $|g(s)|$ at $\sigma = 1$ or $s = 1$. In particular $g(s)$ is bounded in a disk $|s-1| < \frac{1}{2}$.

As an immediate consequence of this and Corollary 4.3 we have

(4.4) (a) $$\log \zeta_K(s) = \sum_{\mathfrak{p}} \frac{1}{\mathcal{N}(\mathfrak{p})^s} + g_1(s),$$

(b) $$\log \zeta_Q(s) = \sum_p \frac{1}{p^s} + g_2(s),$$

with g representing a function bounded at $s = 1$.

To illustrate the kind of reasoning to be used below, we show how the expression 4.4(b) can be used along with Theorem 2.14 to prove the existence of infinitely many primes.

If there were only a finite number of primes then $\log \zeta_Q(s)$ would be bounded near $s = 1$. By Theorem 2.14 the function $(s-1)\zeta_Q(s)$ has a finite positive limit at $s = 1$ so $\log(s-1)\zeta_Q(s)$ is also bounded near $s = 1$. It follows that

$$\log(s-1) = \log(s-1)\zeta_Q(s) - \log \zeta_Q(s)$$

is bounded at $s = 1$. This is impossible, of course, so there exist infinitely many primes.

Since this idea will be used several times we shall use the following notation. For functions $f_1(s)$ and $f_2(s)$ defined for $\operatorname{Re}(s) > 1$ at least, we write.

$$f_1(s) \sim f_2(s)$$

to mean $f_1(s) - f_2(s)$ has a finite limit at $s = 1$.

4.5 Proposition. Let K be an algebraic number field and S the set of primes of K which have relative degree one over Q. Then S is an infinite set.

PROOF. If we exclude the finite number of ramified primes which may be in S, then S is the set of primes $\mathfrak{P}$ of K for which $\mathcal{N}(\mathfrak{P}) = p$ is prime.

By Eq. 4.4(a) it follows that

$$\log \zeta_K(s) \sim \sum_{\mathfrak{P}} \frac{1}{\mathcal{N}(\mathfrak{P})^s}$$

with the sum taken over all primes of K. Estimate this sum by first taking it over those primes outside S. Again ignore the ramified primes since the sum over a finite set is bounded at $s = 1$. For $\mathfrak{P}$ outside S, $\mathcal{N}(\mathfrak{P}) = p^f \geqslant p^2$ and at most $(K:Q)$ possible $\mathfrak{P}$ have their norms equal to a power of the same prime. Thus

$$\left| \sum_{\mathfrak{P} \notin S} \frac{1}{\mathcal{N}(\mathfrak{P})^s} \right| \leqslant (K:Q) \sum_p \frac{1}{p^{2\sigma}}.$$

This sum is bounded at $\sigma = 1$ so it follows that

$$\log \zeta_K(s) \sim \sum_{\mathfrak{P} \in S} \frac{1}{\mathcal{N}(\mathfrak{P})^s}.$$

By Theorem 2.14 it follows that $\log(s-1)\zeta_K(s)$ is bounded at $s = 1$ and since $\log(s-1)$ is not bounded at $s = 1$ it follows that

$$\log \zeta_K(s) \sim -\log(s-1) \sim \sum_{\mathfrak{P} \in S} \frac{1}{\mathcal{N}(\mathfrak{P})^s}$$

and S must be an infinite set which completes the proof.

This idea can be expanded.

Definition. Let S be a set of primes of the algebraic number field K. If there exists a real number δ such that

$$\sum_{\mathfrak{P} \in S} \frac{1}{\mathcal{N}(\mathfrak{P})^s} \sim -\delta \log(s-1)$$

then we say δ is the *Dirichlet density of* S and we write $\delta(S) = \delta$.

We have just proved that $\delta(S) = 1$ when S is the set of primes having relative degree one over Q.

4.6 Properties of the Dirichlet Density

(4.6.1) If S has $\delta(S) \neq 0$ then S is an infinite set.

PROOF. If S is a finite set then

$$\sum_{\mathfrak{P} \in S} \mathcal{N}(\mathfrak{P})^{-s} \sim 0.$$

(4.6.2) Let S_1 denote the set of primes of K having relative degree one over Q. If S is any set which has a Dirichlet density, then $\delta(S) = \delta(S \cap S_1)$.

PROOF. The estimates given in the proof of Proposition 4.5 can be used to show

$$\sum_{\substack{\mathfrak{P} \in S \\ \mathfrak{P} \notin S_1}} \mathcal{N}(\mathfrak{P})^{-s} \sim 0.$$

Then it follows

$$\sum_{\mathfrak{P} \in S} \mathcal{N}(\mathfrak{P})^{-s} \sim \sum_{\mathfrak{P} \in S \cap S_1} \mathcal{N}(\mathfrak{P})^{-s}.$$

(4.6.3) If $S \subset S'$ then $\delta(S) \leqslant \delta(S')$ whenever both densities exist.

This follows from the observation that $\sum \mathcal{N}(\mathfrak{P})^{-s}$ cannot be negative for real s sufficiently close to $s = 1$.

One sees from these remarks that $0 \leqslant \delta(S) \leqslant 1$ whenever S has a density. Thus δ provides a way of measuring the ratio of primes in S to all primes of K.

We shall make a computation of a density that will be important later.

Let $\mathfrak{m}$ be a modulus for K and H a group such that

$$i(K_{\mathfrak{m},1}) \subseteq H \subseteq \mathbf{I}^{\mathfrak{m}}.$$

Let h denote the index of H in $\mathbf{I}^{\mathfrak{m}}$ (finite by Corollary 1.6).

4.7 Theorem. Let S be a set of primes contained in H. If S has density $\delta(S)$, then $\delta(S) \leqslant 1/h$.

PROOF. Let χ be a character of $\mathbf{I}^{\mathfrak{m}}/H$ viewed as a homomorphism on $\mathbf{I}^{\mathfrak{m}}$ with kernel containing H. Then

$$\log L(s,\chi) = \sum_{\mathfrak{P} \nmid \mathfrak{m}} \chi(\mathfrak{P}) \mathcal{N}(\mathfrak{P})^{-s} + g_\chi(s)$$

with $g_\chi(s)$ a convergent Dirichlet series for $\mathrm{Re}(s) > \frac{1}{2}$. In particular $g_\chi(s)$ is bounded at $s = 1$.

For any $\mathfrak{P}$, $\sum \chi(\mathfrak{P})$ taken over all characters of $\mathbf{I}^{\mathfrak{m}}/H$ is zero unless $\mathfrak{P} \in H$ in which case the value is h. Thus

$$\sum_{\mathfrak{P} \in H} h\mathcal{N}(\mathfrak{P})^{-s} = \sum_{\chi \neq \chi_0} \{\log L(s,\chi) - g_\chi(s)\}$$
$$+ \log(s-1) L(s,\chi_0) - \log(s-1) - g_{\chi_0}(s).$$

By assumption we also have

$$\sum_{\mathfrak{P} \in s} \mathcal{N}(\mathfrak{P})^{-s} = -\delta(S) \log(s-1) + g(s)$$

with $g(s)$ bounded at $s = 1$. The assumption $S \subset H$ implies

$$\sum_{\mathfrak{P} \in H} \mathcal{N}(\mathfrak{P})^{-s} - \sum_{\mathfrak{P} \in S} \mathcal{N}(\mathfrak{P})^{-s}$$

is nonnegative when s is real and > 1. This means that

$$-(1/h - \delta(S)) \log(s-1) + \sum_{\chi \neq \chi_0} \{\log L(s,\chi) - g_\chi(s)\}$$
$$+ \log(s-1) L(s,\chi_0) - g_{\chi_0}(s) - g(s)$$

is positive when s is real and > 1. All the functions denoted by gs are bounded at $s = 1$. The same is true of $\log(s-1) L(s,\chi_0)$ by Proposition 4.1.

The terms $\log L(s,\chi)$ are bounded at $s = 1$ unless $L(1,\chi) = 0$. [By Proposition 4.1 one can show that $L(s,\chi)$ is continuous at $s = 1$ provided $\chi \neq \chi_0$. So it is permissible to write $L(1,\chi)$.] In case $L(1,\chi) = 0$ for some χ then the log terms become negatively infinite at $s = 1$. We insist $s > 1$ so $\log(s-1)$ is also negative near $s = 1$. The only way for the expression to be positive is $\delta(S) \leqslant 1/h$ as required.

Remark. It is true that $L(1,\chi) \neq 0$ for $\chi \neq \chi_0$ but this is fairly difficult to prove and will not be done in this generality here. See Section 10, Chapter V for a proof.

If we assume this result one obtains that the set of primes in H has density exactly $1/h$. The above proof does show the following result.

4.8 Proposition. In the above context if $\delta(S) = 1/h$ then $L(1,\chi) \neq 0$ for each nonprincipal character χ of $\mathbf{I}^m/H$.

Proof. The function above now is positive for real s sufficiently close to 1 but the $\log(s-1)$ term has zero coefficient. The remaining terms are bounded at 1 or become negatively infinite when $L(1,\chi) = 0$. Accordingly this cannot occur.

5. FROBENIUS DENSITY THEOREM

In this section we let L denote a Galois extension of K with $G = G(L/K)$. The main idea here is to prove the existence of primes in K having certain prescribed decompositions in L.

We prepare first with a purely group theoretic result.

Definition. Let σ be an element of order n in G. The *division of* σ is the collection of all elements in G which are conjugate to some σ^m with m relatively prime to n.

5.1 Lemma. Let σ be an element of order n in G, H the cyclic group $\langle\sigma\rangle$, and t the number of elements in the division of σ. Then $t = \phi(n)[G : \mathrm{N}_G(H)]$ with $\phi(n)$ the Euler function.

PROOF. For m relatively prime to n we have $\mathrm{C}_G(\sigma) = \mathrm{C}_G(\sigma^m)$. Thus σ^m has $[G : \mathrm{C}_G(\sigma)]$ conjugates. As m ranges over all integers between 1 and n relatively prime to n we count $\phi(n)[G : \mathrm{C}_G(\sigma)]$ conjugates this way. But these are not all different. An element is counted q times if it is conjugate to q distinct powers σ^m. Evaluate q by observing the number of conjugates of σ^m which are also powers of σ is the number of distinct automorphisms of H induced by conjugation; namely the index $[\mathrm{N}_G(H) : \mathrm{C}_G(\sigma)] = q$. It follows that

$$t = \frac{\phi(n)[G : \mathrm{C}_G(\sigma)]}{[\mathrm{N}_G(H) : \mathrm{C}_G(\sigma)]} = \phi(n)[G : \mathrm{N}_G(H)].$$

Now we can state the main result.

5.2 Theorem (Frobenius Density Theorem). Let σ be an element of $G = G(L/K)$ having t elements in its division. Let S_1 denote the set of primes of K which are divisible by a prime of L whose Frobenius automorphism is in the division of σ. Then S_1 has Dirichlet density $t/|G|$.

PROOF. The proof is done by induction on n, the order of σ. Consider the case $n = 1$ first. Then $\sigma = 1$ and S_1 is the set of primes of K which split completely in L. Let S^* denote the set of primes of L dividing some prime in S_1. For each $\mathfrak{p}$ in S_1 there exists exactly $(L : K) = |G|$ distinct primes in S^* dividing $\mathfrak{p}$ and each of these has norm exactly $\mathfrak{p}$. Thus we find

$$\begin{aligned}\sum_{\mathfrak{P} \in S^*} \mathcal{N}_{L/Q}(\mathfrak{P})^{-s} &= \sum_{\mathfrak{P} \in S^*} \mathcal{N}_{K/Q}(\mathcal{N}_{L/K}(\mathfrak{P}))^{-s} \\ &= |G| \sum_{\mathfrak{p} \in S_1} \mathcal{N}_{K/Q}(\mathfrak{p})^{-s}.\end{aligned}$$

The first sum can be evaluated. Let T denote the set of primes of L which have relative degree one over Q. Then $T \subseteq S^*$ and the sum $\sum_{\mathfrak{P} \in S^* - T} \mathcal{N}(\mathfrak{P})^{-s} \sim 0$.

Now it has already been proved that $\delta(T) = 1$ so we have $\delta(S^*) = 1$. This means

$$\sum_{\mathfrak{p} \in S_1} \mathcal{N}(\mathfrak{p})^{-s} \sim -|G|^{-1} \log(s-1)$$

and $\delta(S_1) = 1/|G|$ as required for this case.

Now assume the order of σ is n and $n > 1$. For each divisor d of n let t_d denote the number of elements in the division of σ^d. Let S_d denote the set of primes of K divisible by some prime of L whose Frobenius automorphism belongs to the division of σ^d. By induction we know $\delta(S_d) = t_d/|G|$ if $d \neq 1$.

Let E denote the subfield of L left fixed by $\langle\sigma\rangle$. Let H denote the group $\langle\sigma\rangle$. We use the description given in Chapter III, Corollary 2.9 for the decomposition of primes in E. The primes $\mathfrak{p}$ of K which have at least one prime factor in E having relative degree one are precisely those $\mathfrak{p}$ divisible by a prime $\mathfrak{P}$ of L such that the Frobenius automorphism τ of $\mathfrak{P}$ has a cycle of length one on the cosets of H. This occurs precisely when $\sigma_i\tau\sigma_i^{-1}$ is in H for some σ_i. This means τ is conjugate to some power of σ and so $\mathfrak{p}$ is in S_d for some d.

Let S_E denote the set of primes of E having relative degree one over K. For $\mathfrak{p} \in S_d$ let $n(\mathfrak{p})$ denote the number of primes of E dividing $\mathfrak{p}$ and having relative degree one over K. Thus each $\mathfrak{p}$ in S_d is the norm of $n(\mathfrak{p})$ distinct primes in S_E. Just as before $\delta(S_E) = 1$ because S_E contains the primes of E having relative degree one over Q. Put these facts together to obtain

$$(1) \quad -\log(s-1) \sim \sum_{\mathfrak{P} \in S_E} \mathscr{N}_{K/Q}(\mathrm{N}_{E/K}(\mathfrak{P}))^{-s} = \sum_{d|n} \sum_{\mathfrak{p} \in S_d} n(\mathfrak{p})\mathscr{N}(\mathfrak{p})^{-s}.$$

Next we evaluate $n(\mathfrak{p})$. Suppose $\mathfrak{p} \in S_d$. By Chapter III, Corollary 2.9 this is the number of distinct cosets $H\sigma_i$ such that $H\sigma_i\sigma^d = H\sigma_i$. This holds if and only if

$$\sigma_i\sigma^d\sigma_i^{-1} \in H.$$

Since H is cyclic this only can happen if $\sigma_i \in \mathrm{N}_G(\langle\sigma^d\rangle)$. So we have

$$n(\mathfrak{p}) = [\mathrm{N}_G(\langle\sigma^d\rangle) : H], \qquad \mathfrak{p} \in S_d.$$

Using this equation and the induction hypothesis we write Eq. (1) as

$$[\mathrm{N}_G(H) : H] \sum_{\mathfrak{p} \in S_1} \mathrm{N}(\mathfrak{p})^{-s} \sim \left\{-1 + \sum_{\substack{d|n \\ d \neq 1}} \frac{[\mathrm{N}_G\langle\sigma^d\rangle : H]\, t_d}{|G|}\right\} \log(s-1).$$

Next use Lemma 5.1 to evaluate $t_d = \phi(n/d)[G : \mathrm{N}_G\langle\sigma^d\rangle]$. The coefficient of $\log(s-1)$ on the right becomes

$$-1 + \sum_{\substack{d|n \\ d \neq 1}} \frac{1}{n}\phi\left(\frac{n}{d}\right) = -1 - \phi(n)/n + \frac{1}{n}\sum_{d|n}\phi\left(\frac{n}{d}\right).$$

A well-known formula of elementary number theory asserts the summation remaining here has the value n so the entire expression is just $-\phi(n)/n$. Thus finally

$$\sum_{\mathfrak{p} \in S_1} \mathscr{N}(\mathfrak{p})^{-s} \sim \frac{-\phi(n)}{[\mathrm{N}_G(H) : H]\,n} \log(s-1) = \frac{-t}{|G|} \log(s-1),$$

again using Lemma 5.1 for the value of t. This completes the proof. We can now prove an important property of the Artin map.

5.3 Corollary. Assume L is an abelian extension of K and S is any finite set of primes of K containing all the primes ramified in L. Then the Artin map $\varphi_{L|K}$ carries $\mathbf{I}_K{}^S$ onto $G(L/K)$.

PROOF. Given an element σ in G, the division of σ consists precisely of the elements which generate the cyclic group $\langle\sigma\rangle$. By the Frobenius density theorem there exist infinitely many primes of L whose Frobenius automorphism generates $\langle\sigma\rangle$ so some $\mathfrak{P}$ can be found with $\mathfrak{p} = \mathfrak{P} \cap K$ not in S and $\varphi_{L|K}(\mathfrak{p})$ a generator of $\langle\sigma\rangle$. Thus $\varphi_{L|K}$ maps $\mathbf{I}_K{}^S$ onto G.

5.4 Corollary. Assume $G(L/K)$ is cyclic of order n, and let d be a divisor of n. The set S_d of primes of K having exactly d prime factors in L has Dirichlet density $\phi(n/d)/n$.

In particular there exist infinitely many primes of K which remain prime in L.

PROOF. In Chapter III, Proposition 2.8, take $H = 1$ so $E = L$. A prime $\mathfrak{p}$ has d factors in L if and only if $\sigma = \varphi_{L|K}(\mathfrak{p})$ has d cycles in the representation of G by permutations of its elements (the cosets of H). When G is cyclic σ has this property if and only if it has order n/d. There exist exactly $\phi(n/d)$ such elements in a cyclic group so by the Frobenius density theorem, $\delta(S_d) = \phi(n/d)/n$.

5.5 Corollary. Let L_1, L_2 be normal extensions of K and let S_1, S_2 denote the sets of primes of K which split completely in L_1, L_2, respectively. If $S_1 \subset S_2$ (except possibly for a set of density 0) then $L_2 \subset L_1$ and conversely.

PROOF. One direction is immediate. If $L_2 \subset L_1$ and $\mathfrak{p}$ splits completely in L_1 then it also splits completely in L_2.

Now suppose $S_1 \subset S_2$ except for a set of density 0. Let $L = L_1 \cdot L_2$ be the least extension of K containing both L_1 and L_2. By Chapter III, Corollary 2.7, the primes of K which split completely in L are those in S_1 since they split completely in both L_1 and L_2. Now compute the densities of S_1 and S_2. Use Theorem 5.2 to get

$$(L : K)^{-1} = \delta(S_1) = (L_1 : K)^{-1}.$$

This forces $L = L_1 L_2 = L_1$ and so $L_2 \subset L_1$ as required.

This seems like a theorem which classifies all normal extensions of K in terms of objects in K alone; namely certain sets of primes. Unfortunately it is not known yet which sets of primes can arise as the set of primes which split completely in some normal extension of K. This problem is solved for abelian extensions by class field theory in Chapter V.

Another application of Theorem 5.2 will give us the so-called *first fundamental inequality* of class field theory.

5.6 Theorem. Assume L is normal over K and $\mathfrak{m}$ is a modulus for K. Let $\mathbf{I}_L{}^{\mathfrak{m}}$ denote the subgroup of I_L generated by all primes $\mathfrak{P}$ of L for which $\mathfrak{P} \cap K$ is in $\mathbf{I}_K{}^{\mathfrak{m}}$. Then

$$[\mathbf{I}_K{}^{\mathfrak{m}} : \mathrm{N}_{L|K}(\mathbf{I}_L{}^{\mathfrak{m}})\, i(K_{\mathfrak{m},1})] \leqslant (L:K).$$

PROOF. Except for a finite set of primes, the primes that split completely in L are found in $\mathrm{N}_{L|K}(\mathbf{I}_L{}^{\mathfrak{m}})$. The density of this set is $1/(L:K)$ by Theorem 5.2. If h is the index on the left above, then by Theorem 4.7, we obtain

$$1/(L:K) \leqslant 1/h$$

which is equivalent with the desired conclusion.

Notice this first inequality holds for any modulus and any normal extension. The reverse inequality will be proved later under very restrictive conditions. The Galois group of L over K must be abelian and the modulus must be divisible by all the primes of K which ramify in L. Some restriction is also necessary upon the exponents of these primes. This theorem is called the second inequality and the proof is entirely different from that of the first inequality.

As a final application of the Frobenius density theorem we shall prove Dirichlet's famous theorem on primes in an arithmetic progression.

Let m be a positive integer and $\mathfrak{m}$ the modulus $(m)p_\infty$ for Q. We write $H = i(Q_{\mathfrak{m},1})$ in $\mathbf{I}_Q{}^{\mathfrak{m}}$.

Let β denote a primitive mth root of unity. We have seen already in Proposition 3.3 of Chapter III that the set of primes of Q which split completely in $L = Q(\beta)$ is the set of primes in H. The density of this set is $(L:Q)^{-1}$ by Theorem 5.2. Also by Theorem 4.7 this number is at most $[\mathbf{I}^{\mathfrak{m}} : H]^{-1} = \phi(m)^{-1}$. This index is computed using the results of Section 1. Thus $(L:Q) \geqslant \phi(m)$. On the other hand, β has at most $\phi(m)$ conjugates so its minimum polynomial has degree $\leqslant \phi(m)$. Thus $(L:Q) = \phi(m)$. [We have proved again the cyclotomic polynomials are irreducible.] Now this information along with Proposition 4.8 yields a useful fact.

5.7 Proposition. If χ is a nonprincipal character of $\mathbf{I}^{\mathfrak{m}}/i(Q_{\mathfrak{m},1})$ then $L(1,\chi) \neq 0$.

This is the crucial step in the next theorem.

5.8 Theorem. Let k_0 be any coset of $i(Q_{\mathfrak{m},1})$ in $\mathbf{I}_Q{}^{\mathfrak{m}}$. Then the set of primes in k_0 has density $1/\phi(m)$.

PROOF. Let $\mathbf{C}^{\mathfrak{m}}$ denote $\mathbf{I}_Q{}^{\mathfrak{m}}/i(Q_{\mathfrak{m},1})$. For any character χ of $\mathbf{C}^{\mathfrak{m}}$ we have

$$\log L(s,\chi) \sim \sum_p \frac{\chi(p)}{p^s} = \sum_{k \in \mathbf{C}^{\mathfrak{m}}} \chi(k) \sum_{p \in k} p^{-s}.$$

Now multiply this by $\chi(k_0^{-1})$ and sum over all characters χ of $\mathbf{C}^m$ to obtain,

$$\log L(s,\chi_0) + \sum_{\chi\neq\chi_0} \chi(k_0^{-1}) \log L(s,\chi) = \sum_k \sum_\chi \chi(k_0^{-1}k) \sum_{p\in k} p^{-s}.$$

The orthogonality relations for characters allow the evaluation of the sum over χ as

$$\begin{aligned} \sum_\chi \chi(k_0^{-1}k) &= \phi(m) \qquad \text{if} \quad k = k_0 \\ &= 0 \qquad\quad\ \text{if} \quad k \neq k_0. \end{aligned}$$

Use the fact that $L(1,\chi) \neq 0$ if $\chi \neq \chi_0$ (Proposition 5.7) to see that the sum over nonprincipal characters above is bounded at $s = 1$. We finally obtain

$$\log L(s,\chi_0) \sim \phi(m) \sum_{p\in k_0} p^{-s}.$$

The function $L(s,\chi_0)$ differs from $\zeta(s)$ only by a finite number of factors due to primes dividing m. It follows then

$$\log L(s,\chi_0) \sim \log\zeta(s) \sim -\log(s-1).$$

Finally we combine the last equivalences to get

$$\sum_{p\in k_0} p^{-s} \sim -1/\phi(m)\log(s-1)$$

as required.

Now let m be a positive integer and a an integer relatively prime to m. Suppose p is a prime in the arithmetic progression $mt+a$, $t\in Z$. Then clearly p is in the coset $aQ_{m,1}$ since $mt+a \equiv a$ implies $(mt+a)/a \in Q_{m,1}$. Conversely if p belongs to $aQ_{m,1}$ then $p = ax/y$ with $x \equiv y \bmod m$. It follows $x = mt_1+y$ and $p = mt+a$ for some t. So the primes p which are congruent to $a \bmod m$ are precisely those generating a prime ideal in a fixed coset of $i(Q_{m,1})$. By the above result there exist infinitely many of these so we have recovered Dirichlet's well-known theorem.

5.9 Theorem. For each positive integer m and each integer a relatively prime to m, there exist infinitely many primes of the form $mt+a$.

The proof of this theorem depends upon the nonvanishing of $L(1,\chi)$ which in turn depends upon the existence of a certain field, $Q(\beta)$ in this case. Appropriate generalizations of these facts will be seen later in Section 10 of Chapter V.

In the following exercises let $f(X)$ denote a monic polynomial with integer coefficients, L the splitting field of $f(X)$ over Q, $E = Q(\theta)$ with θ a root of $f(X)$, $G = \mathrm{Gal}(L/Q)$ and $H = \mathrm{Gal}(L/E)$. Regard G as a permutation group on the roots of $f(X)$. Assume $f(X)$ is irreducible so that G is transitive.

EXERCISE 1. For an element σ in G, the cycles of σ on the cosets of H have the same length as the cycles of σ when viewed as a permutation of the roots of $f(X)$. In fact the correspondence $\tau H \to \tau(\theta)$ between cosets of H and roots of $f(X)$ is preserved by G.

EXERCISE 2. Let p be a rational prime, $\mathfrak{P}$ a prime of L dividing p. Assume p is unramified in L. Then $f(X)$ is irreducible modulo p if and only if the Frobenius automorphism of $\mathfrak{P}$ has a single cycle on the cosets of H.

EXERCISE 3. If p is a rational prime which ramifies in L, then p ramifies in E. Conclude $f(X)$ is reducible modulo p.

Procedure. Let $\mathfrak{P}$ be a prime of L dividing p. If p does not ramify in E then the inertia group $T(\mathfrak{P})$ is contained in H. Moreover p is not ramified in $\tau(E)$ so $T(\mathfrak{P})$ is contained in $\tau H \tau^{-1}$. However the intersection of the conjugates $\tau H \tau^{-1}$ is the identity so $T(\mathfrak{P})$ has order one.

EXERCISE 4. Let $f(X)$ have prime degree q. There exist infinitely many primes p for which $f(X)$ is irreducible modulo p.

Procedure. The prime q divides $|G|$ but not $|H|$. Any prime p of Q divisible by a prime $\mathfrak{P}$ of L whose Frobenius automorphism has order q satisfies the requirements.

EXERCISE 5. Let $f(X)$ have arbitrary degree. There exist infinitely many primes p such that $f(X)$ has no root modulo p.

Procedure. Any prime p divisible by a prime $\mathfrak{P}$ of L whose Frobenius automorphism has no cycle of length one will do. To show G contains such an element observe that any element fixing a point must lie in some conjugate $\tau H \tau^{-1}$. The conjugates of H can account for at most

$$[G : H](|H| - 1) + 1 < |G|$$

elements.

EXERCISE 6. Show that an irreducible polynomial $f(X)$ can be reducible modulo every prime.

First Procedure. Let $f(X)$ be a polynomial such that G is abelian but noncyclic.

Second Procedure. Let $f(X)$ be a polynomial with even degree $2n$ whose Galois group is isomorphic to the alternating group A_{2n}. There is no element in A_{2n} consisting of a single cycle on all $2n$ symbols. Apply Problems 1–3.

Chapter V

CLASS FIELD THEORY

1. COHOMOLOGY OF CYCLIC GROUPS

Let G be a finite group. By a *G-module* we mean an abelian group A together with a homomorphism $\sigma \to \bar{\sigma}$ of G into the automorphism group of A. The action of $\bar{\sigma}$ upon an element $a \in A$ will be written as $\sigma(a)$.

The most frequently encountered examples of G-modules include the groups associated with a normal extension field L of K and G is the Galois group. Some G-modules are L^*, $\mathbf{I}_L$, $\mathbf{C}_L$; the multiplicative group of L, the ideal group, and the ideal class group.

Consider now the case with $G = \langle\sigma\rangle$, a cyclic group of order n. The generator σ is fixed throughout the discussion. Let

$$\Delta = 1 - \sigma, \qquad N = 1 + \sigma + \cdots + \sigma^{n-1}.$$

For each G-module A, Δ and N act as endomorphisms on A; $\Delta(a) = a - \sigma(a)$ if A is written additively, $\Delta(a) = a/\sigma(a)$ if A is multiplicative. In case $A = L^*$, then $\mathrm{N} = \mathrm{N}_{L|K}$ is the usual norm.

In general we write $\Delta|A$ and $\mathrm{N}|A$ to emphasize the module upon which these endomorphisms are acting.

Regardless of the module, the equations $\Delta\mathrm{N} = \mathrm{N}\Delta = 0$ always hold. This means

$$\operatorname{Im}\mathrm{N} \subseteq \ker\Delta \qquad \text{and} \qquad \operatorname{Im}\Delta \subseteq \ker\mathrm{N}.$$

Equality need not hold here so we measure the difference by a pair of abelian groups.

Definition. For a G-module A, the *cohomology groups* of A are

$$H^0(A) = \frac{\ker \Delta|A}{\mathrm{N}(A)}, \qquad H^1(A) = \frac{\ker \mathrm{N}|A}{\Delta(A)}.$$

1.1 Lemma. If A and B are G-modules and $f: A \to B$ is a G-homomorphism, then there exist maps f_0, f_1 induced by f such that f_i is a homomorphism from $H^i(A)$ to $H^i(B)$.

PROOF. Since f is a G-homomorphism, it commutes with Δ and N. In particular $f(\ker \Delta|A) \subseteq \ker \Delta|B$ and $f(\mathrm{N}(A)) \subseteq \mathrm{N}(B)$. There is an induced map f_0 from $H^0(A)$ to $H^0(B)$ by $f_0(a+\mathrm{N}(A)) = f(a)+\mathrm{N}(B)$.
Similary one obtains f_1.

1.2 Lemma (The Exact Hexagon). Let

$$0 \longrightarrow A \xrightarrow{f} B \xrightarrow{g} C \longrightarrow 0$$

be an exact sequence of G-modules and G-homomorphisms. There exists maps δ_0, δ_1 such that the hexagon is exact at each group:

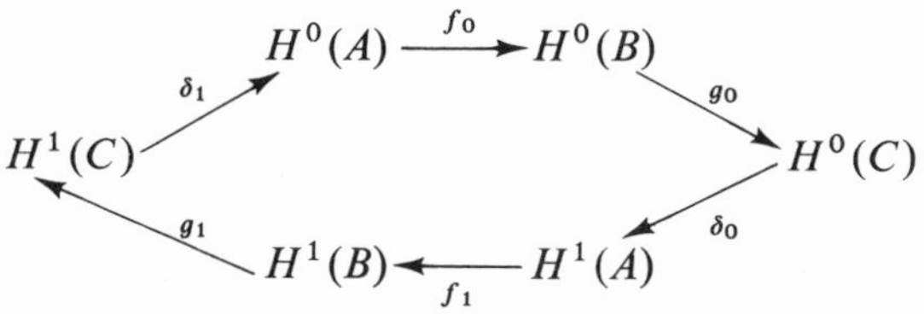

PROOF. We shall define δ_0, δ_1, the so-called connecting homomorphisms. Once this is done, the verification of exactness is an elementary, but tedious exercise using nothing but the exactness of the original sequence. This is left to the reader.

Begin with $c \in \ker \Delta|C$. There exists $b \in B$ such that $g(b) = c$. Then $\Delta g(b) = g(\Delta b) = \Delta c = 0$ implies $\Delta b \in \ker g = \mathrm{Im} f$. So there exists $a \in A$ with $f(a) = \Delta b$. The equation $\mathrm{N}f(a) = f(\mathrm{N}a) = \mathrm{N}\Delta b = 0$ shows $\mathrm{N}a \in \ker f = (0)$. Finally $a \in \ker \mathrm{N}$. The map δ_0 is defined by the equation

$$\delta_0(c+\mathrm{N}(C)) = a + \Delta(A)$$

and thus it is a map from $H^0(C)$ to $H^1(A)$. It must be shown that δ_0 is well defined. Suppose $c+\mathrm{N}(C) = c'+\mathrm{N}(C)$ and $g(b') = c'$. Then $\Delta b' = f(a')$ and it must be shown that $a-a' \in \Delta(A)$. There exists $b'' \in B$ with $c-c' = \mathrm{N}g(b'') = g(\mathrm{N}b'')$. We now have $g(b-b'-\mathrm{N}b'') = 0$ so there is $a'' \in A$ with $b-b'-\mathrm{N}b'' = f(a'')$. It follows that $\Delta b - \Delta b' = f(a-a') = f(\Delta a'')$. Since f is one to one it follows $a-a' \in \Delta(A)$ as required.

That δ_0 is a homomorphism follows easily since at each step an element was selected using the homomorphisms in the sequence.

The map δ_1 is defined analogously. For $c \in \ker \mathrm{N}|C$ we set

$$\delta_1(c+\Delta(C)) = a + \mathrm{N}(A)$$

if $g(b) = c$ and $f(a) = \mathrm{N}b$. The remaining details of the proof are left as an exercise.

Definition. For a G-module A we say the *Herbrand quotient* $q(A)$ is *defined* if $H^0(A)$, $H^1(A)$ are finite groups. We set

$$q(A) = |H^1(A)|/|H^0(A)|$$

and call this the *Herbrand quotient of* A.

1.3 Lemma. If $0 \to A \to B \to C \to 0$ is an exact sequence of G-modules and if two of the objects $q(A)$, $q(B)$, $q(C)$ are defined, then all three are defined and $q(A)q(C) = q(B)$.

PROOF. Use the exact hexagon above. Suppose $q(A)$ and $q(B)$ are defined. Then

$$|H^i(C)| = |\ker \delta_i||\mathrm{Im}\, \delta_i| = |\mathrm{Im}\, g_i||\mathrm{Im}\, \delta_i|.$$

Now g_i is defined on a finite group and $\mathrm{Im}\, \delta_i$ is contained in a finite group so $q(C)$ is defined. Similar arguments work if two others are assumed to be defined.

For the second assertion we prove the equation

$$|H^0(A)||H^0(C)||H^1(B)| = |H^1(A)||H^1(C)||H^0(B)|$$

by observing both sides are equal to

$$|\ker f_0||\mathrm{Im} f_0||\ker \delta_0||\mathrm{Im}\, \delta_0||\ker g_1||\mathrm{Im}\, g_1|.$$

The equality $q(A)q(C) = q(B)$ now follows.

1.4 Corollary. If $A \subset B$ are G-modules and $C = B/A$ is finite, then $q(A) = q(B)$ whenever either one is defined.

PROOF. The result follows immediately if we show $q(C) = 1$ when C is finite. For this case we find

$$q(C) = \frac{[\ker \mathrm{N} : \mathrm{Im}\, \Delta]}{[\ker \Delta : \mathrm{Im}\, \mathrm{N}]} = \frac{|\ker \mathrm{N}||\mathrm{Im}\, \mathrm{N}|}{|\ker \Delta||\mathrm{Im}\, \Delta|} = \frac{|C|}{|C|}$$

so we are done.

To illustrate these ideas as well as for later use we shall consider an example.

Let d be any divisor of $n = |G|$ and let

$$A = \sum_1^d Zu_i$$

be a free Z-module on the d generators u_i. Let G operate on A according to the rules:

$$\sigma u_i = u_{i+1}, \qquad i < d;$$

$$\sigma u_d = u_1.$$

Clearly σ^d generates the subgroup which acts like the identity on A. Let $G_A = \langle \sigma^d \rangle$.

1.5 Proposition. When Z is the ring of integers then $q(A)$ is defined and $q(A) = |G_A|^{-1}$. When Z is a commutative ring with characteristic zero, then $q(A) = [Z : mZ]^{-1}$, $md = n$, provided this is finite.

PROOF. One simply computes all the groups involved. The calculations are not difficult so we shall only state the relevant facts:

(a) $\ker \mathrm{N} = \{\sum a_i u_i | \sum a_i = 0\}$,
(b) $\mathrm{Im}\, \Delta = \ker \mathrm{N}$,
(c) $\ker \Delta = Z(u_1 + \cdots + u_d)$,
(d) $\mathrm{Im}\, \mathrm{N} = mZ(u_1 + \cdots + u_d)$.

This yields $H^0(A) \cong Z/mZ$ and $H^1(A) = 1$ and the result is proved.

2. PREPARATIONS FOR THE SECOND INEQUALITY

Assume L/K is normal with a cyclic Galois group $G = \langle \sigma \rangle$.

Definition. Let $\mathfrak{p}$ be an infinite prime of K. We say $\mathfrak{p}$ is *ramified in* L if $\mathfrak{p}$ is real (on K) but $\mathfrak{p}$ extends to a complex prime of L. In this case we set the ramification index $e_{\mathfrak{p}} = 2$ and formally set $f_{\mathfrak{p}} = 1$. For *unramified* infinite primes $\mathfrak{p}$ we set $e_{\mathfrak{p}} = f_{\mathfrak{p}} = 1$.

Let $\mathfrak{m}$ be a modulus for K containing at least the primes of K which ramify in L (finite and infinite).

Each prime $\mathfrak{p}$ of K may be viewed also as a product of primes in L. In this way $\mathfrak{m}$ is also considered a modulus for L. Thus $\mathbf{I}_L{}^{\mathfrak{m}}$ has meaning and in fact this is a G-module. That is if $\mathfrak{P}$ is a prime of L not dividing $\mathfrak{m}$, then $\sigma(\mathfrak{P})$ does not divide $\mathfrak{m}$ either.

We begin by computing some groups.

2.1 Proposition

(a) $H^1(\mathbf{I}_L{}^{\mathfrak{m}}) = H^1(L^*) = 1$,
(b) $H^0(L^*) = K^*/\mathrm{N}(L^*)$,
(c) $H^0(\mathbf{I}_L{}^{\mathfrak{m}}) = \mathbf{I}_K{}^{\mathfrak{m}}/\mathrm{N}(\mathbf{I}_L{}^{\mathfrak{m}})$.

PROOF. The statement $H^1(L^*) = 1$ is just a restatement of Hilbert's Theorem 90. Statement (b) is clear because $\ker\Delta$ consists of the elements left fixed by G. In the same way $\ker\Delta|\mathbf{I}_L{}^{\mathfrak{m}}$ is the set of fractional ideals $\mathfrak{A}$ of $\mathbf{I}_L{}^{\mathfrak{m}}$ left fixed by G. Suppose $\mathfrak{P}^k$ divides $\mathfrak{A}$ for a prime $\mathfrak{P}$ of L. Then $\sigma^i(\mathfrak{P})^k$ also divides $\mathfrak{A}$ and so the product of the distinct conjugates of $\mathfrak{P}$ divides $\mathfrak{A}$ (with exponent k). Now if $\mathfrak{P} \cap K = \mathfrak{p}$, then $\mathfrak{p} \nmid \mathfrak{m}$ so $\mathfrak{p}$ is unramified in L. Thus $\mathfrak{p}$ is the product of the distinct conjugates of $\mathfrak{P}$ so $\mathfrak{A} = \mathfrak{p}^k \mathfrak{A}_0$ with $\mathfrak{A}_0$ also G-invariant. It follows in this way that $\mathfrak{A} \in \mathbf{I}_K{}^{\mathfrak{m}}$ and then (c) is obvious.

We have left to prove $H^1(\mathbf{I}_L{}^{\mathfrak{m}}) = 1$. Take $\mathfrak{A} \in \mathbf{I}_L{}^{\mathfrak{m}}$ with $\mathrm{N}(\mathfrak{A}) = 1$. Let $\mathfrak{P}_0^{a_0}$ be the exact power of the prime $\mathfrak{P}_0$ appearing in $\mathfrak{A}$. Let $\mathfrak{P}_i = \sigma^i(\mathfrak{P}_0)$ for $1 \leqslant i \leqslant g-1$ and suppose $\sigma^g(\mathfrak{P}_0) = \mathfrak{P}_0$ with g minimal. Let $\mathfrak{P}_i^{a_i}$ be the power of $\mathfrak{P}_i$ dividing $\mathfrak{A}$. Set

$$\mathfrak{B} = \mathfrak{P}_0^{a_0}\mathfrak{P}_1^{a_0+a_1}\cdots\mathfrak{P}_{g-2}^{a_0+\cdots+a_{g-2}}$$

so that

$$\Delta\mathfrak{B} = \mathfrak{P}_0^{a_0}\mathfrak{P}_1^{a_1}\cdots\mathfrak{P}_{g-2}^{a_{g-2}}\mathfrak{P}_{g-1}^{b},$$

where $b = -a_0 - \cdots - a_{g-2}$. Let $\mathrm{N}(\mathfrak{P}_0) = \mathfrak{p}^f$. Since $\mathrm{N}(\mathfrak{A}) = 1$ and since the $\mathfrak{p}$-part of $\mathrm{N}(\mathfrak{A})$ must come from the terms $\mathrm{N}(\mathfrak{P}_i)$ we see

$$\mathrm{N}\left(\prod \mathfrak{P}_i^{a_i}\right) = \mathfrak{p}^{f(a_0+\cdots+a_{g-1})} = 1.$$

It follows that $\sum a_i = 0$ and so $b = a_{g-1}$. Thus $\Delta\mathfrak{B}$ is the part of $\mathfrak{A}$ contributed by the $\mathfrak{P}_i$. After this procedure is repeated with the other primes dividing $\mathfrak{A}$ we are left with $\mathfrak{A} \in \operatorname{Im}\Delta$ and so $\ker \mathrm{N} = \operatorname{Im}\Delta$ as was to be proved.

The next step is to define some maps that will be used frequently in this and the next few sections.

The homomorphism

$$j : \mathbf{I}_L \to \mathbf{I}_L{}^{\mathfrak{m}}$$

is defined on primes by

$$j(\mathfrak{P}) = \begin{cases} \mathfrak{P}, & \mathfrak{P} \nmid \mathfrak{m}, \\ 1, & \mathfrak{P} \mid \mathfrak{m}. \end{cases}$$

Evidently j maps an ideal $\mathfrak{A}$ onto the part of $\mathfrak{A}$ relatively prime to $\mathfrak{m}$.

Recall that the map i carries an element α in L^* to its principal ideal in $\mathbf{I}_L$. We denote by f the composite ji so that

(1) $$f:\ L^* \to \mathbf{I}_L{}^{\mathfrak{m}}.$$

Notice that all the groups mentioned are G-modules and the maps i, j, f are G-homomorphisms.

Now let S denote the set of primes of L which divide $\mathfrak{m}$. Let $L^S = \ker f$. One easily sees

(2) $$L^S = \{\alpha \in L^* \mid i(\alpha) \text{ is divisible only by primes in } S\}.$$

In the case that $S \subseteq S_\infty$ the set of infinite primes, then $L^S = \mathbf{U}_L$ is the group of (absolute) units of L. For this reason one says (for any S) L^S is the group of *S-units*. One must keep in mind that L^S is the set of elements which are units locally *outside* S.

We begin a series of calculations to determine the Herbrand quotient of several G-modules.

2.2 Lemma. If $q(\mathbf{U}_L)$ and $q(\ker j)$ are defined, then $q(L^S) = q(\mathbf{U}_L)q(\ker j)$.

PROOF. The equation $1 = f(L^S) = ji(L^S)$ gives rise to an exact sequence

$$1 \to i(L^S) \to \ker j \to C \to 1$$

for some group C. In fact C is finite because

$$C \cong \frac{\ker j}{i(L^S)} \cong \frac{i(L^*)\ker j}{i(L^*)}$$

which is a subgroup of the class group of L. We are assuming $q(\ker j)$ is defined so by the results in Section 1 we obtain $q(i(L^S)) = q(\ker j)$. Next use the exact sequence

$$1 \to \mathbf{U}_L \to L^S \to i(L^S) \to 1$$

to conclude $q(L^S) = q(\mathbf{U}_L)q(i(L^S)) = q(\mathbf{U}_L)q(\ker j)$.

The next task is the computation of $q(\mathbf{U}_L)$ and $q(\ker j)$. The first is done by finding a suitable subgroup with finite index in $\mathbf{U}_L$ so that Corollary 1.4 can be applied.

Let L have r real primes $\mathfrak{P}_1, \ldots, \mathfrak{P}_r$ and s complex primes $\mathfrak{P}_{r+1}, \ldots, \mathfrak{P}_{r+s}$. Let $|X|_i$ denote a valuation in $\mathfrak{P}_i$. When $\mathfrak{P} = \mathfrak{P}_i$ we also write $|X|_\mathfrak{P}$ for $|X|_i$.

2.3 Theorem. There exists units $w_1, \ldots, w_{r+s}$ in $\mathbf{U}_L$ in one-to-one correspondence with the infinite primes of L such that

(a) G permutes cyclically the w_i corresponding to the $\mathfrak{P}_i$ which extend an infinite prime $\mathfrak{p}$ of K;
(b) $1 = \prod w_i$ is the only relation between them;
(c) the subgroup W generated by all the w_i has finite index in $\mathbf{U}_L$.

PROOF. For each infinite prime $\mathfrak{p}$ of K select one prime $\mathfrak{P}$ of L which extends $\mathfrak{p}$. For each such $\mathfrak{P}$ select a unit $w_\mathfrak{P}$ of $\mathbf{U}_L$ such that $|w_\mathfrak{P}|_i < 1$ whenever $\mathfrak{P} \neq \mathfrak{P}_i$. Such a unit exists as one sees from the proof of the Dirichlet unit theorem (Chapter I, Section 11).

Let $G(\mathfrak{P})$ denote the decomposition group of $\mathfrak{P}$ and set

$$w_\mathfrak{p}' = \prod_{\tau \in G(\mathfrak{P})} \tau(w_\mathfrak{P}).$$

Then $w_\mathfrak{P}'$ is still in $\mathbf{U}_L$ and

$$|w_\mathfrak{P}'|_i = \prod |\tau(w_\mathfrak{P})|_i = \prod |w_\mathfrak{P}|_{\tau(\mathfrak{P}_i)}.$$

When $\tau \in G(\mathfrak{P})$ and $\mathfrak{P}_i = \mathfrak{P}$, then $\tau(\mathfrak{P}_i) = \mathfrak{P}$ so it follows that $|w_{\mathfrak{P}}'|_i < 1$ if $\mathfrak{P}_i \neq \mathfrak{P}$.

The point of changing to $w_{\mathfrak{P}}'$ is that $\mathfrak{P}$ and $w_{\mathfrak{P}}'$ are left fixed by exactly the same subgroup of G. Now if $\mathfrak{Q}$ is a prime of L also extending $\mathfrak{p}$ and if $\tau(\mathfrak{P}) = \mathfrak{Q}$ with $\tau \in G$ then we define

$$w_{\mathfrak{Q}}' = \tau(w_{\mathfrak{P}}').$$

We now have a one to one correspondence between the images of $w_{\mathfrak{P}}'$ and the primes of L above $\mathfrak{p}$. When all the infinite primes are used, we obtain a set of $r+s$ units in $\mathbf{U}_L$. The conditions of (11.17), Chapter I are satisfied by the $w_{\mathfrak{P}}'$ in place of the u_j used there. Set W' equal to the group generated by all $w_{\mathfrak{P}}'$ and apply Proposition 11.18 of Chapter I to conclude $\ell(W')$ is a lattice of rank $r+s-1$ and in fact any $r+s-1$ of the elements $\ell(w_{\mathfrak{P}}')$ give a basis.

For each infinite prime $\mathfrak{p}$ of K let

$$v_{\mathfrak{p}} = \prod_{\mathfrak{P}|\mathfrak{p}} w_{\mathfrak{P}}'.$$

Notice $v_{\mathfrak{p}} \in \mathbf{U}_K$ because $v_{\mathfrak{p}}$ is fixed by G. Let $r'+s'$ be the number of infinite primes of K. There are $r'+s'$ elements $v_{\mathfrak{p}}$ of which $r'+s'-1$ are independent. There exist integers $a_{\mathfrak{p}}$ (not all zero) such that

$$1 = \prod_{\mathfrak{p}} v_{\mathfrak{p}}^{a_{\mathfrak{p}}}.$$

In fact none of the $a_{\mathfrak{p}}$ are zero because there cannot be a relation between fewer than $r+s-1$ of the $\ell(w_{\mathfrak{p}}')$. Finally set $w_i = (w_{\mathfrak{P}_i}')^{a_{\mathfrak{p}}}$ if $\mathfrak{P}_i | \mathfrak{p}$. Then the product of the w_i equals 1 and there is no other relation between them. Let W denote the group generated by the w_i. Then $\ell(W)$ has rank $r+s-1$ and so $[\ell(\mathbf{U}_L):\ell(W)]$ is finite. Since $\ker \ell$ is also finite it follows $[\mathbf{U}_L : W]$ is finite and the proof is complete.

This group W is useful because we are able to evaluate $q(W)$. Proceed as follows. For each infinite prime $\mathfrak{p}$ of K and prime $\mathfrak{P}$ of L extending $\mathfrak{p}$, let $d_{\mathfrak{p}} = [G : G(\mathfrak{P})]$ and set

$$A_{\mathfrak{p}} = \sum_{1}^{d_{\mathfrak{p}}} Z u_{i,\mathfrak{p}}.$$

This is a free Z-module on $d_{\mathfrak{p}}$ generators upon which G acts by permuting the basis elements cyclically. We now have an exact sequence

$$0 \longrightarrow Z \xrightarrow{\ g\ } \sum_{\mathfrak{p}|\infty} A_{\mathfrak{p}} \xrightarrow{\ h\ } W \longrightarrow 1,$$

where $g(z) = z \sum_i \sum_{\mathfrak{p}} u_{i,\mathfrak{p}}$ and h is defined so that the $d_{\mathfrak{p}}$ basis elements $u_{i,\mathfrak{p}}$ map onto those w_j corresponding to the divisors $\mathfrak{P}_j$ of $\mathfrak{p}$ in such a way that h is

G-homomorphism. The exactness follows at once from (b) of Theorem 2.3. Now compute

$$q(W)q(Z) = q(\sum A_{\mathfrak{p}}) = \prod q(A_{\mathfrak{p}}).$$

The action of G upon Z is the trivial action so $q(Z) = 1/|G|$ by Proposition 1.5. The same proposition gives $q(A_{\mathfrak{p}}) = |G(\mathfrak{P})|^{-1}$ if $\mathfrak{P}|\mathfrak{p}$. The groups $G(\mathfrak{P})$ are easily calculated. If $\mathfrak{p}$ is an unramified infinite prime then there exist $(L:K)$ extensions to L so $G(\mathfrak{P}) = 1$. When $\mathfrak{p}$ is ramified there exist $(L:K)/2$ extension to L so $|G(\mathfrak{P})| = 2$.

Now use $q(W) = q(\mathbf{U}_L)$ to get the end of the computation.

2.4 Theorem. Let r_0 be the number of infinite primes of K which ramify in L. Then

$$q(\mathbf{U}_L) = (L:K)/2^{r_0}.$$

To complete the evaluation of $q(L^S)$ we need now to determine $q(\ker j)$.

The group $\ker j$ is the free abelian group on the primes in S. For each prime $\mathfrak{p}$ of K with $\mathfrak{p}$ divisible by primes in S set $I(\mathfrak{p})$ = subgroup of $\ker j$ generated by the divisors of $\mathfrak{p}$. We see at once that

$$\ker j = \prod I(\mathfrak{p}), \qquad q(\ker j) = \prod q(I(\mathfrak{p})).$$

Let $\mathfrak{p} = (\mathfrak{P}_1 \cdots \mathfrak{P}_g)^e$ and $\mathrm{N}_{L|K}(\mathfrak{P}_i) = \mathfrak{p}^f$. We know G is transitive on the $\mathfrak{P}_i$ so one easily verifies

(a) $\ker \mathrm{N}|I(\mathfrak{p}) = \{\prod \mathfrak{P}_i^{a_i} | \sum a_i = 0\}$,
(b) $\mathrm{Im}\,\Delta|I(\mathfrak{p}) = \ker \mathrm{N}|I(\mathfrak{p})$,
(c) $\ker \Delta|I(\mathfrak{p}) = \langle \mathfrak{Q} \rangle, \qquad \mathfrak{Q} = \mathfrak{P}_1 \cdots \mathfrak{P}_g$,
(d) $\mathrm{Im}\,\mathrm{N}|I(\mathfrak{p}) = \langle \mathfrak{p}^f \rangle = \langle \mathfrak{Q}^{ef} \rangle$.

Thus $q(I(\mathfrak{p})) = 1/ef = 1/e_{\mathfrak{p}} f_{\mathfrak{p}}$. Finally apply Lemma 2.2 and these last calculations to compute $q(L^S)$.

2.5 Theorem

$$q(L^S) = \frac{(L:K)}{\prod_{\mathfrak{p}|\mathfrak{m}} e_{\mathfrak{p}} f_{\mathfrak{p}}}.$$

3. A NORM INDEX COMPUTATION

As before L/K has cyclic Galois group $G = \langle \sigma \rangle$. For a modulus $\mathfrak{m}$ for K we set

$$a(\mathfrak{m}) = [K^* : \mathrm{N}(L^*) K_{\mathfrak{m},1}].$$

Our goal is to evaluate this at least under the assumption that the prime divisors of $\mathfrak{m}$ have sufficiently large exponents. This will be made more precise later.

3.1 Lemma. If $\mathfrak{m}$ and $\mathfrak{n}$ are relatively prime moduli, then $a(\mathfrak{m}\mathfrak{n}) = a(\mathfrak{m})\,a(\mathfrak{n})$.

PROOF. The approximation theorem implies that the map

$$\alpha \to (\alpha K_{\mathfrak{m},1}, \alpha K_{\mathfrak{n},1})$$

induces an isomorphism

$$\frac{K^*}{K_{\mathfrak{m}\mathfrak{n},1}} \to \frac{K^*}{K_{\mathfrak{m},1}} \times \frac{K^*}{K_{\mathfrak{n},1}}.$$

This induces a homomorphism

$$\frac{K^*}{K_{\mathfrak{m}\mathfrak{n},1}} \to \frac{K^*}{\mathrm{N}(L^*)\,K_{\mathfrak{m},1}} \times \frac{K^*}{\mathrm{N}(L^*)\,K_{\mathfrak{n},1}}$$

which is onto the direct product. The lemma will be proved if we show the kernel of this map is exactly $\mathrm{N}(L^*)\,K_{\mathfrak{m}\mathfrak{n},1}/K_{\mathfrak{m}\mathfrak{n},1}$.

Suppose $\alpha K_{\mathfrak{m}\mathfrak{n},1}$ is in the kernel. There exist β_1, β_2 in L such that

$$\alpha \equiv \mathrm{N}(\beta_1) \bmod \mathfrak{m}, \qquad \alpha \equiv \mathrm{N}(\beta_2) \bmod \mathfrak{n}.$$

View $\mathfrak{m}$ and $\mathfrak{n}$ as moduli for L. They are still relatively prime and so there is a solution β in L to

$$\beta \equiv \beta_1 \bmod \mathfrak{m}, \qquad \beta \equiv \beta_2 \bmod \mathfrak{n}.$$

From this it follows that

$$\mathrm{N}(\beta)\,\mathrm{N}(\beta_1)^{-1} \in K \cap \mathrm{N}_{L/K}(L_{\mathfrak{m},1})$$

$$\mathrm{N}(\beta)\,\mathrm{N}(\beta_2)^{-1} \in K \cap \mathrm{N}_{L/K}(L_{\mathfrak{n},1}).$$

Next we show $K \cap \mathrm{N}_{L/K}(L_{\mathfrak{m},1}) \subseteq K_{\mathfrak{m},1}$. Let $\mathfrak{p}^a$ be the power of the prime $\mathfrak{p}$ of K dividing $\mathfrak{m}$ and let $\mathfrak{p} = (\mathfrak{P}_1 \cdots \mathfrak{P}_g)^e$ be the factorization of $\mathfrak{p}$ in L. If $\gamma = a/b$ in L with a and b algebraic integers and $\gamma \equiv 1 \bmod \mathfrak{m}$, then

$$a \equiv b \bmod (\mathfrak{P}_1 \cdots \mathfrak{P}_g)^{ae}.$$

Since the product $\mathfrak{P}_1 \cdots \mathfrak{P}_g$ is σ invariant for any σ in $G(L/K)$, we have

$$\sigma(a) \equiv \sigma(b) \bmod (\mathfrak{P}_1 \cdots \mathfrak{P}_g)^{ae}$$

and so

$$\mathrm{N}_{L/K}(a) \equiv \mathrm{N}_{L/K}(b) \bmod (\mathfrak{P}_1 \cdots \mathfrak{P}_g)^{ae}.$$

These norms lie in K so the congruences may be read modulo $K \cap (\mathfrak{P}_1 \cdots \mathfrak{P}_g)^{ae} = \mathfrak{p}^a$. It follows that $\mathrm{N}_{L/K}(a) \equiv \mathrm{N}_{L/K}(b)$ and so $\mathrm{N}_{L/K}(\gamma) \equiv 1 \bmod \mathfrak{p}^a$. Since this holds for each divisor of $\mathfrak{m}$, the stated inclusion follows.

We already have $\alpha \mathrm{N}_{L/K}(\beta_1)^{-1}$ in $K_{\mathfrak{m},1}$ and $\alpha \mathrm{N}_{L/K}(\beta_2)^{-1}$ in $K_{\mathfrak{n},1}$, so

$$\alpha \mathrm{N}_{L/K}(\beta)^{-1} \in K_{\mathfrak{m},1} \cap K_{\mathfrak{n},1} = K_{\mathfrak{mn},1},$$

which is enough to prove the lemma.

This lemma reduces the calculation of $a(\mathfrak{m})$ to the case $\mathfrak{m} = \mathfrak{p}^n$ for a prime $\mathfrak{p}$ and $n \geqslant 1$. The case $\mathfrak{p}$ infinite is easy so we take care of it at once.

3.2 Lemma. If $\mathfrak{p}$ is an infinite prime of K then $a(\mathfrak{p}) = e_{\mathfrak{p}}$ = the ramification index.

PROOF. Suppose $\mathfrak{m} = \mathfrak{p}$ is real and is ramified in L. Let $\mathfrak{P}_1 \cdots \mathfrak{P}_g$ be the (complex) primes of L extending $\mathfrak{p}$. Then for $\alpha \in L^*$,

$$\mathrm{N}_{L|K}(\alpha) = \prod_i \mathrm{N}_{L_{\mathfrak{P}_i}|K_{\mathfrak{p}}}(\alpha).$$

Since $L_{\mathfrak{P}_i}$ is the complex field and $K_{\mathfrak{p}}$ the real field, all the norms on the right are positive. Hence $\mathrm{N}(L^*) = K_{\mathfrak{m},1}$ and $K^*/K_{\mathfrak{m},1}$ has order $2 = e_{\mathfrak{p}}$.

The reader can verify in all other cases $[K^* : \mathrm{N}(L^*)\, K_{\mathfrak{m},1}] = 1 = e_{\mathfrak{p}}$ as required.

For the rest of the section we work with a finite prime $\mathfrak{p}$ and $\mathfrak{m} = \mathfrak{p}^n$. $\mathfrak{P}$ denotes a prime of L dividing $\mathfrak{p}$. As usual, $e_{\mathfrak{p}}, f_{\mathfrak{p}}$ denote the ramification number and relative degree of $\mathfrak{P}$ over $\mathfrak{p}$.

3.3 Lemma

(a) $[K^* : \mathrm{N}(L^*)\, K_{\mathfrak{m}}] = f_{\mathfrak{p}}$;

(b) $a(\mathfrak{m}) = f_{\mathfrak{p}} [K_{\mathfrak{m}} : (K_{\mathfrak{m}} \cap \mathrm{N}(L^*))\, K_{\mathfrak{m},1}]$.

PROOF. Since $\mathfrak{m} = \mathfrak{p}^n$ it follows that $K_{\mathfrak{m}}$ = units in the valuation ring $R_{\mathfrak{p}}$ of $\mathfrak{p}$. Let π be a generator for the maximal ideal so every element in K^* has the form $\pi^a u$, $u \in K_{\mathfrak{m}}$.

We know $\mathrm{N}(\mathfrak{P}) = (\pi^f)$ so elements in $\mathrm{N}(L^*)$ have the form $\pi^{fb} w$ with $w \in K_{\mathfrak{m}}$. It follows

$$K^*/K_{\mathfrak{m}}\, \mathrm{N}(L^*) \cong \langle \pi \rangle / \langle \pi^f \rangle$$

which has order $f = f_{\mathfrak{p}}$ as required for Part (a).

A factorization of $a(\mathfrak{m})$ can be obtained from the successive indices of subgroups in the chain

$$K^* \supset \mathrm{N}(L^*)\, K_{\mathfrak{m}} \supset \mathrm{N}(L^*)\, K_{\mathfrak{m},1}.$$

The first index is $f_{\mathfrak{p}}$ by Part (a). To get the second index in the stated form one need only observe there is a natural isomorphism

$$\frac{K_{\mathfrak{m}}}{(K_{\mathfrak{m}} \cap \mathrm{N}(L^*))\, K_{\mathfrak{m},1}} \cong \frac{\mathrm{N}(L^*)\, K_{\mathfrak{m}}}{\mathrm{N}(L^*)\, K_{\mathfrak{m},1}}$$

induced by the inclusion of $K_{\mathfrak{m}}$ into $\mathrm{N}(L^*)\, K_{\mathfrak{m}}$.

The procedure from this point is to express this last-mentioned group purely in local terms. For this we need some additional notation. As usual $K_{\mathfrak{p}}, L_{\mathfrak{P}}$ denote the completions at the primes $\mathfrak{p}$ and $\mathfrak{P}$. Let $\mathbf{U}_{\mathfrak{p}}$ denote the units in the valuation ring of $K_{\mathfrak{p}}$ and $\mathbf{U}_{\mathfrak{P}}$ the units in the valuation ring of $L_{\mathfrak{P}}$. The maximal ideals in the valuation rings are just denoted by $\mathfrak{p}$ and $\mathfrak{P}$. For each positive integer n we set

$$\mathbf{U}_{\mathfrak{p}}^{(n)} = 1 + \mathfrak{p}^n$$

which is the set of units in $\mathbf{U}_{\mathfrak{p}}$ congruent to 1 modulo $\mathfrak{p}^n$. Finally we write $\mathrm{N}_{\mathfrak{p}}$ for the norm from $L_{\mathfrak{P}}$ to $K_{\mathfrak{p}}$.

3.4 Lemma. For $\mathfrak{m} = \mathfrak{p}^n$ there is an isomorphism

$$\frac{K_{\mathfrak{m}}}{(K_{\mathfrak{m}} \cap \mathrm{N}(L^*))\, K_{\mathfrak{m},1}} \cong \frac{\mathbf{U}_{\mathfrak{p}}}{\mathrm{N}_{\mathfrak{p}}(\mathbf{U}_{\mathfrak{P}})\, \mathbf{U}_{\mathfrak{p}}^{(n)}}.$$

Proof. The group $\mathbf{U}_{\mathfrak{p}}/\mathbf{U}_{\mathfrak{p}}^{(n)}$ is the image of the units in $R_{\mathfrak{p}} = K_{\mathfrak{m}}$ and so the map

$$v : \alpha \to \alpha \mathrm{N}_{\mathfrak{p}}(\mathbf{U}_{\mathfrak{P}})\, \mathbf{U}_{\mathfrak{p}}^{(n)}$$

maps $K_{\mathfrak{m}}$ onto the given group and the kernel contains $K_{\mathfrak{m},1}$ since this is the set mapping onto $\mathbf{U}_{\mathfrak{p}}^{(n)}$. Now suppose α is in the kernel. For some $\beta \in \mathbf{U}_{\mathfrak{P}}$ we have $\alpha \mathrm{N}_{\mathfrak{p}}(\beta)^{-1} \in \mathbf{U}_{\mathfrak{p}}^{(n)}$. Let $\mathfrak{P} = \mathfrak{P}_1, \ldots, \mathfrak{P}_g$ be all the primes of L dividing $\mathfrak{p}$. There is an element $\gamma \in L$ such that

$$\gamma \equiv \beta \bmod \mathfrak{P}_1^{ne}$$

$$\gamma \equiv 1 \bmod \mathfrak{P}_j^{ne}, \qquad j \neq 1, \quad e = e(\mathfrak{P}/\mathfrak{p})$$

Now for $\tau \in G(L/K)$ but $\tau \notin G(\mathfrak{P})$ we have $\tau^{-1}(\mathfrak{P}) = \mathfrak{P}_j \neq \mathfrak{P}_1$ and so

$$\tau(\gamma) \equiv 1 \bmod \mathfrak{P}_1^{ne}.$$

Let $\tau_i G(\mathfrak{P})$ denote the cosets of $G(\mathfrak{P})$ in G. Then

$$\mathrm{N}_{L|K}(\gamma) = \prod_i \prod_{\tau \in G(\mathfrak{P})} \tau_i \tau(\gamma) \equiv \prod_{\tau \in G(\mathfrak{P})} \tau(\gamma) \equiv \mathrm{N}_{\mathfrak{p}}(\gamma) \bmod \mathfrak{P}^{ne}.$$

Thus $\mathrm{N}_{L|K}(\gamma) \equiv \mathrm{N}_{\mathfrak{p}}(\beta) \bmod \mathfrak{P}^{ne}$ and so $\alpha \in \mathrm{N}(\gamma)\, K_{\mathfrak{m},1}$. Since $\alpha \in K_{\mathfrak{m}}$ we see $\mathrm{N}(\gamma)$ is in $(K_{\mathfrak{m}} \cap \mathrm{N}(L^*))\, K_{\mathfrak{m},1}$. We have already seen $K_{\mathfrak{m},1}$ is in the kernel. It is trivial that norms from L are mapped into $\mathrm{N}_{\mathfrak{p}}(\mathbf{U}_{\mathfrak{P}})$ so the stated isomorphism is proved.

The computation of $a(\mathfrak{p}^n)$ is now reduced to a problem about the units in the complete fields. This is carried out in two steps, each of which requires a number of preliminary calculations. Roughly one must show $\mathbf{U}_\mathfrak{p}^{(n)} \subseteq \mathrm{N}_\mathfrak{p}(\mathbf{U}_\mathfrak{P})$ if n is sufficiently large. Then the problem is to compute $[\mathbf{U}_\mathfrak{p} : \mathrm{N}_\mathfrak{p}(\mathbf{U}_\mathfrak{P})]$ which is done using the Herbrand quotient, $q(\mathbf{U}_\mathfrak{P})$.

We work only in the complete fields $K_\mathfrak{p}, L_\mathfrak{p}$ now.

Consider the series

$$\exp(x) = \sum_0^\infty x^n/n!,$$

$$\log(1+x) = \sum_1^\infty (-1)^{n+1} x^n/n.$$

This will be used for x in $L_\mathfrak{P}$ and $K_\mathfrak{p}$. It is necessary to determine the region of convergence. Let $v_\mathfrak{P}$ denote the exponential valuation. An infinite series $\sum a_n x^n$ converges in $L_\mathfrak{P}$ if and only if $v_\mathfrak{P}(a_n x^n) \to \infty$ as $n \to \infty$.

Let q be the prime integer in $\mathfrak{P}$ and set $e_0 = v_\mathfrak{P}(q)$. Suppose $n! = q^r t$ with $(q, t) = 1$. Then

$$r = \left[\frac{n}{q}\right] + \left[\frac{n}{q^2}\right] + \cdots < n \sum_1^\infty q^{-i} = \frac{n}{q-1}.$$

Thus $v_\mathfrak{P}(x^n/n!) = n v_\mathfrak{P}(x) - v_\mathfrak{P}(n!) > n(v_\mathfrak{P}(x) - e_0/(q-1))$. It follows that $\exp(x)$ converges if $v_\mathfrak{P}(x) > e_0/(q-1)$.

Similarly $v_\mathfrak{P}(x^n/n) = n(v_\mathfrak{P}(x) - v_\mathfrak{P}(n)/n)$ so $\log(1+x)$ converges whenever $v_\mathfrak{P}(x) \geqslant 1$. Finally the two familiar relations

$$\log\exp(x) = x, \qquad \exp\log(1+x) = 1 + x$$

can be verified formally. We use the properties at once.

3.5 Proposition. For sufficiently large n, the function log gives an isomorphism of $\mathbf{U}_\mathfrak{p}^{(n)}$ with the additive group $\mathfrak{p}^n$.

Proof. Take n large enough so that $\exp(x)$ converges for x in $\mathfrak{p}^n$. Then $\exp(x) = 1 + x + \cdots$ is in $\mathbf{U}_\mathfrak{p}^{(n)}$ and for $1+y$ in $\mathbf{U}_\mathfrak{p}^{(n)}$, $\log(1+y) = y + \cdots$ is in $\mathfrak{p}^n$. Moreover these are group homomorphisms and are inverses of each other.

3.6 Proposition. Let d be a positive integer. For n sufficiently large, every element of $\mathbf{U}_\mathfrak{p}^{(n)}$ is the dth power of an element in $\mathbf{U}_\mathfrak{p}$. In particular with $d = (L_\mathfrak{P} : K_\mathfrak{p})$ and n sufficiently large, $\mathbf{U}_\mathfrak{p}^{(n)} \subseteq \mathrm{N}_\mathfrak{p}(\mathbf{U}_\mathfrak{P})$.

Proof. Let $v_\mathfrak{p}(d) = k$ and take n large enough so $\exp(x)$ converges for x in $\mathfrak{p}^{n-ke_0}$. Take any $1+x$ in $\mathbf{U}_\mathfrak{p}^{(n)}$ and set $y = \log(1+x)$. Then y is in $\mathfrak{p}^n$ and y/d is in the region of convergence of exp. When we set $z = \exp(y/d)$, then $z \in \mathbf{U}_\mathfrak{p}$ and $z^d = 1 + x$.

To prove the statement about norms notice that for u in $K_\mathfrak{p}$, $\mathrm{N}_\mathfrak{p}(u) = u^d$ and so all dth powers are norms. In particular $\mathbf{U}_\mathfrak{p}^{(n)}$ consists entirely of norms.

3.7 Corollary. For n sufficiently large, we have

$$a(\mathfrak{p}^n) = f_\mathfrak{p}[\mathbf{U}_\mathfrak{p} : \mathrm{N}_\mathfrak{p}(\mathbf{U}_\mathfrak{P})].$$

The index remaining here is equal to $|H^0(\mathbf{U}_\mathfrak{P})|$ which is equal to $q(\mathbf{U}_\mathfrak{P})^{-1}|H^1(\mathbf{U}_\mathfrak{P})|$. We shall evaluate these.

3.8 Lemma. $|H^1(\mathbf{U}_\mathfrak{P})| = e_\mathfrak{p} =$ ramification index of $\mathfrak{p}$.

PROOF. By Hilbert's Theorem 90, $\ker \mathrm{N}|\mathbf{U}_\mathfrak{P} = \mathbf{U}_\mathfrak{P} \cap \Delta(L_\mathfrak{P}{}^*)$. Let π be a generator of the ideal $\mathfrak{P}$ so that any x in $L_\mathfrak{P}$ has the form $\pi^a u$ with u in $\mathbf{U}_\mathfrak{P}$. Let σ be a generator of $G(\mathfrak{P}) = G(L_\mathfrak{P}/K_\mathfrak{p})$. Then $\sigma(\pi) = \pi w$, $w \in \mathbf{U}_\mathfrak{p}$ so $\Delta(x) = \pi^a u/\pi^a w'$ is in $\mathbf{U}_\mathfrak{P}$. Thus $\ker \mathrm{N}|\mathbf{U}_\mathfrak{P} = \Delta(L_\mathfrak{P}{}^*)$. This enables us to write

$$H^1(\mathbf{U}_\mathfrak{P}) \cong \Delta(L_\mathfrak{P}{}^*)/\Delta(\mathbf{U}_\mathfrak{P}) \cong L_\mathfrak{P}{}^*/K_\mathfrak{p}{}^*\mathbf{U}_\mathfrak{P}$$

with the last isomorphism induced by the map $x \to \Delta(x)$ of $L_\mathfrak{P}{}^*$ onto $\Delta(L_\mathfrak{P}{}^*)$.

Let π_0 be a generator for $\mathfrak{p}$ so that $\pi_0 = \pi^e u$ for some unit u and $e = e_\mathfrak{p}$. Then

$$L_\mathfrak{P}{}^* \cong \langle \pi \rangle \times \mathbf{U}_\mathfrak{P},$$

$$K_\mathfrak{p}{}^*\mathbf{U}_\mathfrak{P} \cong \langle \pi^e \rangle \times \mathbf{U}_\mathfrak{P}$$

and so the quotient has order $e_\mathfrak{p}$ as we wished to prove.

Finally we come to the last term remaining.

3.9 Lemma

$$q(\mathbf{U}_\mathfrak{P}) = 1.$$

PROOF. Lemma 1.3 and Corollary 1.4 will be used repeatedly.

For any positive integer n, $\mathbf{U}_\mathfrak{P}/\mathbf{U}_\mathfrak{P}{}^{(n)}$ is finite since it is the unit group of the finite ring $R_\mathfrak{P}/\mathfrak{P}^n$. Thus $q(\mathbf{U}_\mathfrak{P}) = q(\mathbf{U}_\mathfrak{P}{}^{(n)})$.

Take n large enough so that Proposition 3.5 can be used. Then log gives an isomorphism of $\mathbf{U}_\mathfrak{P}{}^{(n)}$ with the additive group $\mathfrak{P}^n$. Moreover this is a $G(\mathfrak{P})$-isomorphism so $q(\mathbf{U}_\mathfrak{P}{}^{(n)}) = q(\mathfrak{P}^n)$. This in turn equals $q(R_\mathfrak{P})$ because $R_\mathfrak{P}/\mathfrak{P}^n$ is finite.

Next apply the normal basis theorem. There is an element α in $R_\mathfrak{P}$ such that the distinct images under $G(\mathfrak{P})$ are linearly independent over $K_\mathfrak{p}$. This means

$$\mathfrak{M} = \sum_{\tau \in G(\mathfrak{P})} R_\mathfrak{p} \tau(\alpha)$$

is a free $R_\mathfrak{p}$-module with the same rank as $R_\mathfrak{P}$ over $R_\mathfrak{p}$. This forces $R_\mathfrak{P}/\mathfrak{M}$ finite and $q(R_\mathfrak{p}) = q(\mathfrak{M})$. Now $\mathfrak{M}$ is a module upon which $G(\mathfrak{P})$ merely permutes the basis elements. By Proposition 1.5 we find $q(\mathfrak{M}) = 1$ because no elements of $G(\mathfrak{P})$ act trivially (except 1).

Now combine the results of this section to finish the original problem.

3.10 Theorem. Let $\mathfrak{m}$ be a modulus for K. If the finite primes dividing $\mathfrak{m}$ have sufficiently high exponents, then

$$[K^* : \mathrm{N}(L^*) K_{\mathfrak{m},1}] = \prod_{\mathfrak{p} | \mathfrak{m}} e_{\mathfrak{p}} f_{\mathfrak{p}}.$$

In the course of obtaining this result some other facts of interest were proved for cyclic extensions.

3.11 Proposition. If $\mathfrak{p}$ is unramified then every element in $\mathbf{U}_{\mathfrak{p}}$ is a norm from $\mathbf{U}_{\mathfrak{P}}$.

PROOF. By Lemmas 3.8 and 3.9 it follows $H^0(\mathbf{U}_{\mathfrak{P}}) = 1$ so $\mathbf{U}_{\mathfrak{p}} = \mathrm{N}_{\mathfrak{p}}(\mathbf{U}_{\mathfrak{P}})$.

3.12 Proposition. The quotient $K_{\mathfrak{p}}^*/\mathrm{N}_{\mathfrak{p}}(L_{\mathfrak{P}}^*)$ has order $(L_{\mathfrak{P}} : K_{\mathfrak{p}})$.

PROOF. We have $G(\mathfrak{P})$-modules $\mathbf{U}_{\mathfrak{P}} \subseteq L_{\mathfrak{P}}^*$ with quotient $\langle \bar{\pi} \rangle$, the infinite cyclic group generated by the image of the prime element π. Then $\langle \bar{\pi} \rangle$ is also a $G(\mathfrak{P})$-module with trivial action.

It follows that $q(\langle \bar{\pi} \rangle) = |G(\mathfrak{P})|^{-1}$ and so $q(L_{\mathfrak{P}}^*) = |G(\mathfrak{P})|^{-1}$ follows also because $q(\mathbf{U}_{\mathfrak{P}}) = 1$. Use the definition of q and Hilbert's Theorem 90 to conclude $H^0(L_{\mathfrak{P}}^*)$ has order $|G(\mathfrak{P})|$.

In the first four exercises below, we sketch a proof of Proposition 3.11 which avoids the logarithmic function. Let $L_{\mathfrak{P}}$ be an unramified finite-dimensional extension of the completion $K_{\mathfrak{p}}$ of an algebraic number field; $R_{\mathfrak{P}}$ and $R_{\mathfrak{p}}$ are the respective valuation rings and bars denote passage to the residue class fields.

EXERCISE 1. Let $\overline{\mathrm{N}}$ denote the norm from $\bar{R}_{\mathfrak{P}}$ to $\bar{R}_{\mathfrak{p}}$ and N the norm from $L_{\mathfrak{P}}$ to $K_{\mathfrak{p}}$. For x in $R_{\mathfrak{P}}$ we have $\overline{\mathrm{N}(x)} = \overline{\mathrm{N}}(\bar{x})$.

EXERCISE 2. An element u of $R_{\mathfrak{p}}$ is a norm of an element $w \in R_{\mathfrak{P}}$ if and only if the characteristic polynomial of w over $K_{\mathfrak{p}}$ has constant term $(-1)^n u$, $n = (L_{\mathfrak{P}} : K_{\mathfrak{p}})$.

EXERCISE 3. Suppose $(L_{\mathfrak{P}} : K_{\mathfrak{p}}) = n$ is prime and u is a unit in $R_{\mathfrak{p}}$. Let $\bar{w}$ be an element of $\bar{R}_{\mathfrak{P}}$ with $\overline{\mathrm{N}}(\bar{w}) = \bar{u}$. The minimum polynomial of $\bar{w}$ over $\bar{R}_{\mathfrak{p}}$ is the reduction mod $\mathfrak{p}$ of a monic polynomial $f(X) \in R_{\mathfrak{p}}[X]$ for which $f(0) = (-1)^n u$. Conclude u is a norm from $\mathbf{U}_{\mathfrak{P}}$.

EXERCISE 4. Use induction and the above problems to prove Proposition 3.11.

EXERCISE 5. The multiplicative group $K_{\mathfrak{p}}^*/\mathrm{N}(L_{\mathfrak{P}}^*)$ is cyclic of order $(L_{\mathfrak{P}} : K_{\mathfrak{p}})$.

4. THE FUNDAMENTAL EQUALITY FOR CYCLIC EXTENSIONS

We assume L/K is normal with cyclic Galois group $G = \langle\sigma\rangle$. For a modulus $\mathfrak{m}$ for K we let

$$\mathbf{C}^{\mathfrak{m}} = \mathbf{I}_K^{\mathfrak{m}}/\mathrm{N}_{L/K}(\mathbf{I}_L^{\mathfrak{m}})\, i(K_{\mathfrak{m},1}).$$

This is a finite group whose order is denoted by $h_{\mathfrak{m}}(L/K)$. Our object is to prove this order equals $(L:K)$ for suitable choice of $\mathfrak{m}$. The calculations of the previous sections play a crucial part.

Start with the exact sequence

$$(1) \qquad 1 \longrightarrow L^S \longrightarrow L^* \xrightarrow{\ f\ } I_L^{\mathfrak{m}} \longrightarrow V \longrightarrow 1,$$

where f is defined in Section 2 and L^S in Eq. (2) of Section 2. From f we obtain the induced map f_0 sending $H^0(L^*)$ into $H^0(\mathbf{I}_L^{\mathfrak{m}})$. These groups are described in Proposition 2.1. Using these one constructs the following diagram:

$$\begin{array}{ccccccccc}
1 \longrightarrow & \ker f_0 & \longrightarrow & K^*/\mathrm{N}(L^*) & \xrightarrow{\ f_0\ } & \mathbf{I}_K^{\mathfrak{m}}/\mathrm{N}(\mathbf{I}_L^{\mathfrak{m}}) & \longrightarrow & \operatorname{cok} f_0 & \longrightarrow 1 \\
& \downarrow & & \downarrow{\scriptstyle \partial} & & \downarrow & & \downarrow & \\
1 \longrightarrow & \ker g & \longrightarrow & K^*/\mathrm{N}(L^*)\,K_{\mathfrak{m},1} & \xrightarrow{\ g\ } & \mathbf{C}^{\mathfrak{m}} & \longrightarrow & \operatorname{cok} g & \longrightarrow 1
\end{array}$$

Here ∂ is the natural projection and all other vertical maps are induced by ∂. The map g is the unique map that makes the square (and the whole diagram) commute. All the groups in the lower row are finite so using the exactness, we obtain a formula for the order of $\mathbf{C}^{\mathfrak{m}}$.

$$(2) \qquad h_{\mathfrak{m}}(L/K) = [K^* : \mathrm{N}(L^*)\,K_{\mathfrak{m},1}]\,|\operatorname{cok} g|/|\ker g|.$$

4.1 Lemma. The groups $\ker f_0$, $\operatorname{cok} f_0$ are finite and

$$\frac{|\operatorname{cok} g|}{|\ker g|} = \frac{|\operatorname{cok} f_0|}{|\ker f_0|}\, n(\mathfrak{m}),$$

where $n(\mathfrak{m}) = [K_{\mathfrak{m},1} \cap i^{-1}(\mathrm{N}(\mathbf{I}_L^{\mathfrak{m}})) :\ K_{\mathfrak{m},1} \cap \mathrm{N}(L^*)]$.

PROOF. By simply applying a standard isomorphism theorem one finds $\operatorname{cok} f_0 \cong \operatorname{cok} g$.

From the exact sequence Eq. (1) and the exact hexagon theorem we find $\ker f_0$ is a subgroup of $H^0(L^S)$. This group is finite since $q(L^S)$ is defined.

Again using exactness we see that

$$|\ker g| = [\ker f_0 :\ \ker f_0 \cap \ker \partial].$$

The elements in the intersection on the right are the cosets $\alpha\mathrm{N}(L^*)$ for which $\alpha \in K_{\mathfrak{m},1}$ and $f(\alpha) \in \mathrm{N}(\mathbf{I}_L^{\mathfrak{m}})$. The condition $\alpha \in K_{\mathfrak{m},1}$ implies $i(\alpha)$ is relatively prime to $\mathfrak{m}$ so $f(\alpha) = i(\alpha)$. The number of such cosets $\alpha\mathrm{N}(L^*)$ equals the number of cosets of $\mathrm{N}(L^*)$ with representatives in $K_{\mathfrak{m},1} \cap i^{-1}(\mathrm{N}(\mathbf{I}_L{}^{\mathfrak{m}}))$. This is the number $n(\mathfrak{m})$ given above.

4.2 Lemma

$$q(L^S) = \frac{|\mathrm{cok} f_0|}{|\ker f_0|}.$$

PROOF. From the exact sequence (Eq. (1)) we extract two short exact sequences,

$$1 \longrightarrow L^S \xrightarrow{\gamma} L^* \xrightarrow{\alpha} f(L^*) \longrightarrow 1,$$

$$1 \longrightarrow f(L^*) \xrightarrow{\beta} \mathbf{I}_L{}^{\mathfrak{m}} \xrightarrow{\lambda} V \longrightarrow 1.$$

We shall write the exact hexagon for each of these sequences. They are simplified by Conditions (a) of Proposition 2.1:

$$\begin{array}{ccccccccccc}
1 \longrightarrow H^1(fL^*) & \xrightarrow{\delta_1} & H^0(L^S) & \xrightarrow{\gamma_0} & H^0(L^*) & \xrightarrow{\alpha_0} & H^0(fL^*) & \xrightarrow{\delta_2} & H^1(L^S) \longrightarrow 1 \\
 & & & & \downarrow f_0 & & & & \\
1 \longrightarrow H^1(V) & \xrightarrow{\delta_3} & H^0(fL^*) & \xrightarrow{\beta_0} & H^0(\mathbf{I}_L{}^{\mathfrak{m}}) & \xrightarrow{\lambda_0} & H^0(V) & \xrightarrow{\delta_4} & H^1(fL^*) \longrightarrow 1
\end{array}$$

We are able to pass from the upper sequence to the lower because they have a common group. Notice that $f_0 = \beta_0\alpha_0$.

We shall make a rather long calculation of orders of groups in these sequences. At each step we use nothing but the exactness of the rows and the following result whose proof is left to the reader.

Lemma. Let β be a homomorphism defined on an abelian group A. Let B be a subgroup of finite index in A. Then

$$[A : B] = [\beta(A) : \beta(B)][\ker\beta : B \cap \ker\beta].$$

Now we begin with

$$\begin{aligned}
|\mathrm{cok} f_0| &= [H^0(\mathbf{I}_L{}^{\mathfrak{m}}) : \mathrm{Im}\,\beta_0\alpha_0] \\
&= |\mathrm{cok}\,\beta_0|\,[\mathrm{Im}\,\beta_0 : \mathrm{Im}\,\beta_0\alpha_0] \\
&= |\mathrm{Im}\,\lambda_0|\,[\mathrm{Im}\,\beta_0 : \mathrm{Im}\,\beta_0\alpha_0] \\
&= \frac{|H^0(V)|}{|H^1(fL^*)|}\frac{|H^1(L^S)|}{|H^1(V)|}\,|\mathrm{Im}\,\alpha_0 \cap \ker\beta_0|,
\end{aligned}$$

where the last step is made with the help of the preceding lemma using $A = H^0(fL^*)$, $B = \mathrm{Im}\,\alpha$, and $\beta = \beta_0$.

Now do a similar calculation:

$$\begin{aligned}|\ker f_0| &= |\ker \beta_0 \alpha_0| = |\ker \alpha_0|\,|\mathrm{Im}\,\alpha_0 \cap \ker \beta_0| \\ &= |\mathrm{Im}\,\gamma_0|\,|\mathrm{Im}\,\alpha_0 \cap \ker \beta_0| \\ &= \frac{|H^0(L^S)|}{|H^1(fL^*)|}|\mathrm{Im}\,\alpha_0 \cap \ker \beta_0|.\end{aligned}$$

Combine these two equations and get

$$\frac{|\mathrm{cok} f_0|}{|\ker f_0|} = q(L^S)/q(V).$$

The group V is finite since it is isomorphic to the class group of L (Section 1 of Chapter IV). Thus $q(V) = 1$ and the lemma is proved.

This finally brings us to the main result.

4.3 Theorem (Fundamental Equality). Let L/K have a cyclic Galois group and let $\mathfrak{m}$ be a modulus for K divisible by sufficiently high powers of every prime of K which ramifies in L. Then $h_{\mathfrak{m}}(L/K) = (L:K)$.

PROOF. The results of this section yield the equation

$$h_{\mathfrak{m}}(L/K) = q(L^S)[K^*:\ \mathrm{N}(L^*)K_{\mathfrak{m},1}]n(\mathfrak{m}).$$

When $\mathfrak{m}$ is sufficiently large Theorem 2.5 and Theorem 3.10 combined give $h_{\mathfrak{m}}(L/K) = (L:K)n(\mathfrak{m})$.

The first fundamental inequality says $h_{\mathfrak{m}}(L/K) \leqslant (L:K)$ so the theorem follows. In addition we obtain $n(\mathfrak{m}) = 1$. We state this separately.

4.4 Corollary. With the same assumptions as in the theorem it holds that

$$K_{\mathfrak{m},1} \cap i^{-1}(\mathrm{N}(\mathbf{I}_L^{\mathfrak{m}})) = K_{\mathfrak{m},1} \cap \mathrm{N}(L^*).$$

This corollary says that any element in $K_{\mathfrak{m},1}$ which generates an ideal that is a norm must itself be a norm. We shall use the corollary to prove an elegant theorem originally proved by Hasse.

An element α in K is a *local norm at* $\mathfrak{p}$ if for some prime $\mathfrak{P}$ of L dividing $\mathfrak{p}$, α is a norm from $L_{\mathfrak{P}}$.

4.5 Theorem (Hasse Norm Theorem). Let L/K be a cyclic extension. An element in K is a norm from L if and only if it is a local norm at every prime of K.

PROOF. It is elementary to see that a norm from L to K is a local norm at all primes of K.

Suppose the element $\alpha \in K^*$ is a local norm at all primes of K. We show first the ideal (α) is the norm of an ideal from L. Let $\mathfrak{p}^a$ be the exact power of the prime $\mathfrak{p}$ dividing (α) and let $\mathfrak{P}$ be a prime of L dividing $\mathfrak{p}$. If $\mathrm{N}_{L/K}(\mathfrak{P}) = \mathfrak{p}^f$

then $\mathfrak{p}^a$ is a norm if and only if f divides a. To see this is the case let $\beta \in L_{\mathfrak{P}}$ be an element such that $N_{\mathfrak{p}}(\beta) = \alpha$. Then $(\beta) = \mathfrak{P}^t$ (in $L_{\mathfrak{P}}$) and

$$N_{\mathfrak{p}}(\mathfrak{P}^t) = \mathfrak{p}^{ft} = (\alpha)_{\mathfrak{p}} = \mathfrak{p}^a.$$

We have written $(\alpha)_{\mathfrak{p}}$ for the ideal in $K_{\mathfrak{p}}$ generated by α. This last equation shows f divides a.

Now let $\mathfrak{m}$ be a modulus such that $n(\mathfrak{m}) = 1$. Suppose

$$\mathfrak{m} = \prod \mathfrak{p}_i^{b_i}.$$

Let $\mathfrak{P}_i$ be one prime of L dividing $\mathfrak{p}_i$ and let e_i denote the ramification index.

There is an element $\beta_i \in L_{\mathfrak{P}_i}$ with $\alpha = N_{\mathfrak{p}_i}(\beta_i)$. Use the approximation theorem in L to obtain an element $\gamma \in L$ which satisfies

$$\gamma \equiv \beta_i \bmod \mathfrak{P}_i^{b_i e_i}$$

$$\gamma \equiv 1 \bmod \mathfrak{P}^{b_i e_i} \qquad \text{if} \quad \mathfrak{P} | \mathfrak{p}_i, \quad \mathfrak{P} \neq \mathfrak{P}_i.$$

For $\sigma \in G(L/K)$ but $\sigma \notin G(\mathfrak{P}_i)$ one has

$$\gamma \equiv 1 \bmod \sigma^{-1}(\mathfrak{P}_i)^{b_i e_i} \qquad \text{and} \qquad \sigma(\gamma) \equiv 1 \bmod \mathfrak{P}_i^{b_i e_i}.$$

This implies

$$N_{L/K}(\gamma) = \prod_{\tau_j} \prod_{\sigma \in G(\mathfrak{P}_i)} \tau_j \sigma(\gamma) \equiv \prod_{\sigma \in G(\mathfrak{P}_i)} \sigma(\gamma) \equiv N_{\mathfrak{p}}(\gamma) \equiv N_{\mathfrak{p}}(\beta_i),$$

all congruences modulo $\mathfrak{P}_i^{b_i e_i}$. After combining these for all i one sees that

$$\alpha \equiv N_{L/K}(\gamma) \bmod \mathfrak{m}$$

and so

$$\alpha N_{L/K}(\gamma)^{-1} \in K_{\mathfrak{m},1}.$$

This implies $\alpha N_{L/K}(\gamma)^{-1}$ is in $K_{\mathfrak{m},1} \cap i^{-1}(N_L(\mathbf{I}_L^{\mathfrak{m}}))$ and so the element is a norm. It follows that α is also a norm.

It will be seen later that $h_{\mathfrak{m}}(L/K) = (L:K)$ for L an abelian extension of K not necessarily cyclic. It is tempting to guess the Hasse norm will also hold in this case. It does not however. Counterexamples are discussed in Cassels and Frohlich [4, p. 360]. For example one takes $K = Q$, $L = Q(\sqrt{13}, \sqrt{17})$ so the Galois group is noncyclic of order four. One can check easily that any prime of Q must split completely in one of the three quadratic subfields and so $(L_{\mathfrak{P}} : K_{\mathfrak{p}}) = 1$ or 2. Thus every square of a rational number is a local norm at all primes. However (the hard part) 5^2 is not a global norm from L.

5. THE RECIPROCITY THEOREM

For a cyclic extension L/K we have seen that the group $N(\mathbf{I}_L^{\mathfrak{m}})\, i(K_{\mathfrak{m},1})$ has index $(L:K)$ in $\mathbf{I}_K^{\mathfrak{m}}$ for suitable $\mathfrak{m}$. Also the Artin map $\varphi_{L/K}$ maps $\mathbf{I}_K^{\mathfrak{m}}$ onto the Galois group $G(L/K)$ so $\ker \varphi_{L/K}$ has index $(L:K)$ in $\mathbf{I}_K^{\mathfrak{m}}$. Our object here is to show these two subgroups are actually equal. In fact this will be the case for L/K any abelian extension.

We say the *reciprocity law* holds for the triple $(L, K, \mathfrak{m})$ if $G(L/K)$ is abelian and $i(K_{\mathfrak{m},1}) \subseteq \ker \varphi_{L/K}$.

5.1 Lemma. If $\mathfrak{m}$ contains all the primes which ramify in L and the reciprocity law holds for $(L, K, \mathfrak{m})$ then $\ker \varphi_{L/K} = \mathrm{N}(\mathbf{I}_L^{\mathfrak{m}})\, i(K_{\mathfrak{m},1})$.

PROOF. By Corollary 3.2 of Chapter III we know

$$\mathrm{N}(\mathbf{I}_L^{\mathfrak{m}}) \subseteq \ker \varphi_{L/K}$$

so the reciprocity law implies

$$\mathrm{N}(\mathbf{I}_L^{\mathfrak{m}})\, i(K_{\mathfrak{m},1}) \subseteq \ker \varphi_{L/K} \subseteq \mathbf{I}_K^{\mathfrak{m}}.$$

By the first inequality the first group has index at most $(L:K)$ in $\mathbf{I}_K^{\mathfrak{m}}$ whereas $\ker \varphi_{L/K}$ has index exactly $(L:K)$ in $\mathbf{I}_K^{\mathfrak{m}}$. This forces the first inclusion to be an equality.

There are several important situations in which the reciprocity law is already known to hold.

(5.1.1) If β is a primitive mth root of unity, and $\mathfrak{m}$ is the modulus $(m)p_\infty$, then the reciprocity law holds for $(Q(\beta), Q, \mathfrak{m})$.
This was proved in Chapter III, Proposition 3.3.

(5.1.2) If the reciprocity law holds for $(L, K, \mathfrak{m})$ and E is any finite-dimensional extension of K, then the reciprocity law holds for $(LE, E, \mathfrak{m})$.

PROOF. We know $\varphi_{EL/E} = \varphi_{L/K} \mathrm{N}_{E/K}$ by Chapter III, Proposition 3.1. Then for $\alpha \in E_{\mathfrak{m},1}$ it follows $\mathrm{N}_{E/K}(\alpha) \in K_{\mathfrak{m},1}$ so

$$\varphi_{EL/E}(\alpha) \in \varphi_{L/K}(i(K_{\mathfrak{m},1})) = 1.$$

Thus $i(E_{\mathfrak{m},1}) \subseteq \ker \varphi_{EL/E}$.

(5.1.3) If the reciprocity law holds for $(L, K, \mathfrak{m})$ then it also holds for $(L, K, \mathfrak{m}\mathfrak{n})$.
This is evident because $K_{\mathfrak{m}\mathfrak{n},1} \subseteq K_{\mathfrak{m},1}$.

(5.1.4) If β is a primitive nth root of unity and $\mathfrak{m}$ is a modulus for K divisible by $(n)p_\infty$ (the modulus on Q extended to K) then the reciprocity law holds for $(K(\beta), K, \mathfrak{m})$.
This just combines the last three assertions.

(5.1.5) With the same assumptions as in (5.1.4) and with $K(\beta) \supseteq E \supseteq K$, the reciprocity law holds for $(E, K, \mathfrak{m})$.

PROOF. By Property 2.4, Chapter III of the Artin map we see $\varphi_{E|K} = \operatorname{res} \varphi_{K(\beta)|K}$ with res denoting the restriction of automorphisms to E. If $i(K_{\mathfrak{m},1}) \subseteq \ker \varphi_{K(\beta)|K}$ then also $i(K_{\mathfrak{m},1}) \subseteq \ker \varphi_{E|K}$ and so the reciprocity law holds.

Any field extension $K(\beta)$ of K with β a root of unity is called a cyclotomic extension. These fields seem to be crucial to the proof of the reciprocity law. It is necessary to construct cyclotomic extensions having some very delicate properties. We begin with two numerical lemmas.

5.2 Lemma. Let a and r be integers $\geqslant 2$ and q a prime integer. There exists a prime p such that a has order q^r modulo p.

PROOF. We shall use the polynomial

$$(1) \qquad g(X) = \frac{X^q - 1}{X - 1} = X^{q-1} + X^{q-2} + \cdots + X + 1$$

$$= (X-1)^{q-1} + \cdots + \binom{q}{t}(X-1)^{t-1} + \cdots + q.$$

Let p be a prime divisor of $g(a^{q^{r-1}}) = g$. If p does not divide the denominator

$$a^{q^{r-1}} - 1,$$

then r must be the least integer such that

$$a^{q^r} \equiv 1 \bmod p,$$

so this choice of p works.

Now suppose $p \mid a^{q^{r-1}} - 1$. Then (with $X = a^{q^{r-1}}$) we see that $p = q$ by Eq. (1). We shall prove g is not a power of q so that some choice of p can be made covered by the first case.

Suppose first $q > 2$. Then every term

$$(X-1)^{q-1}, \qquad \binom{q}{t}(X-1)^{t-1}, \qquad t \neq 1$$

is divisible by q^2 since q divides the binomial coefficient. It follows from Eq. (1) that $q^2 \nmid g$. But also $a \geqslant 2$ implies $q \neq g$ so some suitable p can be selected.

Finally suppose $q = 2$. Then

$$g = (a^{2^{r-1}} - 1) + 2 = a^{2^{r-1}} + 1.$$

It is necessary to show this is not a power of 2. Clearly a cannot be even if g is a power of 2. But with $a = 2k + 1$ we see

$$g \equiv 2 \bmod 4$$

because $r-1 \geqslant 1$. So g is not a power of 2 and all cases have been considered.

Two elements σ, τ is an abelian group are *independent* if $\langle\sigma\rangle \cap \langle\tau\rangle = 1$.

Two integers a, b relatively prime to m are *independent* $\bmod m$ if they are independent in the multiplicative group of integers $\bmod m$.

5.3 Lemma. Let $n = q_1^{r_1} \cdots q_s^{r_s}$ be the factorization of n as a product of distinct primes q_i, and let $a > 1$ be an integer. There exist infinitely many square free integers

$$m = p_1 \cdots p_s p_1' \cdots p_s'$$

such that the order of $a \bmod m$ is divisible by n. Also there exists an integer b whose order $\bmod m$ is divisible by n and such that a and b are independent $\bmod m$. Furthermore the smallest divisor of m can be selected arbitrarily large.

PROOF. For any $r \geqslant r_i$ and $r \geqslant 2$ there is a prime p_i such that a has order $q_i^r \bmod p_i$. As r increases, also p_i increases and the order of a is still divisible by q^{r_i}.

Now find (large) distinct primes $p_1, \ldots, p_s$ such that a has order $q_i^{r_i'} \bmod p_i$ with $r_i' \geqslant r_i$. Find still larger primes p_i' such that a has order $q_i^{r_i''} \bmod p_i'$ with $r_i'' > r_i'$. Then

$$m = p_1 \cdots p_s p_1' \cdots p_s'$$

is square free and n divides the order of $a \bmod m$. Select b an integer >1 such that

$$b \equiv a \bmod p_1 \cdots p_s, \qquad b \equiv 1 \bmod p_1' \cdots p_s'.$$

Also n divides the order of $b \bmod m$. To show a and b are independent suppose u and v are positive integers for which

$$a^u b^v \equiv 1 \bmod m.$$

Then $1 \equiv a^u b^v \equiv a^u \bmod p_1' \cdots p_s'$ so that $q_i^{r_i''} | u$. This forces $a^u \equiv 1 \bmod m$ and so also $b^v \equiv 1 \bmod m$. Thus a and b are independent.

Now we consider an *abelian* extension L/K of algebraic number fields and translate these lemmas into results about cyclotomic extensions of K.

5.4 Proposition. Let $n = (L:K)$ and $s =$ positive integer. Select a prime $\mathfrak{p}$ of K which is unramified in L. There exists a positive integer m relatively prime to $\mathfrak{p}$ and s with the following properties:

(i) For a primitive mth root of unity, β, and $E = K(\beta)$, the element $\varphi_{E|K}(\mathfrak{p})$ has order divisible by n;

(ii) $L \cap E = K$;

(iii) there is an automorphism τ in $G(E/K)$ whose order is divisible by n and which is independent of $\varphi_{E|K}(\mathfrak{p})$.

Proof. We shall apply Lemma 5.3 using $a = \mathcal{N}_{K|Q}(\mathfrak{p})$. The field L has only a finite number of subfields so there exists an Mth root of unity β_M such that $Q(\beta_M)$ contains every cyclotomic subfield of L. In Lemma 5.3 arrange that m has no prime divisor less than Ms. Then $Q(\beta_M) \cap Q(\beta_m) = Q$ and $L \cap Q(\beta_m) = Q$. With $E = K(\beta_m)$ it follows that Property (ii) holds.

Let $\sigma = \varphi_{E|K}(\mathfrak{p})$. The defining property of the Frobenius automorphism insures

$$\sigma(\beta_m) = \beta_m^{\mathcal{N}(\mathfrak{p})} = \beta_m{}^a,$$

so Property (i) holds. Finally take b as in the lemma and set

$$\tau(\beta_m) = \beta_m{}^b.$$

Then Property (iii) holds and the lemma is proved.

5.5 Artin's Lemma. Let L/K be a cyclic extension, s an integer, $\mathfrak{p}$ a prime of K unramified in L. There exists an extension field F of K and an integer m such that

(i) $L \cap F = K$,
(ii) $L \cap K(\beta_m) = K$,
(iii) $L(\beta_m) = F(\beta_m)$,
(iv) $\mathfrak{p}$ splits completely in F.

Proof. Select m and $\beta = \beta_m$ as in the last proposition. Then $L(\beta) = LE$ and $L \cap E = K$ so

$$G(L(\beta)/K) = G(L/K) \times G(E/K).$$

Let σ be a generator for $G(L/K)$ and τ the element in $G(E/K)$ defined in Proposition 5.4 (iii).

Let H be the subgroup generated by

$$\sigma \times \tau \qquad \text{and} \qquad \varphi_{L|K}(\mathfrak{p}) \times \varphi_{E|K}(\mathfrak{p}),$$

and let F be the subfield of LE fixed by H:

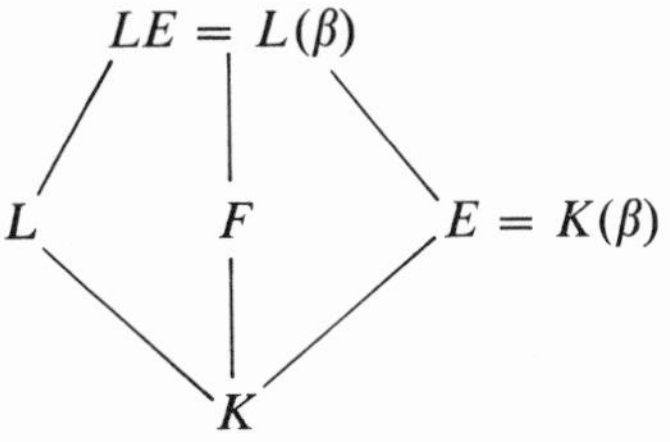

The Property 2.4, Chapter III, of the Frobenius automorphism yields

$$\varphi_{LE|K}(\mathfrak{p}) = \varphi_{L|K}(\mathfrak{p}) \times \varphi_{E|K}(\mathfrak{p}).$$

This element generates the decomposition group of $\mathfrak{p}$ in $G(LE/K)$ (that is the decomposition group of a prime of LE over $\mathfrak{p}$) and so the decomposition group is in H. Now Proposition 2.7, Chapter III and the fact that $G(LE/K)$ is abelian implies $\mathfrak{p}$ splits completely in the fixed field F of H.

The field $F(\beta) = FE$ is the fixed field of

$$H \cap (G(L/K) \times 1).$$

We shall argue this is the identity to prove Property (iii). Suppose

$$(\sigma \times \tau)^u (\varphi_{L|K}(\mathfrak{p}) \times \varphi_{E|K}(\mathfrak{p}))^v \in G(L/K) \times 1.$$

Then τ^u is in $\langle \varphi_{E|K}(\mathfrak{p}) \rangle$ and so $\tau^u = 1$ by the independence. This means n divides u and so $\sigma^u = 1$ because σ has order $(L:K) = n$. This implies also $\varphi_{E|K}(\mathfrak{p})^v = 1$ and n divides v. Once again $\varphi_{L|K}(\mathfrak{p})^v = 1$ to complete the proof of this point.

To finish the lemma we observe $L \cap F$ is the subfield of L fixed by H. Since $\sigma \times \tau$ is in H, this is the field fixed by σ, namely K.

5.6 Theorem. Let L/K be a cyclic extension with group G and let $\mathfrak{m}$ be a modulus for K divisible by all the ramified primes. Suppose also the fundamental equality $h_{\mathfrak{m}}(L/K) = (L:K)$ holds. Then the reciprocity law holds for $(L, K, \mathfrak{m})$.

PROOF. We shall prove $\ker \varphi_{L|K} | \mathbf{I}_K^{\mathfrak{m}} \subseteq i(K_{\mathfrak{m},1}) \, \mathrm{N}(\mathbf{I}_L^{\mathfrak{m}})$. Then equality must hold here because both groups have index $(L:K)$ in $\mathbf{I}_K^{\mathfrak{m}}$.

Take an ideal $\mathfrak{A}$ in $\mathbf{I}_K^{\mathfrak{m}}$ and suppose $\varphi_{L|K}(\mathfrak{A}) = 1$. Factor $\mathfrak{A}$ as

$$\mathfrak{A} = \prod_1^r \mathfrak{p}_i^{a_i}.$$

The primes $\mathfrak{p}_i$ are unramified in L because all ramified primes divide $\mathfrak{m}$. Apply Artin's lemma to each prime $\mathfrak{p}_i$ in turn to obtain for each a root of unity, β_{m_i}, such that the integers m_i are relatively prime in pairs. According to the choices made in Proposition 5.4 we can also insure $K \cap Q(\beta_{m_i}) = Q$. Then the group

$$G_i = G(K(\beta_{m_i})/K) \cong G(Q(\beta_{m_i})/Q).$$

Furthermore the group of $L(\beta_{m_1}, \cdots, \beta_{m_r})$ over K is the direct product

$$G \times G_1 \times \cdots \times G_r.$$

Let $G = \langle \sigma \rangle$ and let τ_i be the element in G_i selected as above and let H_i be the subgroup of $G \times G_i$ generated by

$$\sigma \times \tau_i \qquad \text{and} \qquad \varphi_{L|K}(\mathfrak{p}_i) \times \varphi_{K(\beta_{mi})|K}(\mathfrak{p}_i).$$

Let F_i denote the fixed field under $H_i \times \prod_{j \neq i} G_j$ and $F = F_1 F_2 \cdots F_r$.

Assertion. $L \cap F = K$ and $G(L/K) = G(LF/F)$.

The intersection of the groups $G(LF/F_i)$ fixes F and contains

$$\sigma \times \tau_1 \times \cdots \times \tau_r.$$

The field $L \cap F$ is fixed also by this element and also by $1 \times \tau_1 \times \cdots \times \tau_r$. Thus $L \cap F$ is fixed by σ and $L \cap F = K$ which proves the assertion.

Now let $\varphi_{L|K}(\mathfrak{p}_i^{a_i}) = \sigma^{d_i}$ for some integer $d_i \geqslant 0$. Then $\varphi_{L|K}(\mathfrak{A}) = \sigma^d = 1$ with

$$d = d_1 + \cdots + d_r.$$

Necessarily $n|d$, $n = (L : K) = |G|$.

The Artin map $\varphi_{LF|F}$ maps $\mathbf{I}_F{}^{\mathfrak{m}'}$ onto $G(LF/F)$ for a sufficiently large modulus $\mathfrak{m}'$. So there is an ideal $\mathfrak{B}_0$ relatively prime to $\mathfrak{m}$ and to all the integers $m_1, \ldots, m_r$ such that

$$\varphi_{LF|F}(\mathfrak{B}_0) = \sigma.$$

Let $N_{F|K}(\mathfrak{B}_0) = \mathfrak{B} \in \mathbf{I}_K{}^{\mathfrak{m}}$. By the proof of Statement 5.1.2 we see $\varphi_{L|K}(\mathfrak{B}) = \sigma$. Each prime $\mathfrak{p}_i$ splits completely in F_i so is a norm. There is then an ideal $\mathfrak{C}_i$ prime to $\mathfrak{m}$ and all m_i such that

$$N_{F_i|K}(\mathfrak{C}_i) = \mathfrak{p}_i^{a_i} \mathfrak{B}^{-d_i}.$$

By the choice of d_i one obtains

$$\varphi_{LF_i|F_i}(\mathfrak{C}_i) = \varphi_{L|K}(N_{F_i|K}(\mathfrak{C}_i)) = 1.$$

The extension LF_i of F_i satisfies

$$F_i \subseteq LF_i \subseteq F_i(\beta_{m_i})$$

by Property (iii) of Artin's lemma. By Statement 5.1.5 the reciprocity law holds for $(LF_i, F_i, \mathfrak{m}')$ so long as $\mathfrak{m}'$ is divisible by $m_i \mathfrak{p}_\infty$. By choice $\mathfrak{C}_i$ is prime to m_i and so we may select $\mathfrak{m}'$ in such a way that $\mathfrak{C}_i \in \mathbf{I}_{F_i}{}^{\mathfrak{m}'}$. Thus there exists $\gamma_i \in F_i$, $\gamma_i \equiv 1 \bmod \mathfrak{m}'$ and an ideal $\mathfrak{D}_i \in \mathbf{I}_{LF_i}{}^{\mathfrak{m}'}$ such that

$$\mathfrak{C}_i = (\gamma_i) N_{LF_i|F_i}(\mathfrak{D}_i).$$

Take norms into K to obtain

$$\mathfrak{p}_i^{a_i} \mathfrak{B}^{-d_i} = (N_{F_i|K}(\gamma_i)) N_{LF_i|K}(\mathfrak{D}_i).$$

The modulus $\mathfrak{m}'$ could have been selected so that $\mathfrak{m}|\mathfrak{m}'$ because $\mathfrak{C}_i$ is prime to $\mathfrak{m}$. Thus with this choice

$$\alpha_i = N_{F_i|K}(\gamma_i) \in K_{\mathfrak{m},1}.$$

Now take products over all subscripts i to get

$$\mathfrak{A}\mathfrak{B}^{-d} = \prod \mathfrak{p}_i^{a_i} \mathfrak{B}^{-d_i} = \prod \alpha_i \prod N_{LF_i|K}(\mathfrak{D}_i).$$

Write $\mathfrak{D}_i' = \mathrm{N}_{LF_i|L}(\mathfrak{D}_i)$ and observe that $\mathfrak{D}_i'$ is prime to $\mathfrak{m}$. Finally

$$\mathfrak{A} = \mathfrak{B}^d(\alpha_1 \cdots \alpha_s)\,\mathrm{N}_{L|K}(\mathfrak{D}_1' \cdots \mathfrak{D}_s').$$

We saw above that $n|d$ so $\mathfrak{B}^d$ is a norm from L; finally $\mathfrak{A} \in i(K_{\mathfrak{m},1})\,\mathrm{N}_{L|K}(\mathbf{I}_L^{\mathfrak{m}})$ as required.

We now have good results for cyclic extensions. The step up to abelian extensions is not difficult.

5.7 Theorem (Artin Reciprocity Theorem). Let L/K be an extension with abelian Galois group G. Let $\mathfrak{m}$ be a modulus for K divisible at least by all primes which ramify in L and assume the exponents of the prime divisors of $\mathfrak{m}$ are sufficiently large. Then the Artin map $\varphi_{L|K}$ maps $\mathbf{I}_K^{\mathfrak{m}}$ onto G and the kernel is $\mathrm{N}_{L|K}(\mathbf{I}_L^{\mathfrak{m}})\,i(K_{\mathfrak{m},1})$.

PROOF. Express the group G as a direct product

$$G = C_1 \times \cdots \times C_s, \qquad C_i \text{ cyclic.}$$

Let H_j be the direct product of all C_i with $i \neq j$ so that $G = C_j \times H_j$. Let E_j denote the subfield fixed by H_j. Then E_j is a cyclic extension of K with group C_j. There is a modulus $\mathfrak{m}_j$ such that the reciprocity law holds for $(E_j, K, \mathfrak{m}_j)$. It is possible to arrange $\mathfrak{m}_i|\mathfrak{m}$ so the reciprocity law holds for $(E_j, K, \mathfrak{m})$. This means

$$i(K_{\mathfrak{m},1}) \subseteq \bigcap_j \ker \varphi_{E_j|K}.$$

For any ideal $\mathfrak{A}$ we have from Property 2.4, Chapter III of the Frobenius automorphism

$$\varphi_{L|K}(\mathfrak{A})|E_j = \varphi_{E_j|K}(\mathfrak{A})$$

and so for $\mathfrak{A} \in i(K_{\mathfrak{m},1})$ we see

$$\varphi_{L|K}(\mathfrak{A})|E_j = 1 \qquad \text{for each } j.$$

But $E_1 \cdots E_s = L$ because the group fixing all the E_j is the intersection of the H_j which is trivial. Any automorphism trivial on each E_j is the identity on all of L. Thus $i(K_{\mathfrak{m},1}) \subseteq \ker \varphi_{L|K}$ and the reciprocity law holds for $(L, K, \mathfrak{m})$. This implies the desired conclusion because of Lemma 5.1.

As an illustration of the scope of this result one obtains the following results as corollaries.

5.8 Theorem. Let L/K be an abelian extension and $\mathfrak{m}$ a modulus such that the reciprocity law holds for $(L, K, \mathfrak{m})$. Let E/K be a normal extension such that

$$\mathrm{N}_{E|K}(\mathbf{I}_E^{\mathfrak{m}}) \subseteq \mathrm{N}_{L|K}(\mathbf{I}_L^{\mathfrak{m}})\,i(K_{\mathfrak{m},1}).$$

Then $L \subseteq E$.

PROOF. Except for a finite number of primes dividing $\mathfrak{m}$, the primes of K which split completely in E are in $N_{E|K}(\mathbf{I}_K^{\mathfrak{m}})$. These must also split completely in L because they lie in $\ker \varphi_{L|K}$. Thus $L \subseteq E$ by Corollary 5.5, Chapter IV.

5.9 Theorem (Kronecker–Weber). If L is an abelian extension of Q, then $L \subseteq Q(\beta)$ for some root of unity β.

PROOF. The reciprocity law holds for $(L, Q, \mathfrak{m})$ for some modulus $\mathfrak{m}$. We may suppose $\mathfrak{m} = (m)p_\infty$ for some positive integer m. Take β a primitive mth root of unity. Then from the calculation in Proposition 3.3 of Chapter III we have $E = Q(\beta)$ and $i(Q_{\mathfrak{m},1}) = \ker \varphi_{E/Q}$ so

$$i(Q_{\mathfrak{m},1}) = i(Q_{\mathfrak{m},1}) N_{E|Q}(\mathbf{I}_E^{\mathfrak{m}}) \subseteq i(Q_{\mathfrak{m},1}) N_{L|Q}(\mathbf{I}_L^{\mathfrak{m}}) = \ker \varphi_{L|Q}.$$

The previous theorem applies to give $L \subseteq E$.

Because of its scope and the simplicity of the statement, this must be regarded as one of the really elegant theorems in mathematics. The result was first stated by Kronecker and a complete proof was given by Weber in 1886. There are proofs which are more elementary than the one given here. For example Speiser [10] gives a proof using only ramification theory.

An explicit classification of all abelian extensions of $Q(\sqrt{D})$, $D < 0$, can also be given. This requires the study of "complex multiplications" and is discussed by J. P. Serre in [4, Chapter XIII]. This description, as well as Kronecker–Weber, classifies the abelian extensions by explicitly giving generating elements for the extension fields. This is not possible for general algebraic number fields with the present state of knowledge. We shall give a classification of all the abelian extensions of a number field K in other terms; they are classified by certain subgroups of the ideal group $\mathbf{I}_K$. This is the main goal of the rest of the chapter.

EXERCISE 1. Let $K = Q(\sqrt{d}) \neq Q$. Show K has an abelian extension which is not contained in $K(\theta)$ for any root of unity, θ.

Procedure. Find u in K such that $K(\sqrt{u})$ is not normal over Q. Since $K(\theta)$ is abelian over Q, every subfield must be normal over Q; thus $\sqrt{u}$ is not in $K(\theta)$.

EXERCISE 2. An element θ in the Galois extension L of Q is said to give an *integral normal basis* if the ring of algebraic integers in L is $\sum Z\sigma(\theta)$ with σ in $G(L/Q)$.

a. Let n be a square free positive integer and θ a primitive nth root of unity. Show θ gives an integral normal basis for $Q(\theta)$.

b. Let K be any intermediate field between Q and $Q(\theta)$. Show $T_{Q(\theta)/K}(\theta)$ gives an integral normal basis for K.

c. Let K be an abelian extension of Q which is tamely ramified at every prime; that is, for each prime p of Q, p does not divide the ramification number $e(\mathfrak{P}/p)$ in K. Show that K has an integral normal basis.

6. IDEAL GROUPS, CONDUCTORS, AND CLASS FIELDS

In this section we consider the groups which eventually will be the objects by which all abelian extensions of a number field are classified.

A subgroup $\mathbf{H}$ of $\mathbf{I}_K$ is called a *congruence subgroup* if there is a modulus $\mathfrak{m}$ such that

$$i(K_{\mathfrak{m},1}) \subseteq \mathbf{H} \subseteq \mathbf{I}^{\mathfrak{m}}.$$

We will say $\mathbf{H}$ is defined mod $\mathfrak{m}$ in this context. Suppose $\mathfrak{n}$ is a modulus and $\mathfrak{n}|\mathfrak{m}$. Then $\mathbf{I}^{\mathfrak{m}}$ is a subgroup of $\mathbf{I}^{\mathfrak{n}}$. There may (or may not) be a congruence subgroup $\mathbf{H}^{\mathfrak{n}}$ defined mod $\mathfrak{n}$ such that $\mathbf{H} = \mathbf{I}^{\mathfrak{m}} \cap \mathbf{H}^{\mathfrak{n}}$. When this does hold we say $\mathbf{H}$ is the *restriction* of $\mathbf{H}^{\mathfrak{n}}$ to $\mathbf{I}^{\mathfrak{m}}$. The first lemma shows $\mathbf{H}^{\mathfrak{n}}$ is uniquely determined by $\mathbf{H}$ and $\mathfrak{n}$.

6.1 Lemma. Let $\mathfrak{n}|\mathfrak{m}$ and $\mathbf{H}^{\mathfrak{m}}$, $\mathbf{H}^{\mathfrak{n}}$ be congruence subgroups defined mod $\mathfrak{m}$ and $\mathfrak{n}$. Suppose $\mathbf{H}^{\mathfrak{m}} = \mathbf{I}^{\mathfrak{m}} \cap \mathbf{H}^{\mathfrak{n}}$. Then

(a) $\mathbf{I}^{\mathfrak{m}}/\mathbf{H}^{\mathfrak{m}} \cong \mathbf{I}^{\mathfrak{n}}/\mathbf{H}^{\mathfrak{n}}$, (b) $\mathbf{H}^{\mathfrak{n}} = \mathbf{H}^{\mathfrak{m}} i(K_{\mathfrak{n},1})$.

PROOF. We show first that $\mathbf{I}^{\mathfrak{n}} = \mathbf{I}^{\mathfrak{m}} i(K_{\mathfrak{n},1})$. Take an ideal $\mathfrak{A}_0$ in $\mathbf{I}^{\mathfrak{n}}$ and write $\mathfrak{A}_0 = \mathfrak{A}\mathfrak{A}_1$ with $\mathfrak{A}_1$ in $\mathbf{I}^{\mathfrak{m}}$ and

$$\mathfrak{A} = \prod \mathfrak{p}_i^{a_i}$$

with $\mathfrak{p}_i$ a divisor of $\mathfrak{m}$ but not $\mathfrak{n}$. By the CRT we can find elements π_i which satisfy

$$\pi_i \in \mathfrak{p}_i - \mathfrak{p}_i^2,$$

$$\pi_i \equiv 1 \bmod \mathfrak{n}.$$

Then the element $\alpha = \pi_1^{a_1} \cdots$ belongs to $K_{\mathfrak{n},1}$ and $\mathfrak{A}\alpha^{-1}$ is relatively prime to $\mathfrak{m}$. It follows that $\mathfrak{A}_0 \in \mathbf{I}^{\mathfrak{m}} i(K_{\mathfrak{n},1})$ as claimed.

Now $\mathbf{H}^{\mathfrak{n}}$ is a congruence subgroup mod $\mathfrak{n}$ so $\mathbf{I}^{\mathfrak{n}} = \mathbf{I}^{\mathfrak{m}}\mathbf{H}^{\mathfrak{n}}$. Now use the assumption about $\mathbf{H}^{\mathfrak{m}}$ to obtain

$$\frac{\mathbf{I}^{\mathfrak{m}}}{\mathbf{H}^{\mathfrak{m}}} \cong \frac{\mathbf{I}^{\mathfrak{m}}}{\mathbf{H}^{\mathfrak{n}} \cap \mathbf{I}^{\mathfrak{m}}} \cong \frac{\mathbf{I}^{\mathfrak{m}}\mathbf{H}^{\mathfrak{n}}}{\mathbf{H}^{\mathfrak{n}}} \cong \frac{\mathbf{I}^{\mathfrak{n}}}{\mathbf{H}^{\mathfrak{n}}}.$$

This proves (a).

It is clear from definitions that $\mathbf{H}^{\mathfrak{n}} \supseteq \mathbf{H}^{\mathfrak{m}}$ and $\mathbf{H}^{\mathfrak{n}} \supseteq i(K_{\mathfrak{n},1})$ so $\mathbf{H}^{\mathfrak{n}} \supseteq$

$\mathbf{H}^{\mathfrak{m}}i(K_{\mathfrak{n},1})$. This implies $\mathbf{H}^{\mathfrak{m}}i(K_{\mathfrak{n},1}) \cap \mathbf{I}^{\mathfrak{m}} = \mathbf{H}^{\mathfrak{m}}$.

$$\frac{\mathbf{I}^{\mathfrak{m}}}{\mathbf{H}^{\mathfrak{m}}} \cong \frac{\mathbf{I}^{\mathfrak{n}}}{\mathbf{H}^{\mathfrak{m}}i(K_{\mathfrak{n},1})}$$

and so $\mathbf{H}^{\mathfrak{n}} = \mathbf{H}^{\mathfrak{m}}i(K_{\mathfrak{n},1})$ because both have the same index in $\mathbf{I}^{\mathfrak{n}}$ and one is contained in the other. This proves (b).

We now define a relation between congruence subgroups. Say $\mathbf{H}_1 \sim \mathbf{H}_2$ if there is a modulus $\mathfrak{m}$ such that

$$\mathbf{H}_1 \cap \mathbf{I}^{\mathfrak{m}} = \mathbf{H}_2 \cap \mathbf{I}^{\mathfrak{m}}.$$

Observe that whenever this condition holds for a modulus $\mathfrak{m}$, it also holds for any modulus divisible by $\mathfrak{m}$.

It is easy to see this relation is an equivalence relation. The next lemma shows that equivalent congruence subgroups are obtained as the restriction of a single congruence subgroup.

6.2 Lemma. Let $\mathbf{H}_1, \mathbf{H}_2$ be congruence subgroups defined mod $\mathfrak{m}_1, \mathfrak{m}_2$, respectively, which have a common restriction $\mathbf{H}_3 = \mathbf{H}_i \cap \mathbf{I}^{\mathfrak{m}_3}$, $i = 1, 2$. Let $\mathfrak{m}$ be the greatest common divisor of $\mathfrak{m}_1, \mathfrak{m}_2$. Then there is a congruence subgroup $\mathbf{H}$ defined mod $\mathfrak{m}$ such that $\mathbf{H} \cap \mathbf{I}^{\mathfrak{m}_i} = \mathbf{H}_i$, $i = 1, 2$.

PROOF. Since $\mathbf{H}_1$ and $\mathbf{H}_2$ have equal restrictions in $\mathbf{I}^{\mathfrak{m}_3}$ the same holds for any larger modulus. There is no harm then in assuming $\mathfrak{m}_3$ is divisible by $\mathfrak{m}_1$ and $\mathfrak{m}_2$. If the lemma is correct, then $\mathbf{H}$ is uniquely determined so we set

$$\text{(1)} \qquad \mathbf{H} = \mathbf{H}_3\, i(K_{\mathfrak{m},1})$$

and try to verify it has the required properties. If we show

$$\text{(2)} \qquad \mathbf{H} \cap \mathbf{I}^{\mathfrak{m}_3} = \mathbf{H}_3$$

then $\mathbf{H}_1$ and $\mathbf{H} \cap \mathbf{I}^{\mathfrak{m}_1}$ both restrict to $\mathbf{H}_3$ so by uniqueness $\mathbf{H}_1 = \mathbf{H} \cap \mathbf{I}^{\mathfrak{m}_1}$. Same reasoning applies to $\mathbf{H}_2$. The lemma will be proved then if equation (2) is verified.

Suppose $\mathfrak{A}(\alpha) \in \mathbf{H} \cap \mathbf{I}^{\mathfrak{m}_3}$. By (1) we may suppose $\mathfrak{A} \in \mathbf{H}_3$, $\alpha \in K_{\mathfrak{m},1}$. Then $\mathfrak{A}$ and $\mathfrak{A}(\alpha)$ are prime to $\mathfrak{m}_3$ so (α) is also. Now find an element $\beta \in K$ which satisfies

$$\beta \in K_{\mathfrak{m}_1,1}, \qquad \alpha\beta^{-1} \in K_{\mathfrak{m}_2,1}, \qquad (\beta) \in \mathbf{I}^{\mathfrak{m}_3}.$$

To show such an element exists we shall list a set of congruences, one for each prime divisor of $\mathfrak{m}_1 \mathfrak{m}_2 \mathfrak{m}_3$, to be satisfied by β.

Let $\mathfrak{p}^{a_1} | \mathfrak{m}_1$ and $\mathfrak{p}^{a_2} | \mathfrak{m}_2$ and assume these are the exact powers of $\mathfrak{p}$ dividing the moduli.

If $a_1 \geqslant a_2$ the congruence at $\mathfrak{p}$ for β is

$$\beta \equiv 1 \bmod \mathfrak{p}^{a_1}.$$

In this case $\mathfrak{p}^{a_2}$ is the power of $\mathfrak{p}$ dividing $\mathfrak{m}$ so $\beta \equiv \alpha \equiv 1 \bmod \mathfrak{p}^{a_2}$.

If $a_2 \geqslant a_1$ then the congruence at $\mathfrak{p}$ for β is

$$\beta \equiv \alpha \bmod \mathfrak{p}^{a_2}.$$

This time $\mathfrak{p}^{a_1}$ divides $\mathfrak{m}$ and $\beta \equiv \alpha \equiv 1 \bmod \mathfrak{p}^{a_1}$.

For a prime $\mathfrak{p}$ dividing $\mathfrak{m}_3$ but not $\mathfrak{m}_1 \mathfrak{m}_2$ we require

$$\beta \equiv 1 \bmod \mathfrak{p}.$$

This β satisfies our requirements.

Thus

$$\mathfrak{A}(\beta) \in \mathbf{H}_3\, i(K_{\mathfrak{m}_1, 1}) = \mathbf{H}_1.$$

But $\mathfrak{A}(\beta)$ is prime to $\mathfrak{m}_3$ so

$$\mathfrak{A}(\beta) \in \mathbf{H}_1 \cap \mathbf{I}^{\mathfrak{m}_3} = \mathbf{H}_3.$$

Then $\mathfrak{A}(\beta)(\beta^{-1}\alpha) = \mathfrak{A}(\alpha) \in \mathbf{H}_3\, i(K_{\mathfrak{m}_2, 1}) = \mathbf{H}_2$ as well as $\mathfrak{A}(\alpha) \in \mathbf{I}^{\mathfrak{m}_3}$. Thus

$$\mathfrak{A}(\alpha) \in \mathbf{H}_2 \cap \mathbf{I}^{\mathfrak{m}_3} = \mathbf{H}_3$$

as we wished to prove.

An equivalence class of congruence subgroups is called an *ideal group*. If $\mathbf{H}$ denotes an ideal group and $\mathfrak{m}$ a modulus for which some congruence subgroup mod $\mathfrak{m}$ belongs to $\mathbf{H}$, we shall denote that (unique) subgroup by $\mathbf{H}^{\mathfrak{m}}$.

Lemma 6.2 shows us whenever $\mathbf{H}^{\mathfrak{m}}$ and $\mathbf{H}^{\mathfrak{n}}$ belong to the ideal group $\mathbf{H}$, then also $\mathbf{H}^{\mathfrak{m}'} \in \mathbf{H}$ for $\mathfrak{m}'$ = greatest common divisor of $\mathfrak{m}$ and $\mathfrak{n}$. This implies there is a unique modulus $\mathfrak{f}$ such that

$$\mathbf{H}^{\mathfrak{f}} \in \mathbf{H} \quad \text{and} \quad \mathbf{H}^{\mathfrak{m}} \in \mathbf{H} \quad \text{implies} \quad \mathfrak{f} \mid \mathfrak{m}.$$

Clearly $\mathfrak{f}$ is the g.c.d. of all $\mathfrak{m}$ for which $\mathbf{H}^{\mathfrak{m}} \in \mathbf{H}$. This modulus is called the *conductor* of $\mathbf{H}$.

Now suppose L/K is an abelian extension. Let $\mathfrak{m}$ be a modulus such that the reciprocity law holds for $(L, K, \mathfrak{m})$. Then the kernel of $\varphi_{L/K}$ acting on $\mathbf{I}^{\mathfrak{m}}$ is a congruence subgroup which we shall denote by $\mathbf{H}^{\mathfrak{m}}(L/K)$.

If $\mathfrak{m}'$ is another modulus such that the reciprocity law holds for $(L, K, \mathfrak{m}')$ then $\mathbf{H}^{\mathfrak{m}'}(L/K)$ and $\mathbf{H}^{\mathfrak{m}}(L/K)$ have a common restriction in $\mathbf{I}^{\mathfrak{m}\mathfrak{m}'}$. This is immediate because $\ker(\varphi_{1|K}|\mathbf{I}^{\mathfrak{m}}) \cap \mathbf{I}^{\mathfrak{m}\mathfrak{m}'} = \ker(\varphi_{L|K}|\mathbf{I}^{\mathfrak{m}\mathfrak{m}'}) = \ker(\varphi_{L|K}|\mathbf{I}^{\mathfrak{m}'}) \cap \mathbf{I}^{\mathfrak{m}\mathfrak{m}'}$.

This implies there is a unique ideal group—denoted by $\mathbf{H}(L/K)$—containing $\mathbf{H}^{\mathfrak{m}}(L/K)$. This ideal group is called the *class group* to L and L is called the *class field* to $\mathbf{H}(L/K)$. The conductor of $\mathbf{H}(L/K)$ is denoted by $\mathfrak{f}(L/K)$.

The main goal of the next few sections will be to show the correspondence between ideal groups and abelian extensions of K is in fact a one-to-one correspondence between all ideal groups and all abelian extensions of K. In other words we shall obtain a classification of all abelian extensions of K purely in terms of objects defined directly in terms of K.

With regard to the conductor we see at once that $\mathfrak{f}(L/K)|\mathfrak{m}$ whenever the reciprocity law holds for $\mathfrak{m}$ but we do not know at this point that the reciprocity law holds for $(L, K, \mathfrak{f}(L/K))$. The main point of concern is that a ramified prime might fail to divide the conductor. However this does not actually happen and in fact all ramified primes divide $\mathfrak{f}$ as we shall see later.

7. REDUCTION STEPS TOWARD THE EXISTENCE THEOREM

Consider an ideal group $\mathbf{H}$. Our object is to produce an abelian extension L/K which is class field to $\mathbf{H}$. The construction of L is indirect and is performed by a series of reductions. The first shows it is enough to produce a class field for a subgroup between $i(K_{\mathfrak{m},1})$ and $\mathbf{H}^{\mathfrak{m}}$.

7.1 Proposition. Suppose there is a chain of groups

$$i(K_{\mathfrak{m},1}) \subseteq \mathbf{H}_0 \subseteq \mathbf{H}_1 \subseteq \mathbf{I}^{\mathfrak{m}}$$

such that the ideal group containing $\mathbf{H}_0$ is class group to the abelian extension L/K. Suppose $\mathfrak{m}$ contains all primes of K which ramify in L. Then the ideal group containing $\mathbf{H}_1$ is class group to the subfield of L fixed by the subgroup $\varphi_{L|K}(\mathbf{H}_1)$ of $G(L/K)$.

Remark. The assumption about the ramified primes is necessary so that $\varphi_{L|K}$ is defined on $\mathbf{I}^{\mathfrak{m}}$. If the first choice of $\mathfrak{m}$ does not contain all the ramified primes, enlarge $\mathfrak{m}$ so that it does and then take restrictions of the groups involved. This does not change the ideal groups so the conclusion is unchanged.

Proof. Let $G_1 = \varphi_{L|K}(\mathbf{H}_1)$ and E the subfield of L fixed by G_1. Let res denote the restriction map from $G(L/K)$ to $G(E/K)$ so that $\operatorname{res} G_1 = 1$. For any $\mathfrak{A}$ in $\mathbf{I}^{\mathfrak{m}}$ we have

$$\varphi_{E|K}(\mathfrak{A}) = \operatorname{res} \varphi_{L|K}(\mathfrak{A})$$

so in particular $\varphi_{E|K}(\mathfrak{A}) = 1$ when $\mathfrak{A} \in \mathbf{H}_1$. Thus $\mathbf{H}_1 \subseteq \ker \varphi_{E|K}$. Since $\mathbf{H}_1$ is a congruence subgroup we see the reciprocity law holds for the triple $(E, K, \mathfrak{m})$ and so

$$[\mathbf{I}^{\mathfrak{m}} : \ker \varphi_{E|K}] = [G(L/K) : G_1] = [\mathbf{I}^{\mathfrak{m}} : \mathbf{H}_1].$$

It follows that $\mathbf{H}_1 = \ker \varphi_{E|K}$. The reciprocity theorem describes this and so E is class field to the ideal group of $\mathbf{H}_1$.

The next reduction shows how the ground field can be changed.

7.2 Proposition. Suppose E/K is a cyclic extension and $\mathbf{H}^{\mathfrak{m}}$ is a congruence subgroup mod $\mathfrak{m}$. Let $\mathbf{H}_E$ denote the subgroup $\{\mathfrak{A} \in \mathbf{I}_E{}^{\mathfrak{m}} | N_{E|K}(\mathfrak{A}) \in \mathbf{H}^{\mathfrak{m}}\}$. If the ideal group of $\mathbf{H}_E$ has a class field over E, then the ideal group for $\mathbf{H}^{\mathfrak{m}}$ has a class field over K.

PROOF. Let $L|E$ be an abelian extension such that

$$\mathbf{H}_E = N_{L|E}(\mathbf{I}_L{}^{\mathfrak{m}})\, i(E_{\mathfrak{m},1}).$$

We first show L is normal over K. Let σ be an isomorphism (over K) of L into some normal field over K. Then $\sigma(E) = E$ because E/K is normal. From the equations

$$\sigma(\mathbf{H}^{\mathfrak{m}}) = \mathbf{H}^{\mathfrak{m}}, \qquad \sigma(\mathbf{H}_E) = \mathbf{H}_E$$

we deduce $\sigma(L)$ and L have the same norm groups, $\mathbf{H}_E$. By Theorem 5.8 we have $L = \sigma(L)$ and so L is normal over K.

Next we prove L/K has an abelian Galois group. Let σ be an element in $G(L/K)$ whose restriction to E generates $G(E/K)$. To prove $G(L/K)$ is abelian it is sufficient to show $\tau\sigma = \sigma\tau$ for each τ in $G(L/E)$. For such an element τ take an ideal $\mathfrak{A}$ in $\mathbf{I}_E{}^{\mathfrak{m}}$ such that $\varphi_{L|E}(\mathfrak{A}) = \tau$. Then $\sigma\tau\sigma^{-1} = \sigma\varphi_{L|E}(\mathfrak{A})\sigma^{-1} = \varphi_{\sigma L|\sigma E}(\sigma\mathfrak{A}) = \varphi_{L|E}(\sigma\mathfrak{A})$. The ideal $\mathfrak{A}/\sigma(\mathfrak{A})$ has norm 1 so in particular

$$\mathrm{N}_{E|K}(\mathfrak{A}\sigma(\mathfrak{A})^{-1}) \in \mathbf{H}^{\mathfrak{m}} \qquad \text{and} \quad \mathfrak{A}/\sigma(\mathfrak{A}) \in \mathbf{H}_E.$$

But $\mathbf{H}_E = \ker \varphi_{L|E}$ so $\varphi_{L|E}(\mathfrak{A}/\sigma(\mathfrak{A})) = 1$. It follows that

$$\sigma\tau\sigma^{-1} = \varphi_{L|E}(\sigma\mathfrak{A}) = \varphi_{L|E}(\mathfrak{A}) = \tau.$$

This shows L/K is abelian. Now we have

$$\mathrm{N}_{L|E}(\mathbf{I}_L{}^{\mathfrak{m}}) \subseteq \mathbf{H}_E \qquad \text{and so} \quad \mathrm{N}_{L|K}(\mathbf{I}_L{}^{\mathfrak{m}}) \subseteq \mathbf{H}^{\mathfrak{m}}$$

by transitivity of the norm. Since $\mathbf{H}^{\mathfrak{m}}$ is a congruence subgroup we have

$$\mathrm{N}_{L|K}(\mathbf{I}_L{}^{\mathfrak{m}})\, i(K_{\mathfrak{m},1}) \subseteq \mathbf{H}^{\mathfrak{m}} \subseteq \mathbf{I}^{\mathfrak{m}}.$$

It follows from Proposition 7.1 that the ideal group for $\mathbf{H}^{\mathfrak{m}}$ has a class field which is a subfield of L.

These two reductions are used in the following way. Select an ideal group $\mathbf{H}$ and let $\mathbf{H}^{\mathfrak{m}} \in \mathbf{H}$. Suppose $\mathbf{I}^{\mathfrak{m}}/\mathbf{H}^{\mathfrak{m}}$ has exponent n. Select a primitive nth root of unity β and form a chain

$$K = K^{(1)} \subset K^{(2)} \subset \cdots \subset K^{(t)} = K(\beta).$$

Let $\mathbf{H}_1 = \mathbf{H}^{\mathfrak{m}}$ and inductively define

$$\mathbf{H}_{i+1} = \{\mathfrak{A} \in \mathbf{I}^{\mathfrak{m}}_{K^{(i+1)}} | \mathrm{N}(\mathfrak{A}) \in \mathbf{H}_i\}$$

where N means the norm from $K^{(i+1)}$ to $K^{(i)}$.

Assume $K^{(i+1)}/K^{(i)}$ is cyclic. If we show the ideal group of $\mathbf{H}_t$ has a class field over $K(\beta)$ then the same is true of $\mathbf{H}_1$ over K by repeated application of Proposition 7.2.

Furthermore the quotient $\mathbf{I}^{\mathfrak{m}}_{K(\beta)}/\mathbf{H}_t$ has exponent n. This can be seen by induction in the chain. It was given that $\mathbf{I}_K{}^{\mathfrak{m}}/\mathbf{H}_1$ has exponent n. For $\mathfrak{A} \in \mathbf{I}^{\mathfrak{m}}_{K(2)}$ we see

$$\mathrm{N}_{K(2)|K}(\mathfrak{A}^n) \in (\mathbf{I}_K{}^{\mathfrak{m}})^n \subseteq \mathbf{H}_1$$

so $\mathfrak{A}^n \in \mathbf{H}_2$. This shows $\mathbf{I}^{\mathfrak{m}}_{K(2)}/\mathbf{H}_2$ has exponent n and the procedure can be repeated.

We summarize now.

Reduction Step. Let $\mathbf{H}$ be an ideal group for K. The assertion that $\mathbf{H}$ is class group to some abelian extension of K is true provided it can be proved under the additional assumption that K contains a primitive nth root of unity where n is the exponent of $\mathbf{I}^{\mathfrak{m}}/\mathbf{H}^{\mathfrak{m}}$ for $\mathbf{H}^{\mathfrak{m}} \in \mathbf{H}$.

This result will be proved in the next sections along with some facts about abelian extensions of K when K contains the nth roots of unity.

8. KUMMER EXTENSIONS AND THE *S*-UNIT THEOREM

A finite, abelian extension L of K is called a *Kummer n-extension* if $G(L/K)$ has exponent n and K contains the nth roots of unity. The word exponent is used in the sense that $\sigma^n = 1$ for all σ in $G(L/K)$ but n need not be the least such positive integer.

The Kummer n-extensions are classified by the finite subgroups of K^*/K^n where K^n is the multiplicative group of nonzero nth powers of elements of K^*.

8.1 Theorem. There is a one-to-one correspondence between the Kummer n-extensions L of K and the subgroups W such that $K^n \subseteq W \subset K^*$ and W/K^n is finite. The correspondence associates W with the field $L = K(\sqrt[n]{W})$. In this case the Galois group $G(L/K)$ is isomorphic to W/K^n.

PROOF. Let L be a Kummer n-extension of K with group G. We show first L has the form asserted in the theorem. Let

$$M = \{\alpha \in L^* \mid \alpha^n \in K^*\}.$$

Each α in M is a root of an equation $X^n - a$ with a in K so for $\sigma \in G$ it follows that $\sigma(\alpha) = \beta\alpha$ with β an nth root of unity in K. For each α in M let ψ_α be the function from G to K^* defined by

$$\psi_\alpha(\sigma) = \sigma(\alpha)/\alpha.$$

The values of ψ_α are nth roots of unity and ψ_α is the principal character precisely when α is fixed by all of G; that is when α is in K^*. One easily checks that $\alpha \to \psi_\alpha$ is a homomorphism of M into the character group $\hat{G}$ and so we have an imbedding

$$\psi : M/K^* \to \hat{G}.$$

We shall prove this is an isomorphism. If ψ_M is a proper subgroup of $\hat{G}$ then there must exist a non-principal character of $\hat{G}$ with ψ_M in the kernel—that is, a character of $\hat{G}/\psi_M$. Since the character group of $\hat{G}$ is known to consist of the evaluation maps at elements of G (Chapter IV, Corollary 3.2) there must exist an element $\sigma \neq 1$ in G such that $\psi_\alpha(\sigma) = 1$ for all $\alpha \in M$.

We will obtain a contradiction to this by using Hilbert's Theorem 90. By the decomposition theory for abelian groups we may write

$$G = \langle\tau\rangle \times G_0, \qquad \sigma \notin G_0$$

and $\sigma = \tau^a \gamma$ with $\tau^a \neq 1$ and $\gamma \in G_0$. Let E denote the subfield of L fixed by G_0 so that $G(E/K) \cong \langle\tau\rangle$ is cyclic of order d with $d|n$. Let β be a primitive dth root of unity in K. Then $\mathrm{N}_{E|K}(\beta) = \beta^d = 1$ so by Theorem 90 there is an element α in E with $\beta = \tau(\alpha)/\alpha$. It follows that α^d is invariant under $\langle\tau\rangle$ and so α^d is in K. Also then α is in M. Our assumption about σ implies $\psi_\alpha(\sigma) = 1$. But ψ_α maps $\langle\tau\rangle$ onto $\langle\beta\rangle$ is a one-to-one way and maps G_0 onto $\langle 1\rangle$. Thus $\tau^a \neq 1$ and $\gamma \in G_0$ imply $1 \neq \psi_\alpha(\tau^a) = \psi_\alpha(\tau^a\gamma) = \psi_\alpha(\sigma)$ which gives the required contradiction. This proves M/K^* is isomorphic to $\hat{G}$ and this in turn is isomorphic to G.

Since no element of G fixes all of M, it must be that $L = K(M)$.

Now let $W = M^n$. The nth power mapping gives a homomorphism of M/K^* onto W/K^n. This is in fact an isomorphism because two elements of M having equal nth powers must differ by a factor from K—an nth root of unity.

This proves the part of the theorem showing all Kummer n-extensions arise as indicated. Now suppose W is a subgroup of K^* containing K^n and with W/K^n finite. Let $\alpha_1, \ldots, \alpha_s$ be elements such that the cosets $\alpha_i K^n$ independently generate W/K^n. Then the field $L = K(\sqrt[n]{W})$ is equal to the finite extension $K(\sqrt[n]{\alpha_1}, \ldots, \sqrt[n]{\alpha_s})$ so $(L:K)$ is finite. Let σ belong to the Galois group of L/K. Then $\sigma\sqrt[n]{\alpha_i} = \beta_i\sqrt[n]{\alpha_i}$ with β_i an nth root of unity. Since β_i is in K, one easily computes that $\sigma^n = 1$ and $G(L/K)$ is abelian. Thus L/K is a Kummer n-extension.

All that remains is to show the group W is uniquely determined by $L = K(\sqrt[n]{W})$. Let M be the group defined above for this field L. Then $M^n \supseteq W$ and M^n/K^n has order equal to $(L:K)$ by what has been proved above. If we prove $(L:K) \leqslant [W:K^n]$ then the equality $M^n = W$ must follow.

Let $\alpha_i K^n$ have order d_i in W/K^n so that $[W:K^n] = d_1 \cdots d_s$. Then $\alpha_i^{d_i}$ is an

nth power so $(K(\sqrt[n]{\alpha_i}):K) \leqslant d_i$ and it follows that:

$$(K(\sqrt[n]{\alpha_1}, \ldots, \sqrt[n]{\alpha_s}):K) \leqslant d_1 \cdots d_s = [W:K^n].$$

This completes the proof.

Before we take up the existence theorem, a result about S-units will be proved. (See the definition in Section 2 of this chapter.) This next result makes no assumptions about the roots of unity in K and is valid for any algebraic number field.

8.2 Theorem. (Dirichlet–Chevalley–Hasse Unit Theorem). Let S be a finite set of primes containing all the infinite primes and let K^S denote the group of S units of K. Then K^S is the direct product of the finite cyclic group of roots of unity in K and a free abelian group of rank $|S|-1$.

PROOF. Let S_0 denote the collection of finite primes in S and $\mathbf{I}(S_0)$ the group of ideals divisible only by primes in S_0. There is an exact sequence

$$1 \longrightarrow \mathbf{U}_K \longrightarrow K^S \xrightarrow{\ i\ } \mathbf{I}(S_0).$$

in which i maps an element to the principal ideal it generates. Let $h = h_K$ denote the class number of K. For each $\mathfrak{p} \in S_0$, $\mathfrak{p}^h$ is in the image of i so

$$\mathbf{I}(S_0)^h \subseteq i(K^S) \subseteq \mathbf{I}(S_0).$$

The first and last groups in this chain are free abelian of rank $|S_0|$ and so $i(K^S)$ is also free of rank $|S_0|$. By the elementary properties of free abelian groups (projective modules) it follows that

$$K^S \cong \mathbf{U}_K \times i(K^S).$$

Since the structure of the units is known by Dirichlet's Theorem, the result follows.

8.3 Corollary. Assume $\sqrt[n]{1} \in K$ and that S is a finite set of primes containing all the infinite primes of K. Then

$$[K^S : (K^S)^n] = n^{|S|}.$$

PROOF. We may write $K^S = \langle w \rangle \times T$ with $\langle w \rangle$ a finite group of order divisible by n and T of free abelian of rank $|S|-1$. Then $\langle w \rangle / \langle w^n \rangle$ has order n and T/T^n has order $n^{|S|-1}$ so the result is true.

9. THE EXISTENCE THEOREM (proof completed)

Assume K contains the nth roots of unity.

Let S_1 be a finite set of primes of K and $\mathfrak{m}_1$ a modulus divisible by the primes in S_1 (and no others) to sufficiently high powers. The proof of the existence of a class field to a given ideal group has been reduced to showing the ideal group containing

$$i(K_{\mathfrak{m}_1,1})(\mathbf{I}^{\mathfrak{m}_1})^n$$

has a class field. Rather than prove this directly, a slightly more general result will be proved for later use when the conductor is considered.

Keep S_1 and $\mathfrak{m}_1$ fixed and let S_2 be a finite set of primes (possibly empty) such that S_1 and S_2 are disjoint and the union $S = S_1 \cup S_2$ contains all primes $\mathfrak{p}$ such that

(i) $\mathfrak{p}|n$,
(ii) $\mathfrak{p}|\infty$,
(iii) $\mathfrak{p}|\mathfrak{A}_i$ where $\{\mathfrak{A}_i\}$ is a finite set of ideals of K whose images cover the class group $\mathbf{C}_K$.

The significance of (iii) is that any ideal $\mathfrak{A}$ can be expressed $\mathfrak{A} = \mathfrak{A}_i(\alpha)$ for some $\alpha \in K$ and $\mathfrak{A}_i$ divisible only by primes in S.

Let $\mathfrak{m}_2$ be a modulus divisible by the primes in S_2 (and no others) to sufficiently high powers.

Consider the congruence subgroups

$$\mathbf{H}_1 = i(K_{\mathfrak{m}_1,1})(\mathbf{I}^{\mathfrak{m}_1})^n\mathbf{I}(S_2),$$
$$\mathbf{H}_2 = i(K_{\mathfrak{m}_2,1})(\mathbf{I}^{\mathfrak{m}_2})^n\mathbf{I}(S_1)$$

where $\mathbf{I}(S_j)$ is the group of ideals divisible only by finite primes in S_j. Because S_1 and S_2 are disjoint we have

$$\mathbf{H}_j \subseteq \mathbf{I}^{\mathfrak{m}_j} \qquad j = 1, 2.$$

Next we consider the subgroups of K^* defined by

$$W_1 = K^S K^n \cap K_{\mathfrak{m}_2,1}$$
$$W_2 = K^S K^n \cap K_{\mathfrak{m}_1,1}$$

and let

$$L_j = K(\sqrt[n]{W_j}) \qquad j = 1, 2.$$

Notice that $(L_j : K)$ is finite because

$$W_j K^n/K^n \subseteq K^S K^n/K^n$$

which is finite by Corollary 8.3. Our object is to prove the following theorem.

9.1 Theorem. The field L_j is class field over K to the ideal group containing $\mathbf{H}_j$.

The proof will require quite a few steps. We begin by recording some properties of the fields L_j.

9.2 Property. The primes in S_1 split completely in L_2; the primes in S_2 split completely in L_1.

PROOF. Let $\mathfrak{p} \in S_1$ and $\mathfrak{P}$ a prime of L_2 which divides $\mathfrak{p}$. We can show $\mathfrak{p}$ splits completely by proving $(L_{2\mathfrak{P}} : K_\mathfrak{p}) = ef = 1$. For any $\alpha \in W_2$, we have also $\alpha \in K_{\mathfrak{m}_1, 1}$. This means $\alpha \in U_\mathfrak{p}^{(t)}$ where $\mathfrak{p}^t$ divides $\mathfrak{m}_1$. For sufficiently large t we know by Proposition 3.6 that

$$U_\mathfrak{p}^{(t)} \subseteq (U_\mathfrak{p})^n$$

so α is an nth power. Thus $K_\mathfrak{p}(\sqrt[n]{\alpha}) = K_\mathfrak{p}$. It follows then

$$K_\mathfrak{p}(\sqrt[n]{W_2}) = K_\mathfrak{p},$$

which implies the desired conclusion.

9.3 Property. L_i/K is unramified outside S_i.

PROOF. We first argue that any prime ramified in L_i is in S. Let $\mathfrak{p}$ be a prime of K with $\mathfrak{p} \notin S$.

The field L_j is obtained by adjoining to K the roots of certain equations $X^n - \alpha$ with α in K^S. The extension of $K_\mathfrak{p}$ determined by this polynomial is seen to be unramified by the following reasoning. The conditions (1) and (2) upon S imply n and α are $\mathfrak{p}$-units so that modulo $\mathfrak{p}$, $X^n - \alpha$ has distinct roots. Thus the irreducible factors of $X^n - \alpha$ over $K_\mathfrak{p}$ have the same degrees as the irreducible factors modulo $\mathfrak{p}$. This uses Hensel's Lemma—Chapter II, Proposition 3.5. It follows that $\mathfrak{p}$ is unramified in the splitting field of $X^n - \alpha$ over $K_\mathfrak{p}$ and so $\mathfrak{p}$ is unramified in L_j. The primes of $S_1 \cup S_2$ are the only primes which can ramify in L_1, but those in S_1 split completely in L_2. Hence only primes in S_2 can ramify in L_2. The same argument applies to the primes in S_1 and the field L_1 so Property 9.3 is proved.

9.4 Property $G(L_j/K) \cong W_j/W_j \cap K^n$.

This is a consequence of Theorem 8.1.

Now let

$$\mathbf{H}_j^* = \mathrm{N}_{L_j|K}(\mathbf{I}_{L_j}^{\mathfrak{m}_j})\, i(K_{\mathfrak{m}_j, 1}).$$

The modulus $\mathfrak{m}_j$ contains all the primes ramified in L_j to high powers so the reciprocity theorem applies to show

9.5 Property $\mathbf{I}^{\mathfrak{m}_j}/\mathbf{H}_j^* \cong G(L_j/K) \cong W_j/W_j \cap K^n$.

Our object is the proof of the equality $\mathbf{H}_j = \mathbf{H}_j^*$. An inclusion can be easily proved.

The groups in Property 9.5 have exponent n so $\mathbf{H}_j^*$ contains the nth powers of $\mathbf{I}^{\mathfrak{m}_j}$. The primes in S_i, $i \neq j$, split completely in L_j so the finite primes in

S_i are norms from L_j. This shows $\mathbf{I}(S_i) \subseteq \mathbf{H}_j^*$. Both $\mathbf{H}_j$ and $\mathbf{H}_j^*$ are congruence subgroups mod $\mathfrak{m}_j$ so we have

(9.6) $$\mathbf{H}_j \subseteq \mathbf{H}_j^* \qquad j = 1 \text{ or } 2.$$

The equality will be proved by showing both groups have the same index in $\mathbf{I}^{\mathfrak{m}_j}$. From the last two statements it follows

$$[\mathbf{I}^{\mathfrak{m}_j} : \mathbf{H}_j] \geqslant [W_j : W_j \cap K^n]$$

with equality only if there is equality in Eq. (9.6). The actual method of proof is to show

(9.7) $$\frac{[\mathbf{I}^{\mathfrak{m}_1} : \mathbf{H}_1][\mathbf{I}^{\mathfrak{m}_2} : \mathbf{H}_2]}{[W_1 : W_1 \cap K^n][W_2 : W_2 \cap K^n]} = 1.$$

9.8 Lemma

$$\mathbf{I}^{\mathfrak{m}_j}/\mathbf{H}_j \cong \mathbf{I}^S/\mathbf{I}^S \cap \mathbf{H}_j.$$

PROOF. Since all the primes dividing $\mathfrak{m}_j$ are in S, this follows from the property of congruence subgroups proved in Lemma 6.1.

9.9 Lemma

$$\frac{K^*}{K^n K^S K_{\mathfrak{m}_j,1}} \cong \frac{\mathbf{I}^S}{\mathbf{I}^S \cap \mathbf{H}_j}.$$

PROOF. Let f denote the composite of the maps

$$K^* \xrightarrow{\ i\ } \mathbf{I}_K \xrightarrow{\ j\ } \mathbf{I}^S$$

where the second map is the identity on primes outside S and maps primes in S to 1. We first show f is onto. Let $\mathfrak{A}$ be any ideal relatively prime to S. Property 3 defining S allows us to write $\mathfrak{A} = \mathfrak{B}(\alpha)$ with $\mathfrak{B}$ divisible only by primes in S. Then $f(\alpha) = j(\mathfrak{A}\mathfrak{B}^{-1}) = \mathfrak{A}$ as required. Next we determine the subgroup of K^* mapped by f onto $\mathbf{I}^S \cap \mathbf{H}_j$. We do this for $\mathbf{H}_1$.

Suppose $\alpha \in K$, and $f(\alpha) \in \mathbf{H}_1$. Write $i(\alpha) = \mathfrak{A}_0 \mathfrak{A}_1$ with $\mathfrak{A}_0$ divisible only by primes in S and $\mathfrak{A}_1 \in \mathbf{I}^S$. Then $f(\alpha) = \mathfrak{A}_1$. Since this is in $\mathbf{H}_1$ we may write

$$\mathfrak{A}_1 = \mathfrak{B}^n(\beta)\mathfrak{C}, \qquad \mathfrak{B} \in \mathbf{I}^{\mathfrak{m}_1}, \qquad \beta \in K_{\mathfrak{m}_1,1}, \qquad \mathfrak{C} \in \mathbf{I}(S_2).$$

We can also write $\mathfrak{B}$ in the form:

$$\mathfrak{B} = \mathfrak{B}_0(\theta), \qquad \mathfrak{B}_0 \text{ divisible only by primes in } S.$$

Then $(\alpha\theta^{-n}\beta^{-1}) = \mathfrak{A}_0 \mathfrak{B}_0{}^n \mathfrak{C}$ is divisible only by primes in S so $\alpha\theta^{-n}\beta^{-1} \in K^S$ and thus $\alpha \in K^n K^S K_{\mathfrak{m}_1,1}$. Now let $\alpha \in K_{\mathfrak{m}_1,1}$ and

$$(\alpha) = \mathfrak{A}\mathfrak{B}, \qquad \mathfrak{B} \in \mathbf{I}(S_2) \qquad \mathfrak{A} \in \mathbf{I}^{S_2}.$$

Since (α) is prime to $\mathfrak{m}_1$, no primes of S_1 divide $\mathfrak{A}$. Thus $\mathfrak{A} \in \mathbf{I}^S$ and $f(\alpha) = \mathfrak{A}$. Now $(\alpha) \in \mathbf{H}_1$ and $\mathfrak{B} \in \mathbf{H}_1$ so $f(\alpha) = (\alpha)\mathfrak{B}^{-1} = \mathfrak{A}$ is in $\mathbf{H}_1$ and also in $\mathbf{I}^S$. This shows $f(K_{\mathfrak{m}_1,1}) \subseteq \mathbf{H}_1$. It is now easy to see that

$$f(K^n K^S K_{\mathfrak{m}_1,1}) = \mathbf{I}^S \cap \mathbf{H}_1$$

and the isomorphism of the lemma now follows.

We begin the index computation now. At one point the following fact is used.

9.10. Lemma. Suppose A, B, and C are abelian subgroups of some larger group and $A \supset B$. Then

$$[A : B] = [AC : BC][A \cap C : B \cap C]$$

if these numbers are finite.

The proof is straightforward and is left to the reader.

In the next few lines we shall write K_1 for $K_{\mathfrak{m}_1,1}$. Now we find

$$\begin{aligned} [K^* : K^n K^S K_1] &= \frac{[K^* : K^n K_1]}{[K^n K^S K_1 : K^n K_1]} \\ &= \frac{[K^* : K^n K_1]}{[K^n K^S : K^n]}[K^n K^S \cap K_1 : K^n \cap K_1] \\ &= \frac{[K^* : K^n K_1]}{n^{|S|}}[W_2 : W_2 \cap K^n]. \end{aligned}$$

To obtain the second equality use Lemma 9.10 with $A = K^n K^S$, $B = K^n$, and $C = K_1$.

Now combine Lemmas 9.8 and 9.9 with the symmetric version of this calculation to obtain

$$\text{(9.11)} \qquad \frac{[\mathbf{I}^{\mathfrak{m}_1} : H_1][\mathbf{I}^{\mathfrak{m}_2} : H_2]}{[W_1 : W_1 \cap K^n][W_2 : W_2 \cap K^n]} = \frac{[K^* : K^n K_{\mathfrak{m}_1,1}][K^* : K^n K_{\mathfrak{m}_2,1}]}{n^{2|S|}}$$

It is necessary to show the numerator on the right is $n^{2|S|}$.

For any modulus $\mathfrak{m}$ let

$$c(\mathfrak{m}) = [K^* : K^n K_{\mathfrak{m},1}].$$

By making trivial changes in the proof of Lemma 3.1 one shows $c(\mathfrak{m})$ is multiplicative on relatively prime moduli. This reduces the problem to calculating $c(\mathfrak{m})$ when $\mathfrak{m}$ is a prime power.

The case for an infinite prime is easy. If $\mathfrak{m}$ is a *complex* prime then $K_{\mathfrak{m},1} = K^*$ so $c(\mathfrak{m}) = 1$.

If $\mathfrak{m}$ is a *real* prime then our assumption that $\sqrt[n]{1} \in K$ forces $n = 2$. Then $K^n K_{\mathfrak{m},1}$ contains only elements positive at $\mathfrak{m}$ so $c(\mathfrak{m}) = 2$.

Now let $\mathfrak{m} = \mathfrak{p}^t$, $\mathfrak{p}$ a finite prime. The computation of $c(\mathfrak{m})$ is very similar to the computation of $a(\mathfrak{m})$ in Section 3. We shall provide the statements of the results needed and leave to the reader the relatively simple exercise of translating the proofs from Section 3 to the present situation.

(9.12) (a) $[K^* : K^n K_{\mathfrak{m}}] = n$

(b) $c(\mathfrak{m}) = n[K_{\mathfrak{m}} : (K_{\mathfrak{m}})^n K_{\mathfrak{m},1}]$

(9.13) $$\frac{K_{\mathfrak{m}}}{(K_{\mathfrak{m}})^n K_{\mathfrak{m},1}} \cong \frac{\mathbf{U}_{\mathfrak{p}}}{\mathbf{U}_{\mathfrak{p}}{}^n \mathbf{U}_{\mathfrak{p}}{}^{(t)}}$$

Here of course $\mathbf{U}_{\mathfrak{p}}$ is the unit group in the completion $K_{\mathfrak{p}}$.

For t sufficiently large we have $\mathbf{U}_{\mathfrak{p}}{}^{(t)} \subseteq \mathbf{U}_{\mathfrak{p}}{}^n$ and so

$$c(\mathfrak{m}) = n[\mathbf{U}_{\mathfrak{p}} : \mathbf{U}_{\mathfrak{p}}{}^n].$$

To keep the analogy with Section 3 we let G denote the cyclic group of order n which operates trivially on all the groups associated with K. For M a G-module we now have for $x \in M$

$$N(x) = x^n, \quad \Delta(x) = 1$$

and

$$H^0(M) = M/M^n, \qquad H^1(M) = n\text{th roots of } 1 \text{ in } M.$$

In particular the computation of $c(\mathfrak{m})$ depends upon

$$|H^0(\mathbf{U}_{\mathfrak{p}})| = [\mathbf{U}_{\mathfrak{p}} : \mathbf{U}_{\mathfrak{p}}{}^n].$$

Notice that $\sqrt[n]{1} \in K$ implies

$$|H^1(\mathbf{U}_{\mathfrak{p}})| = n.$$

We now have

(9.14) $$c(\mathfrak{m}) = n[\mathbf{U}_{\mathfrak{p}} : \mathbf{U}_{\mathfrak{p}}{}^n] = n^2/q(\mathbf{U}_{\mathfrak{p}}).$$

The Herbrand quotient $q(\mathbf{U}_{\mathfrak{p}})$ is determined by making successive reductions as in Lemma 3.9. With $R_{\mathfrak{p}}$ the valuation ring in $K_{\mathfrak{p}}$ we obtain $q(\mathbf{U}_{\mathfrak{p}}) = q(R_{\mathfrak{p}})$ with G operating trivially upon the additive group of $R_{\mathfrak{p}}$. The additive version of the cohomology groups above now yields

$$H^0(R_{\mathfrak{p}}) = R_{\mathfrak{p}}/nR_{\mathfrak{p}}, \qquad H^1(R_{\mathfrak{p}}) = 0.$$

Since the quotient $R_{\mathfrak{p}}/nR_{\mathfrak{p}}$ is isomorphic to the same quotient with $R_{\mathfrak{p}}$

interpreted as the localization at $\mathfrak{p}$ of the algebraic integers in K, we may make this replacement. We then have

(9.15) $c(\mathfrak{m}) = n^2[R_\mathfrak{p} : nR_\mathfrak{p}]$, $\quad R_\mathfrak{p}$ = localization at $\mathfrak{p}$ of the algebraic integers in K.

Now return to the computation of Eq. (9.11). The term of interest is $c(\mathfrak{m}_1)c(\mathfrak{m}_2) = c(\mathfrak{m}_1\mathfrak{m}_2)$. Let S_0 denote the finite primes in $S = S_1 \cup S_2$. We calculate first the contribution from S_0; namely

$$\prod_{\mathfrak{p} \in S_0} n^2[R_\mathfrak{p} : nR_\mathfrak{p}] = n^{2|S_0|}[R : nR].$$

We have used here the fact that all prime divisors of n are in S and the isomorphism

$$R/nR \cong \prod_{\mathfrak{p}|n} R_\mathfrak{p}/nR_\mathfrak{p}.$$

Now by Chapter I, Proposition 8.6, we find

$$[R : nR] = \mathcal{N}_{K|Q}(nR) = n^{(K:Q)}.$$

Let r and s denote the number of real and complex infinite primes of K so $r+2s = (K:Q)$. Also $r+s+|S_0| = |S|$.

Consider the case with $r = 0$. Then

$$c(\mathfrak{m}_1\mathfrak{m}_2) = n^{2|S_0|}n^{2s} = n^{2|S|}.$$

If $r > 0$ then as observed earlier, $n = 2$. In this case

$$c(\mathfrak{m}_1\mathfrak{m}_2) = 2^r 2^{2|S_0|}2^{r+2s} = 2^{2|S|} = n^{2|S|}.$$

Thus in all cases we have proved the index quotient in Eq. (9.11) is equal to 1 and this is enough to complete the proof of Theorem 9.1.

As a corollary we also obtain the rest of the existence theorem which we shall state completely as the classification theorem.

Let K be any number field and $\mathbf{H}, \mathbf{J}$ two ideal groups for K. We say $\mathbf{H} \subseteq \mathbf{J}$ if for some modulus $\mathfrak{m}$, we have $\mathbf{H}^\mathfrak{m} \subseteq \mathbf{J}^\mathfrak{m}$. Notice that this inclusion for one modulus $\mathfrak{m}$ implies the same inclusion for any modulus divisible by the conductor of $\mathbf{H}$ and the conductor of $\mathbf{J}$.

9.16 Theorem (The Classification Theorem). Let K be any algebraic number field. The correspondence $L \to \mathbf{H}(L/K)$ is a one-to-one inclusion reversing correspondence between the collection of finite dimensional abelian extensions L/K and the collection of ideal groups of K.

Proof. We show first that any ideal group $\mathbf{H}$ is a class group to some abelian extension. Take $\mathbf{H}^\mathfrak{m} \in \mathbf{H}$ and suppose $\mathbf{I}^\mathfrak{m}/\mathbf{H}^\mathfrak{m}$ has exponent n. The reduction step in Section 7 allows us to assume K contains the nth roots of unity.

Let S_1 be a finite set of primes containing all the primes dividing $\mathfrak{m}$ and all primes satisfying the conditions (i)–(iii) at the beginning of this section for S. We take S_2 empty so $S = S_1$. We also take the modulus $\mathfrak{m}_1$ large enough so that $\mathfrak{m}|\mathfrak{m}_1$. Now the group $\mathbf{H}_1$ defined at the beginning of this section is contained in

$$\mathbf{H}^{\mathfrak{m}_1} = \mathbf{H}^{\mathfrak{m}} \cap \mathbf{I}^{\mathfrak{m}_1}.$$

By Theorem 9.1 there is an abelian extension L_1 with $\mathbf{H}_1 = \ker(\varphi_{L_1|K}|\mathbf{I}^{\mathfrak{m}_1})$. By Proposition 7.1 there is a subfield of L_1 which is class field to the ideal group $\mathbf{H}$ containing $\mathbf{H}^{\mathfrak{m}_1}$.

Thus every ideal group is a class group. If $\mathbf{H}(L/K) \subseteq \mathbf{H}(E/K)$ for abelian extensions L, E of K, then $E \subseteq L$ by Theorem 5.8. On the other hand $E \subseteq L$ implies

$$\ker \varphi_{L|K} \subseteq \ker \varphi_{E|K}$$

so $\mathbf{H}(L/K) \subseteq \mathbf{H}(E/K)$. This shows the correspondence is one-to-one and completes the proof.

This is the main theorem in class field theory. It gives the classification of all abelian extensions of K in terms of objects defined by the internal structure of K.

10. SOME CONSEQUENCES OF THE CLASSIFICATION THEOREM

We consider now a normal extension E/K whose Galois group G is not necessarily abelian. The theory in the preceding sections can be applied to give information in the nonabelian case.

Begin with a modulus $\mathfrak{m}$ divisible by high powers of all the primes of K which ramify in E. The group

$$\mathbf{H}^{\mathfrak{m}}(E/K) = \mathrm{N}_{E|K}(\mathbf{I}_E^{\mathfrak{m}})\, i(K_{\mathfrak{m},1})$$

is a congruence subgroup and the ideal group containing it has a class field. Let L/K be an abelian extension such that $\mathbf{H}^{\mathfrak{m}}(E/K) \in \mathbf{H}(L/K)$. Then for a suitable modulus $\mathfrak{n}$ divisible by $\mathfrak{m}$ (and containing all the primes ramified in L) we have

$$\mathrm{N}_{E|K}(\mathbf{I}_E^{\mathfrak{n}}) \subseteq \mathbf{H}^{\mathfrak{n}}(L/K).$$

By Theorem 5.8 we obtain $L \subseteq E$. (So the modulus $\mathfrak{m}$ could have been used after all in place of $\mathfrak{n}$.) We know then $\mathbf{H}^{\mathfrak{m}}(E/K) = \mathbf{H}^{\mathfrak{m}}(L/K)$ and so

$$\text{(1)} \qquad \mathbf{I}^{\mathfrak{m}}/\mathbf{H}^{\mathfrak{m}}(E/K) \cong \mathbf{I}^{\mathfrak{m}}/\mathbf{H}^{\mathfrak{m}}(L/K) \cong G(L/K).$$

It is necessary to identify these groups in a more direct way with $G(E/K)$.

10.1 Theorem. The field L is the largest subfield of E with an abelian Galois group over K. Thus $G(L/K) \cong G/G'$ with G' denoting the commutator subgroup of G. Moreover, the isomorphism of $\mathbf{I}^{\mathfrak{m}}/\mathbf{H}^{\mathfrak{m}}(E/K)$ with G/G' is induced by the "Artin-map" $\varphi_{E|K}$. (The Artin map for this nonabelian extension is defined in the course of the proof.)

PROOF. Suppose $L \subseteq L_1 \subseteq E$ with L_1/K an abelian extension. Then by the transitivity of the norm one finds

$$\begin{aligned} \mathrm{N}_{E|K}(\mathbf{I}_E{}^{\mathfrak{m}})\, i(K_{\mathfrak{m},1}) &\subseteq \mathrm{N}_{L_1|K}(\mathbf{I}_{L_1}{}^{\mathfrak{m}})\, i(K_{\mathfrak{m},1}) \\ &\subseteq \mathrm{N}_{L|K}(\mathbf{I}_L{}^{\mathfrak{m}})\, i(K_{\mathfrak{m},1}). \end{aligned}$$

However, the first and third groups are equal and so the ideal groups for L and L_1 are equal and so $L = L_1$ since both are abelian extensions. This means $G(L/K)$ is the largest homomorphic image of G which is abelian; necessarily this is G/G'.

We can give a rather explicit description of the isomorphism (1) with G/G'. Let $\mathfrak{P}$ be a prime in $\mathbf{I}_E{}^{\mathfrak{m}}$ and $\mathfrak{p} = \mathfrak{P} \cap K$. We know that the primes above $\mathfrak{p}$ in E are all conjugate under G and that $\mathfrak{p}$ determines the conjugacy class containing the Frobenius automorphism of $\mathfrak{P}$. Now we observe that since G/G' is abelian, conjugate elements in G have the same image in G/G'. Thus $\mathfrak{p}$ uniquely determines an element in G/G'. We define the Artin map by

$$\varphi_{E|K}(\mathfrak{p}) = \left[\frac{E|K}{\mathfrak{P}}\right] G'.$$

This map extends to a homomorphism of $\mathbf{I}^{\mathfrak{m}}$ to G/G'. We can also determine the kernel. By a property of the Frobenius automorphism we see

$$\left[\frac{E|K}{\mathfrak{P}}\right]\Bigg| L = \left[\frac{L|K}{\mathfrak{P}_L}\right], \qquad \mathfrak{P}_L = \mathfrak{P} \cap L$$

and so

$$\varphi_{E|K}(\mathfrak{p}) = \varphi_{L|K}(\mathfrak{p})\, G'.$$

Thus, anything in the kernel of $\varphi_{L|K}$ is in the kernel of $\varphi_{E|K}$. But $\ker \varphi_{L|K} = \mathbf{H}^{\mathfrak{m}}(E|K)$ which has index in $\mathbf{I}^{\mathfrak{m}}$ equal to $[G : G']$. It follows that $\mathbf{H}^{\mathfrak{m}}(E|K) = \ker \varphi_{E|K}$ and the proof is complete.

The first inequality insured that $[\mathbf{I}^{\mathfrak{m}} : \mathbf{H}^{\mathfrak{m}}(E|K)]$ was at most $(E : K) = |G|$. We have now shown this index to be exactly $[G : G']$ so the first inequality is always a strict inequality for nonabelian normal extensions.

We turn to another consequence of the existence theorem which generalizes the earlier results proved for $K = Q$ in Chapter IV, Section 6.

10.2 Proposition. Let χ be a nonprincipal character of $\mathbf{I}^{\mathfrak{m}}/i(K_{\mathfrak{m},1})$. Then $L(1, \chi) \neq 0$.

PROOF. There is a class field L to the ideal group containing $i(K_{\mathfrak{m},1})$ and the set of primes which split completely in L are in $i(K_{\mathfrak{m},1})$ except for a finite number. Thus the density of the set of primes in $i(K_{\mathfrak{m},1})$ is $(L : K)$ by the Frobenius density theorem and this number equals $[\mathbf{I}^{\mathfrak{m}} : i(K_{\mathfrak{m},1})]$ by the reciprocity theorem. Thus $L(1, \chi) \neq 0$ by Chapter IV, Proposition 4.8.

This is the main step in the proof of the generalized theorem of Dirichlet on primes in arithmetic progressions.

10.3 Theorem. Let $\mathbf{I}_K{}^{\mathfrak{m}} \supseteq \mathbf{H}^{\mathfrak{m}} \supseteq i(K_{\mathfrak{m},1})$ be a chain of subgroups. Then any coset of $\mathbf{H}^{\mathfrak{m}}$ in $\mathbf{I}_K{}^{\mathfrak{m}}$ contains infinitely many primes. In fact this set of primes has density $[\mathbf{I}^{\mathfrak{m}} : \mathbf{H}^{\mathfrak{m}}]^{-1}$.

The proof is almost word for word the same as that of Chapter IV, Theorem 5.8 and so will not be repeated.

One can refine the Frobenius density theorem now. We deal with an abelian extension L/K having Galois group G. Any element σ in G is the image of a unique coset of $\mathbf{H}^{\mathfrak{m}}(L/K)$ under $\varphi_{L|K}$ and so by Theorem 10.3 the density of the primes $\mathfrak{p}$ in this coset for which $\varphi_{L|K}(\mathfrak{p}) = \sigma$ is $1/|G|$. This improves the earlier result for abelian extensions because previously we only knew the density of the primes which mapped onto some σ^d for d relatively prime to $|\sigma|$.

There is also a generalization to the nonabelian case.

10.4 Theorem (Tchebotarev Density Theorem). Let E/K be a normal extension with Galois group G. Let $\sigma \in G$ and suppose σ has c conjugates in G. The set of primes $\mathfrak{p}$ of K which have a prime divisor $\mathfrak{P}$ in E whose Frobenius automorphism is σ has a density $c/|G|$.

PROOF. Let L denote the subfield fixed by $\langle\sigma\rangle$ and S' the set of primes $\mathfrak{P}$ of L for which $\varphi_{E|L}(\mathfrak{P}) = \sigma$. By the remarks just above, S' has density $1/|\sigma|$ because $G(E/L) = \langle\sigma\rangle$ is abelian. When densities are considered we may restrict our attention to primes with relative degree one over K (or over Q even). Let S denote the subset of primes in S' having relative degree one over K.

For $\mathfrak{P} \in S$ and $\mathfrak{p} = \mathfrak{P} \cap K$ we now count the number of $\mathfrak{P}_i \in S$ containing $\mathfrak{p}$. First take a prime $\mathfrak{Q}$ of E which divides $\mathfrak{P}$ and has σ as Frobenius automorphism over L. Let $\langle\sigma\rangle\tau_j$ be the distinct cosets of $\langle\sigma\rangle$ in G. The primes dividing $\mathfrak{p}$ in E are $\tau_i(\mathfrak{Q})$ (and these are distinct) and the primes dividing $\mathfrak{p}$ in L are $\mathfrak{P}_j = \tau_j(\mathfrak{Q}) \cap L$. By Chapter III, Corollary 2.8, we see $\mathfrak{P}_j$ has relative degree one over K if and only if $\langle\sigma\rangle\tau_j\sigma = \langle\sigma\rangle\tau_j$. Assuming this to be the case then

$$\varphi_{E|L}(\mathfrak{P}_j) = \left[\frac{E|L}{\tau_j \mathfrak{Q}}\right] = \tau_j \left[\frac{E|L}{\mathfrak{Q}}\right] \tau_j{}^{-1} = \tau_j \sigma \tau_j{}^{-1}.$$

It follows that $\mathfrak{P}_j \in S$ if and only if

$$\sigma = \tau_j \sigma \tau_j^{-1}.$$

The distinct $\mathfrak{P}_j$ which arise this way correspond to the distinct cosets $\langle\sigma\rangle\tau_j$ so the number of primes in S dividing $\mathfrak{p}$ is

$$d = [C_G(\sigma) : \langle\sigma\rangle].$$

If T denotes the set of primes of K divisible by a prime in S then there arise exactly d primes $\mathfrak{P} \in S$ for which $\mathrm{N}_{L|K}(\mathfrak{P}) = \mathfrak{p}$ for each $\mathfrak{p} \in T$. This means $d\delta(T) = \delta(S) = 1/|\sigma|$. From the form of d it is immediate that

$$\delta(T) = 1/|C_G(\sigma)| = c/|G|.$$

11. PRELIMINARIES FOR THE NORM RESIDUE MAP AND THE CONDUCTOR THEOREM

Our object in the next two sections is to get precise information about the conductor $\mathfrak{f}(L/K)$. We take a rather general approach that will yield considerable information in other directions also. The abelian extension L/K is fixed for this section.

We begin with a simple but useful result.

11.1 Theorem (Translation theorem). Let E/K be any finite dimensional extension and $\mathfrak{m}$ a modulus for K divisible by $\mathfrak{f}(L/K)$. Then the class group to the abelian extension LE/E is the ideal group for E which contains the congruence subgroup

$$\{\mathfrak{A} \in \mathbf{I}_E^{\mathfrak{m}} \mid \mathrm{N}_{E/K}(\mathfrak{A}) \in \mathbf{H}^{\mathfrak{m}}(L/K)\}.$$

PROOF. By Chapter III, Lemma 3.1, we have

$$\varphi_{LE/E} = \varphi_{L/K}\mathrm{N}_{E/K}.$$

Thus the kernel of $\varphi_{LE/E}$ acting on $\mathbf{I}_E^{\mathfrak{m}}$ is the subgroup mapped by $\mathrm{N}_{E/K}$ into the kernel of $\varphi_{L/K}$ acting upon $\mathbf{I}_K^{\mathfrak{m}}$. This kernel is $\mathbf{H}^{\mathfrak{m}}(L/K)$ since the conductor divides $\mathfrak{m}$. The result follows.

Now we consider a prime $\mathfrak{p}$ of K and a modulus

$$\mathfrak{n} = \mathfrak{p}^a\mathfrak{m}, \qquad \mathfrak{p} \nmid \mathfrak{m}.$$

Suppose $\mathfrak{n}$ is divisible by $\mathfrak{f}(L/K)$ and all ramified primes. Eventually we will see that all the ramified primes already divide $\mathfrak{f}(L/K)$.

Let θ denote the composite of the maps

$$(1) \qquad K_{\mathfrak{m},1} \xrightarrow{i} \mathbf{I}_K \xrightarrow{j} \mathbf{I}_K^{\mathfrak{n}} \xrightarrow{\varphi_{L|K}} G(L/K)$$

where j is the identity on primes not dividing $\mathfrak{n}$ and sends to 1 any prime dividing $\mathfrak{n}$. It would be more precise to write $\theta_{L/K}$ but we shall only use the simpler notation.

In the following, we write $G(\mathfrak{p})$ for the decomposition group of a prime of L dividing $\mathfrak{p}$.

11.2 Theorem

$$\theta(K_{\mathfrak{m},1}) = G(\mathfrak{p}).$$

PROOF. The first step is to show $\theta(K_{\mathfrak{m},1}) \subseteq G(\mathfrak{p})$. Let Z be the fixed field under $G(\mathfrak{p})$. The prime $\mathfrak{p}$ splits completely in Z so in particular it is not ramified. This means $\mathfrak{p}$ does not divide $\mathfrak{f}(Z/K)$. The conductor of Z/K divides $\mathfrak{n}$ and so it divides $\mathfrak{m}$. In particular this means

$$\ker \varphi_{Z|K}|\mathbf{I}_K^{\mathfrak{m}} = \mathrm{N}_{Z|K}(\mathbf{I}_Z^{\mathfrak{m}})\, i(K_{\mathfrak{m},1}).$$

Suppose $\mathfrak{p}$ is a finite prime. Then this kernel contains $\mathfrak{p}$ because $\mathfrak{p}$ splits. For $\alpha \in K_{\mathfrak{m},1}$ we write $(\alpha) = \mathfrak{p}^t \mathfrak{A}$ with $\mathfrak{A}$ prime to $\mathfrak{p}$. Then

$$ji(\alpha) = \mathfrak{A} = \mathfrak{p}^{-t}(\alpha) \in \ker \varphi_{Z|K}.$$

If $\mathfrak{p}$ is an infinite prime then $ji(\alpha) = i(\alpha)$ so again $ji(\alpha) \in \ker \varphi_{Z|K}$. In either case

$$1 = \varphi_{Z|K}(ji(K_{\mathfrak{m},1})) = \varphi_{L|K}(ji(K_{\mathfrak{m},1}))\,|Z = \theta(K_{\mathfrak{m},1})|\,Z.$$

This shows $\theta(K_{\mathfrak{m},1})$ is trivial on Z and so is a subgroup of $G(\mathfrak{p})$.

The proof of equality for these two groups requires some subtle moves. We suppose on the contrary that $\theta(K_{\mathfrak{m},1})$ is not all of $G(\mathfrak{p})$. Then there is a subgroup G_0 with

$$\theta(K_{\mathfrak{m},1}) \subseteq G_0 < G(\mathfrak{p})$$

and

$$[G(\mathfrak{p}) : G_0] = q = \text{a prime}.$$

Let E be the field fixed by G_0. A contradiction will be obtained by examining the extension E/Z. Let β be a primitive qth root of unity, $Z(\beta) = Z'$, $E(\beta) = E'$. The procedure is to apply the Existence Theorem 9.1 (which required roots of unity) to Z' in order to deduce certain primes are split in E'. This information is translated down to deduce $\mathfrak{p}$ splits completely in E. This will be contradictory to properties of the decomposition field.

Let S_1 be the set of primes of Z' which divide $\mathfrak{p}$; S_2 is a sufficiently large set of primes of Z' so that $S = S_1 \cup S_2$ satisfies the conditions of Section 9 (for Z' in place of K). Let $\mathfrak{m}_1 = \mathfrak{p}^a$ be viewed as a modulus for Z' and $\mathfrak{m}_2$ a product of the primes in S_2 with sufficiently high exponents. We may assume

$$\mathfrak{n}|\mathfrak{m}_1\mathfrak{m}_2, \qquad \mathfrak{m}|\mathfrak{m}_2.$$

Let $\mathbf{H}_2 = (\mathbf{I}_{Z'}^{\mathfrak{m}_2})^q i(Z'_{\mathfrak{m}_2,1})\mathbf{I}(S_1)$ and L_2 the class field over Z' to $\mathbf{H}_2$.

ASSERTION: $\mathbf{H}_2 \cap \mathbf{I}_{Z'}{}^{\mathfrak{n}} \subseteq \mathbf{H}^{\mathfrak{n}}(E'/Z')$.

To prove the assertion we first need a description of the group on the right. Apply the translation theorem to the extension E/K translated by Z'. It follows

$$\mathbf{H}^{\mathfrak{n}}(E'/Z') = \{\mathfrak{A} \in \mathbf{I}_{Z'}{}^{\mathfrak{n}} | \mathrm{N}_{Z'|K}(\mathfrak{A}) \in \mathbf{H}^{\mathfrak{n}}(E/K)\}.$$

We shall also use the fact that $ji(K_{\mathfrak{m},1})$ is contained in $\mathbf{H}^{\mathfrak{n}}(E/K)$. This inclusion follows because $\varphi_{L|K}$ maps $ji(K_{\mathfrak{m},1})$ into the group G_0 fixing E.

Now suppose $\mathfrak{X} = \mathfrak{B}^q(\alpha)\mathfrak{C}$ is in $\mathbf{H}_2$ and also prime to $\mathfrak{n}$. We may assume

$$\mathfrak{B} \in \mathbf{I}_{Z'}^{\mathfrak{m}_2}, \qquad \alpha \in Z'_{\mathfrak{m}_2,1}, \qquad \mathfrak{C} \in \mathbf{I}(S_1).$$

To prove the assertion we must show $\mathrm{N}_{Z'|K}(\mathfrak{X})$ is $\mathbf{H}^{\mathfrak{n}}(E/K)$.

Suppose $\mathfrak{p}$ is *finite*. Perform factorizations:

$$\mathfrak{B} = \mathfrak{B}_0\mathfrak{B}_1, \quad \mathfrak{B}_0 \text{ prime to } \mathfrak{n}, \qquad \mathrm{N}_{Z'|K}(\mathfrak{B}_1) = \mathfrak{p}^r,$$

$$(\alpha) = \mathfrak{A}_0\mathfrak{A}_1, \quad \mathfrak{A}_0 \text{ prime to } \mathfrak{n}, \qquad \mathrm{N}_{Z'|K}(\mathfrak{A}_1) = \mathfrak{p}^s.$$

We know also $\mathrm{N}_{Z'|K}(\mathfrak{C}) = \mathfrak{p}^t$. Since $\mathfrak{X}$ is prime to $\mathfrak{n}$, $\mathrm{N}_{Z'|K}(\mathfrak{X})$ is prime to $\mathfrak{p}$. This implies $qr+s+t = 0$ and

$$\mathrm{N}_{Z'|K}(\mathfrak{X}) = \mathrm{N}_{Z'|K}(\mathfrak{B}_0)^q\,\mathrm{N}_{Z'|K}(\mathfrak{A}_0).$$

We can prove each factor here is in $\mathbf{H}^{\mathfrak{n}}(E/K)$. Firstly observe that $\mathbf{H}^{\mathfrak{n}}(Z/K) \supseteq \mathbf{H}^{\mathfrak{n}}(E/K)$ and the index is q. Since

$$\mathrm{N}_{Z|K}(\mathrm{N}_{Z'|Z}(\mathfrak{B}_0)) \in \mathbf{H}^{\mathfrak{n}}(Z/K)$$

it follows that $\mathrm{N}_{Z'|K}(\mathfrak{B}_0)^q \in \mathbf{H}^{\mathfrak{n}}(E/K)$. Next we note that $\mathrm{N}_{Z'|K}(\alpha) \in K_{\mathfrak{m},1}$ because $\mathfrak{m}|\mathfrak{m}_2$. Thus

$$\mathrm{N}_{Z'|K}(\mathfrak{A}_0) = \mathrm{N}_{Z'|K}(\alpha)\,\mathfrak{P}^{-s} = ji\,\mathrm{N}_{Z'|K}(\alpha) \in ji(K_{\mathfrak{m},1}).$$

This is known to be in $\mathbf{H}^{\mathfrak{n}}(E/K)$ so the assertion is proved in the case $\mathfrak{p}$ is finite.

Now for $\mathfrak{p}$ *infinite*, the assumption $G(\mathfrak{p}) \neq 1$ forces $\mathfrak{p}$ to be real and ramified in L and moreover $|G(\mathfrak{p})| = 2$. Thus Z already contains a qth root of unity since $q = 2$ and $Z = Z'$, $E = E'$. With $\mathfrak{X}$ as above we have $\mathfrak{B}^q \in \mathbf{H}^{\mathfrak{n}}(E/Z)$ and $\mathfrak{C} = 1$ because $\mathbf{I}(S_1) = 1$. We need only show $(\alpha) \in \mathbf{H}^{\mathfrak{n}}(E/Z)$. It is enough to show $i\mathrm{N}_{Z|K}(\alpha) \in \mathbf{H}^{\mathfrak{n}}(E|K)$. We already know $\mathrm{N}_{Z|K}(\alpha) \in K_{\mathfrak{m},1}$ because $\mathfrak{m}|\mathfrak{m}_2$. This means $\theta(\mathrm{N}_{Z|K}(\alpha)) = 1$ for $\theta(K_{\mathfrak{m},1}) = 1$ when $|G(\mathfrak{p})| = 2$. However this assertion implies

$$1 = \varphi_{L|K}(i\mathrm{N}_{Z|K}(\alpha)) = \varphi_{E|K}(i\mathrm{N}_{Z|K}(\alpha)).$$

We have used the equality $E = L$ which follows because $(L:Z) = 2$. This shows $i\mathrm{N}_{Z|K}(\alpha)$ belongs to $\mathbf{H}^{\mathfrak{n}}(E/K)$ and completes the proof of the assertion.

As an immediate consequence we have

$$\mathbf{H}^{\mathfrak{m}_1\mathfrak{m}_2}(L_2/Z') = \mathbf{H}_2 \cap \mathbf{I}_{Z'}^{\mathfrak{m}_1\mathfrak{m}_2} = (\mathbf{H}_2 \cap \mathbf{I}_{Z'}{}^{\mathfrak{n}}) \cap \mathbf{I}_{Z'}^{\mathfrak{m}_1\mathfrak{m}_2} \subseteq \mathbf{H}^{\mathfrak{m}_1\mathfrak{m}_2}(E'/Z').$$

By the Classification Theorem 9.16 it follows that $E' \subseteq L_2$. Then by Property 9.2 we know every prime in S_1 splits completely in L_2. Let $\mathfrak{p}_Z$ be a prime of Z dividing $\mathfrak{p}$ and $\mathfrak{P}$ a prime of Z' dividing $\mathfrak{p}_Z$. We have $\mathfrak{P} \in S_1$ so $\mathfrak{P}$ splits completely in L_2. Let e, f denote the ramification index and relative degree of $\mathfrak{P}$ over $\mathfrak{p}_Z$ because there is no change when passing from Z' to E' since $\mathfrak{P}$ splits completely in $E' \subseteq L_2$. Now $(E' : Z')$ divides $q-1$ and so

$$e|q-1, \qquad f|q-1.$$

Now consider the ramification and change of relative degree of $\mathfrak{p}_Z$ in E. The ramification number e_0 must divide e, the ramification number in E' and also e_0 divides $(E:Z) = q$. This forces $e_0 = 1$ since $q-1$ and q are relatively prime. In the same way the relative degree is one so $\mathfrak{p}_Z$ splits completely in E.

Now there can be only one prime divisor of $\mathfrak{p}_Z$ in L by Proposition 11.2 and so there is only one prime divisor of $\mathfrak{p}_Z$ in E. In view of the complete splitting this forces $E = Z$ contrary to our assumption. This completes the proof.

We shall describe how the map θ is related to the computation of the conductor. Some additional notation will make the description easier. Let

$$\begin{aligned} V(\mathfrak{p}^b, \mathfrak{m}) &= K_{\mathfrak{p}^b\mathfrak{m},1} \quad \text{if} \quad b > 0, \\ &= \{\alpha \in K_{\mathfrak{m},1} | (\alpha) \text{ prime to } \mathfrak{p}\} \quad \text{if} \quad b = 0 \quad \text{and} \quad \mathfrak{p} \quad \text{finite}, \\ &= K_{\mathfrak{m},1} \quad \text{if} \quad b = 0 \quad \text{and} \quad \mathfrak{p} \quad \text{infinite}. \end{aligned}$$

11.3 Proposition. The power of $\mathfrak{p}$ dividing $\mathfrak{f}(L/K)$ is $\mathfrak{p}^b$ if b is the smallest nonnegative integer such that $V(\mathfrak{p}^b, \mathfrak{m}) \subseteq \ker\theta$.

PROOF. First notice that $V = V(\mathfrak{p}^b, \mathfrak{m})$ is a subgroup of $K_{\mathfrak{m},1}$ and $ji(V) = i(V)$ because the elements of V are prime to $\mathfrak{p}$. It follows that $V \subseteq \ker\theta$ if and only if $i(V) \subseteq \ker\varphi_{L|K}$. Now use the observation that $\mathfrak{f}(L/K)$ divides a modulus $\mathfrak{n}$ if and only if $i(K_{\mathfrak{n},1}) \subseteq \ker\varphi_{L|K}$. An examination of the cases easily gives the result.

We shall refine this further in the next section and compute the exponent of $\mathfrak{p}$ in terms of local data.

12. NORM RESIDUE SYMBOL

This is a continuation of the last section and the notation there carries over. The main object here is to define a map $\theta_\mathfrak{p}$ on the completion $K_\mathfrak{p}$ which serves as the "local Artin map". We use the following notation:

$$\begin{aligned}\mathbf{U}_\mathfrak{p}^{(b)} &= 1 + \mathfrak{p}^b \text{ in } K_\mathfrak{p} \quad \text{if } b > 0 \text{ and } \mathfrak{p} \text{ is finite;}\\ &= \mathbf{U}_\mathfrak{p} \qquad \text{if } b = 0 \text{ and } \mathfrak{p} \text{ is finite;}\\ &= \text{positive reals if } b > 0 \text{ and } \mathfrak{p} \text{ is real;}\\ &= K_\mathfrak{p}^* \qquad \text{if } b = 0 \text{ and } \mathfrak{p} \text{ is real or if } \mathfrak{p} \text{ is complex.}\end{aligned}$$

12.1 Lemma. $K_\mathfrak{p}^*/\mathbf{U}_\mathfrak{p}^{(b)} \cong K_{\mathfrak{m},1}/V(\mathfrak{p}^b, \mathfrak{m})$.

Proof. The natural inclusion of $K_{\mathfrak{m},1}$ into the completion $K_\mathfrak{p}^*$ induces a map of $K_{\mathfrak{m},1}$ onto the quotient $K_\mathfrak{p}^*/\mathbf{U}_\mathfrak{p}^{(b)}$. This is onto because any element of $K_\mathfrak{p}^*$ can be approximated modulo $\mathfrak{p}^b$ by an element in K^* which is congruent to 1 modulo $\mathfrak{m}$. One now compares the definitions of the V's and $\mathbf{U}$'s to see that $V(\mathfrak{p}^b, \mathfrak{m})$ is the kernel.

Now select the integer b so that $V(\mathfrak{p}^b, \mathfrak{m}) \subseteq \ker \theta$. Then θ induces a map (also denoted by θ) of $K_{\mathfrak{m},1}/V(\mathfrak{p}^b, \mathfrak{m})$ onto $G(\mathfrak{p})$. Now compose several maps:

$$\theta_\mathfrak{p} : K_\mathfrak{p}^* \longrightarrow \frac{K_\mathfrak{p}^*}{\mathbf{U}_p^{(b)}} \cong \frac{K_{\mathfrak{m},1}}{V(\mathfrak{p}^b, \mathfrak{m})} \xrightarrow{\ \theta\ } G(\mathfrak{p})$$

The composite $\theta_\mathfrak{p}$ is a homomorphism of $K_\mathfrak{p}^*$ onto $G(\mathfrak{p})$. We call $\theta_\mathfrak{p}$ the $\mathfrak{p}$-*local Artin map*. We shall see later that the kernel of $\theta_\mathfrak{p}$ is the group of $\mathfrak{p}$-local norms from $L_\mathfrak{p}^*$. For this reason $\theta_\mathfrak{p}$ is sometimes called the *norm residue symbol* at $\mathfrak{p}$.

The actual computation of $\theta_\mathfrak{p}(x)$ for x in $K_\mathfrak{p}^*$ is often very difficult. For emphasis we point out the procedure. We approximate the given x by an element y in K^* such that

$$y \equiv 1 \bmod \mathfrak{m}, \qquad y \equiv x \bmod \mathfrak{p}^b.$$

Then $\theta_\mathfrak{p}(x) = \theta(y) = \varphi_{L|K}\, ji(y)$.

12.2 Lemma. $\mathrm{N}_\mathfrak{p}(L_\mathfrak{P}^*) \subseteq \ker \theta_\mathfrak{p}$.

Proof. For $\beta \in L_\mathfrak{P}^*$, there is an element $\gamma \in L^*$ such that

$$\mathrm{N}_{L|K}(\gamma) \equiv 1 \bmod \mathfrak{m}, \qquad \mathrm{N}_{L|K}(\gamma) \equiv \mathrm{N}_\mathfrak{p}(\beta) \bmod \mathfrak{p}^b.$$

For example, this was done in the proof of Hasse's Norm Theorem 4.5. This allows the equation

$$\theta_\mathfrak{p}\, \mathrm{N}_\mathfrak{p}(\beta) = \theta \mathrm{N}_{L|K}(\gamma) = \varphi_{L|K}\, ji \mathrm{N}_{L|K}(\gamma).$$

When $i\mathrm{N}_{L|K}(\gamma)$ is factored as a product of prime powers one sees that each prime power is the norm of an ideal from L. In particular $ji\,\mathrm{N}_{L|K}(\gamma)$ is the norm of an ideal and so belongs to $\ker \varphi_{L|K}$. Thus $\theta_\mathfrak{p}\, \mathrm{N}_\mathfrak{p}(\beta) = 1$ as required.

We know now a part of the kernel of $\theta_{\mathfrak{p}}$. The group $\mathbf{U}_{\mathfrak{p}}^{(b)}$ is in the kernel as seen from the definition so the local Artin map induces a homomorphism

$$\frac{K_{\mathfrak{p}}}{\mathrm{N}_{\mathfrak{p}}(L_{\mathfrak{P}}^*)\,\mathbf{U}_{\mathfrak{p}}^{(b)}} \to G(\mathfrak{p}) \quad \text{(onto)}. \tag{12.3}$$

We can calculate the order of the first group.

12.4 Lemma

$$[K_{\mathfrak{p}}^* : \mathrm{N}_{\mathfrak{p}}(L_{\mathfrak{P}}^*)] \leqslant (L_{\mathfrak{P}} : K_{\mathfrak{p}}).$$

PROOF. Use induction on the dimension $(L : K)$. If this dimension is prime then $G(L/K)$ is cyclic. By Proposition 3.12 the lemma is true and equality holds. If $(L : K)$ is not prime there is a chain

$$K \subset E \subset L$$

with proper inclusions. Let $\mathfrak{p}'$ be a prime of E divisible by $\mathfrak{P}$ in L. Thus

$$K_{\mathfrak{p}} \subseteq E_{\mathfrak{p}}' \subseteq L_{\mathfrak{P}}.$$

Let $\mathrm{N}_{L_{\mathfrak{p}}|E_{\mathfrak{p}'}} = \mathrm{N}_1$, $\mathrm{N}_{E_{\mathfrak{p}'}|K_{\mathfrak{p}}} = \mathrm{N}_2$ so $\mathrm{N}_2\,\mathrm{N}_1 = \mathrm{N}_{\mathfrak{p}}$. Then

$$[K_{\mathfrak{p}}^* : \mathrm{N}_{\mathfrak{p}}(L_{\mathfrak{P}}^*)] = [K_{\mathfrak{p}}^* : \mathrm{N}_2(E_{\mathfrak{p}}^{*\prime})][\mathrm{N}_2(E_{\mathfrak{p}}^{*\prime}) : \mathrm{N}_2\,\mathrm{N}_1(L_{\mathfrak{P}}^*)].$$

By induction $[K_{\mathfrak{p}}^* : \mathrm{N}_2(E_{\mathfrak{p}'}^*)] \leqslant (E_{\mathfrak{p}}' : K_{\mathfrak{p}})$. The second factor is at most $(L_{\mathfrak{P}} : E_{\mathfrak{p}'})$ because the group

$$\mathrm{N}_2(E_{\mathfrak{p}'}^*)/\mathrm{N}_2\,\mathrm{N}_1(L_{\mathfrak{P}}^*)$$

is a homomorphic image (under N_2) of

$$E_{\mathfrak{p}'}^*/\mathrm{N}_1(L_{\mathfrak{P}}^*).$$

After combining the two estimates, the lemma follows.

This inequality tightens the situation. We now have from (12.3)

$$[K_{\mathfrak{p}}^* : \mathrm{N}_{\mathfrak{p}}(L_{\mathfrak{P}}^*)\,\mathbf{U}_{\mathfrak{p}}^{(b)}] \geqslant |G(\mathfrak{p})| = (L_{\mathfrak{P}} : K_{\mathfrak{p}}) \geqslant [K_{\mathfrak{p}}^* : \mathrm{N}_{\mathfrak{p}}(L_{\mathfrak{P}}^*)].$$

Clearly the first index is no larger than the last so in fact equality holds throughout. We can draw two conclusions at once.

12.5 Corollary. If L/K is an abelian extension and $\mathfrak{p}$ a prime of K, then the $\mathfrak{p}$-local Artin map $\theta_{\mathfrak{p}}$ maps $K_{\mathfrak{p}}^*$ onto $G(\mathfrak{p})$ and the kernel is the group of $\mathfrak{p}$-local norms from $L_{\mathfrak{P}}$. (See comments at the end of this section.)

12.6 Corollary. The power of the prime $\mathfrak{p}$ dividing $\mathfrak{f}(L/K)$ is the least non-negative power $\mathfrak{p}^b$ such that $\mathbf{U}_{\mathfrak{p}}^{(b)} \subseteq \mathrm{N}_{\mathfrak{p}}(L_{\mathfrak{P}}^*)$.

PROOF. It is easy to check that $\mathbf{U}_\mathfrak{p}^{(b)} \subseteq \ker \theta_\mathfrak{p}$ if and only if $V(\mathfrak{p}^b, \mathfrak{m}) \subseteq \ker \theta$. We have just seen these conditions are equivalent to

$$\mathrm{N}_\mathfrak{p}(L_\mathfrak{P}^*)\,\mathbf{U}_\mathfrak{p}^{(b)} = \mathrm{N}_\mathfrak{p}(L_\mathfrak{P}^*)$$

so the result follows from Proposition 11.3.

In the classical terminology $\mathfrak{p}^b$ is called the $\mathfrak{p}$-*local conductor*, written $\mathfrak{f}_\mathfrak{p}$, if b is the least non-negative integer such that $\mathbf{U}_\mathfrak{p}^{(b)} \subseteq \mathrm{N}_\mathfrak{p}(L_\mathfrak{P}^*)$. The last corollary can be restated as

$$\mathfrak{f}(L/K) = \prod_\mathfrak{p} \mathfrak{f}_\mathfrak{p}.$$

We shall prove now $\mathfrak{f}_\mathfrak{p} = 1$ if and only if $\mathfrak{p}$ is unramified. This will prove finally that every ramified prime divides the conductor.

If $\mathfrak{f}_\mathfrak{p} = 1$ then $\mathbf{U}_\mathfrak{p} \subseteq \mathrm{N}_\mathfrak{p}(L_\mathfrak{P}^*)$. For finite primes $\mathfrak{p}$ we then have $\mathfrak{p}$ is unramified.

On the other hand suppose $\mathfrak{p}$ is finite and unramified. In case L/K is cyclic we have from Proposition 3.11 that $\mathbf{U}_\mathfrak{p} \subseteq \mathrm{N}_\mathfrak{p}(L_\mathfrak{P}^*)$ so in fact

$$\mathbf{U}_\mathfrak{p} = \mathrm{N}_\mathfrak{p}(\mathbf{U}_\mathfrak{P}).$$

One now uses induction on $(L:K)$ to see this equality holds when L/K is an abelian extension. Thus

$$\mathbf{U}_\mathfrak{p} = \mathbf{U}_\mathfrak{p}^{(0)} \subseteq \mathrm{N}_\mathfrak{p}(L_\mathfrak{P}^*)$$

and $\mathfrak{f}_\mathfrak{p} = 1$.

For $\mathfrak{p}$ real and unramified $L_\mathfrak{P} = K_\mathfrak{p} =$ reals and so

$$\mathbf{U}_\mathfrak{p}^{(0)} = K_\mathfrak{p}^* = \mathrm{N}_\mathfrak{p}(L_\mathfrak{P}^*)$$

and again $\mathfrak{f}_\mathfrak{p} = 1$. This proves the result.

12.7 Theorem. The conductor $\mathfrak{f}(L/K)$ is the product of all the ramified primes with positive exponents determined by the local conductors.

As the last theorem of this section we prove a property of the local Artin maps which is often called the Artin-reciprocity law.

If $\alpha \in K^*$ and $\mathfrak{p}$ is a prime we shall write $\theta_\mathfrak{p}(\alpha)$ for $\theta_\mathfrak{p}(\alpha_\mathfrak{p})$ where $\alpha_\mathfrak{p}$ is the image of α under the imbedding $K \to K_\mathfrak{p}$,

12.8 Theorem (Product formula for the local Artin maps). For each $\alpha \in K^*$ we have

$$\prod_\mathfrak{p} \theta_\mathfrak{p}(\alpha) = 1.$$

PROOF. First note the product is actually a finite one because $\theta_\mathfrak{p}(\alpha) = 1$ whenever $\mathfrak{p}$ is unramified and $\alpha \in \mathbf{U}_\mathfrak{p}$. That is every element of $\mathbf{U}_\mathfrak{p}$ is a norm from $L_\mathfrak{P}^*$ if $\mathfrak{p}$ unramified. Secondly the values $\theta_\mathfrak{p}(\alpha)$ lie in an abelian group so the product may be taken in any order.

Let S denote the set of primes $\mathfrak{p}$ such that either $\mathfrak{p}$ is ramified in L or $\mathfrak{p}$ divides (α). Suppose $\mathfrak{p}_1, \ldots, \mathfrak{p}_s$ are the primes in S. Take

$$\mathfrak{m} = \prod_i \mathfrak{p}_i^{t_i}$$

with the t_i large enough so that $\mathfrak{f}(L/K)$ divides $\mathfrak{m}$.

Suppose also $(\alpha) = \prod \mathfrak{p}_i^{a_i}$ and that t_i is selected so that $t_i \geqslant a_i$. Let $\mathfrak{m}_i$ be the modulus so that $\mathfrak{m} = \mathfrak{p}_i^{t_i}\mathfrak{m}_i$. For each i select an element ω_i in K^* to satisfy

$$\omega_i \equiv \alpha \bmod \mathfrak{p}_i^{t_i}$$

$$\omega_i \equiv 1 \bmod \mathfrak{m}_i.$$

Then for each j, $\omega_1 \cdots \omega_s \equiv \alpha \bmod \mathfrak{p}_j^{t_j}$ and so

$$\text{(1)} \qquad \alpha^{-1}\omega_1 \cdots \omega_s \in K_{\mathfrak{m},1}.$$

We use these elements to evaluate the $\theta_{\mathfrak{p}_i}(\alpha)$. Write

$$(\omega_i) = \mathfrak{p}_i^{a_i}\mathfrak{B}_i$$

with $\mathfrak{B}_i$ not divisible by any prime in S. Then

$$\theta_{\mathfrak{p}_i}(\alpha) = \theta(\omega_i) = \varphi_{L|K}j(\mathfrak{p}_i^{a_i}\mathfrak{B}_i) = \varphi_{L|K}(\mathfrak{B}_i).$$

Now

$$\prod_{\mathfrak{p}} \theta_{\mathfrak{p}}(\alpha) = \prod_i \theta_{\mathfrak{p}_i}(\alpha) = \varphi_{L|K}(\mathfrak{B}_1 \cdots \mathfrak{B}_s).$$

Since $\mathfrak{B}_1 \cdots \mathfrak{B}_s = (\alpha^{-1}\omega_1 \cdots \omega_s) \in i(K_{\mathfrak{m},1})$ it follows that $\varphi_{L|K}(\mathfrak{B}_1 \cdots \mathfrak{B}_s) = 1$ and the theorem is proved.

This form of the reciprocity law has the following consequence.

12.9 Proposition. If L/K is an abelian extension and an element α in K^* is a $\mathfrak{p}$-local norm from L for all primes $\mathfrak{p}$ with the possible exception of one particular prime, then α is a local norm at that prime also.

PROOF. Let $\mathfrak{p}_0$ be the exceptional prime. Then $\theta_{\mathfrak{p}}(\alpha) = 1$ for every $\mathfrak{p}$ except $\mathfrak{p}_0$ because $\theta_{\mathfrak{p}}$ is trivial on $\mathfrak{p}$-local norms. Thus

$$1 = \prod_{\mathfrak{p}} \theta_{\mathfrak{p}}(\alpha) = \theta_{\mathfrak{p}_0}(\alpha)$$

and so α is in the kernel of $\theta_{\mathfrak{p}_0}$; α is a local norm at $\mathfrak{p}_0$ also.

We close this section with a few comments on Corollary 12.5. This is the local version of the Artin reciprocity theorem and it can be made the basis of *local class field theory*. We shall not carry out this project but it seems worth the time to at least state the relevant facts. The main theorem classifies all abelian extensions of the local field $K_{\mathfrak{p}}$.

12.10 Theorem. The abelian extensions of $K_\mathfrak{p}$ are in one-to-one (inclusion reversing) correspondence with the subgroups of $K_\mathfrak{p}{}^*$ which contain some $\mathrm{U}_\mathfrak{p}{}^{(a)}$. The abelian extension $L_\mathfrak{P}/K_\mathfrak{p}$ corresponds to the subgroup $\mathrm{N}_\mathfrak{p}(L_\mathfrak{P}{}^*)$. In this case we have $G(L_\mathfrak{P}/K_\mathfrak{p}) \cong K_\mathfrak{p}{}^*/\mathrm{N}_\mathfrak{p}(L_\mathfrak{P}{}^*)$.

The interested reader may want to provide a proof using the global theory. It is necessary to show that the local map $\theta_\mathfrak{p}$ depends only on the extension $L_\mathfrak{P}/K_\mathfrak{p}$ and not upon any algebraic number fields L/K used to obtain them. Further one shows that any abelian extension $L_\mathfrak{P}/K_\mathfrak{p}$ can be obtained by completion of an abelian extension of number fields L/K.

The local theory can be developed on its own without reference to the global results. Consult Serre's article in Cassels–Frohlich, Algebraic Number Theory.

13. THE HILBERT CLASS FIELD

The group $i(K^*)$ of principal ideals is a congruence subgroup defined for the modulus 1 containing no prime divisors. The class field to the ideal group containing $i(K^*)$ is called the *Hilbert class field* (or absolute class field) of K. We shall denote it by $K^{(1)}$.

13.1 Theorem. The Hilbert class field $K^{(1)}$ of K is an abelian unramified extension of K which contains every abelian unramified extension of K. The Galois group $G(K^{(1)}/K)$ is isomorphic to the class group $\mathbf{C}_K$. In particular $(K^{(1)} : K)$ is the class number h_K of K.

PROOF. The conductor $\mathfrak{f}(K^{(1)}/K) = 1$ so no prime of K ramifies in $K^{(1)}$ by Theorem 12.7. Conversely if L/K is abelian and unramified, then $\mathfrak{f}(L/K) = 1$ so

$$\mathbf{H}(K^{(1)}/K) \subseteq \mathbf{H}(L/K).$$

By the Classification Theorem 9.16 $L \subseteq K^{(1)}$. The remaining statements follow from the reciprocity isomorphism of $G(K^{(1)}/K)$ with $\mathbf{I}_K/i(K^*) = \mathbf{C}_K$.

This shows, for example, that a field K with class number 1 cannot have an abelian unramified extension larger than K itself. Of course the case $K = Q$ is covered here but we already know that Q has no unramified extensions abelian or otherwise. There do exist examples of fields with class number 1 which have unramified extensions (non-abelian of course). If L/K is such an example then there cannot be a subfield not equal to K which is abelian over K. This means $G(L/K)$ has no abelian homomorphic image not equal to 1 and $G(L/K)$ is non-solvable. From finite group theory it follows $(L: K) \geqslant 60$.

Natural unramified extensions occur by considering the *class field tower* of

K. Namely $K^{(1)}$ is the Hilbert class field of K and inductively $K^{(i+1)}$ is the Hilbert class field of $K^{(i)}$. Then the chain

$$K \subseteq K^{(1)} \subseteq K^{(2)} \subseteq \cdots$$

is the *class field tower over K*.

It was not known until 1965 that the class field tower could be infinite. The work of Golod and Shafarevich proved that when "enough" primes of Q ramified in K, the class field tower of K would be infinite. A specific example occurs with $K = Q(\sqrt{d})$, where d is the product of eight or more distinct primes.

Notice that $K^{(i)}/K$ is always unramified. It is also easy to see that $K^{(i)}/K$ is a normal extension. In fact a slightly more general statement holds.

13.2 Proposition. Let F be algebraic number field and K/F a normal extension. Then every term $K^{(i)}$ in the class field tower of K is normal over F.

PROOF. If we prove $K^{(1)}/F$ is normal the general result follows. Let σ be isomorphism of $K^{(1)}$ into a field normal over F and assume σ is the identity on F. Then $\sigma(K) = K$ because K/F is normal. Also $\sigma(K^{(1)})$ is an abelian unramified extension of $\sigma(K) = K$ and so $\sigma(K^{(1)}) \subseteq K^{(1)}$ by Theorem 13.1. This proves normality.

One of the most elegant properties of the Hilbert class field is the following.

Principal Ideal Theorem. Every ideal in K becomes principal when extended to an ideal in $K^{(1)}$.

This theorem was conjectured by Hilbert and its proof was reduced to a purely group theoretic problem by Artin after he proved the reciprocity theorem. The group theoretic question was then resolved by Furtwangler.

We shall describe the group theoretic result and then show Artin's reduction.

Let G be a finite group and H a subgroup. Let

$$H\tau_1, \ldots, H\tau_n$$

be the distinct cosets of H in G. For $\sigma \in G$ and each index i, there is a unique element $\phi_i(\sigma)$ in H which satisfies

$$\tau_i \sigma = \phi_i(\sigma)\tau_j \qquad \text{for some} \quad j.$$

The *transfer* from G into H is the function V defined on G with values in H/H' which satisfies

$$V(\sigma) = H'\phi_1(\sigma) \cdots \phi_n(\sigma).$$

Of course H' is the commutator subgroup of H and so H/H' is an abelian group. The product then is independent of the order of the factors. One shows that $V(\sigma)$ is independent of the choice of coset representatives and is a homomorphism from G into H/H'. In case H is abelian, V is a homomorphism into H.

The transfer map is often used to prove the existence of normal subgroups of G. Here one selects a suitable H and shows $V(G) \neq 1$. Then $\ker V \neq G$. The application to the Hilbert class field however is different. We shall select $H = G'$ and it is required to show $V(G) = 1$. Because of the connection to the principal ideal theorem this result has been called the *Principal Ideal Theorem of Group Theory*.

Theorem. If G is a finite group then the transfer from G into G' is the trivial map.

For a proof see Artin–Tate, Class Field Theory, Chapter 13.

Now to apply this result to the Hilbert class field we use the first two terms above K in the class field tower:

$$K \subseteq K^{(1)} \subseteq K^{(2)}.$$

Let $G = G(K^{(2)}/K)$. Since $K^{(2)}/K$ is unramified, every abelian extension of K inside $K^{(2)}$ must in fact be inside $K^{(1)}$. Thus $K^{(1)}/K$ is the maximal abelian extension of K and

$$G(K^{(2)}/K^{(1)}) = G', \qquad G(K^{(1)}/K) = G/G'.$$

We shall use the following notation: If $\mathfrak{A}$ is an ideal in $\mathbf{I}_K$, then $\mathfrak{A}^{(1)}$ denotes the ideal $\mathfrak{A}$ extended to $\mathbf{I}_{K^{(1)}}$; $[\mathfrak{A}]$ denotes the image of $\mathfrak{A}$ in

$$\mathbf{I}_K/i(K^*) = \mathbf{C}_K$$

and $[\mathfrak{A}^{(1)}]$ is the class of $\mathfrak{A}^{(1)}$ in $\mathbf{C}_{K^{(1)}}$.

The principal ideal theorem states in this notation $[\mathfrak{A}^{(1)}] = 1$ for every $\mathfrak{A}$ in $\mathbf{I}_K$.

Now bring in the Artin maps: $\varphi_{K^{(1)}|K}$ maps $\mathbf{I}_K$ onto G/G' with $\ker i(K^*)$ and so induces an isomorphism

$$\varphi_1 : \mathbf{C}_K \cong G/G'.$$

$\varphi_{K^{(2)}|K^{(1)}}$ in the same way induces an isomorphism

$$\varphi_2 : \mathbf{C}_{K^{(1)}} \cong G'.$$

This gives rise to a chain of maps

$$(*) \qquad G \longrightarrow \frac{G}{G'} \xrightarrow{\varphi_1{}^{-1}} \mathbf{C}_K \longrightarrow \mathbf{C}_{K^{(1)}} \xrightarrow{\varphi_2} G'$$

where the first arrow is the natural projection of G onto G/G' and the third arrow is the map $[\mathfrak{A}] \to [\mathfrak{A}^{(1)}]$.

We now verify that the composite map is the transfer from G into G'.

For any σ in G there is a prime $\mathfrak{p}$ of K such that $\varphi_1[\mathfrak{p}] = G'\sigma$ (by Theorem 10.3). Let the factorization of $\mathfrak{p}$ in $K^{(2)}$ be

$$\mathfrak{p} = \mathfrak{P}_1 \cdots \mathfrak{P}_g.$$

We may suppose

$$\sigma = \left[\frac{K^{(2)}|K}{\mathfrak{P}_1}\right].$$

Now use Proposition 2.8 of Chapter III to describe the factorization of $\mathfrak{p}$ in $K^{(1)}$. The factors of $\mathfrak{p}^{(1)}$ are in one-to-one correspondence with the cycles of σ on the cosets of G'. The length of the cycle is the relative degree of the corresponding factor. In this case all the relative degrees are equal (to f, say) so the cosets of G' can be described as

$$\begin{matrix} G'\tau_1, G'\tau_1\sigma, \ldots, G'\tau_1\sigma^{f-1} \\ \vdots \quad\quad \vdots \quad\quad\quad \vdots \\ G'\tau_t, G'\tau_t\sigma, \ldots, G'\tau_t\sigma^{f-1} \end{matrix}$$

The prime factors of $\mathfrak{p}^{(1)}$ are $\mathfrak{q}_j = \tau_j(\mathfrak{P}_1) \cap K^{(1)}$ and

$$\mathfrak{p}^{(1)} = \mathfrak{q}_1 \cdots \mathfrak{q}_t.$$

Notice that

$$\left[\frac{K^{(2)}|K^{(1)}}{\mathfrak{p}_1}\right] = \left[\frac{K^{(2)}|K}{\mathfrak{p}_1}\right]^f = \sigma^f$$

so $\varphi_{K^{(2)}|K^{(1)}}(\mathfrak{q}_j) = \tau_j \varphi_{K^{(2)}|K^{(1)}}(\mathfrak{q}_1)\tau_j^{-1} = \tau_j\sigma^f\tau_j^{-1}$. Finally we can evaluate the composition of the maps above:

$$\sigma \to G'\sigma \to [\mathfrak{p}] \to [\mathfrak{p}^{(1)}] = [\mathfrak{q}_1 \cdots \mathfrak{q}_t] \to \prod \tau_j\sigma^f\tau_j^{-1}.$$

A brief check of the coset representatives shows this product is precisely the value $V(\sigma)$ with V the transfer into G'. By the principal ideal theorem of group theory this value $V(\sigma) = 1$. An inspection of the maps in (*) shows the first two maps are onto the indicated groups and the last is an isomorphism. Since the composite is trivial, the map $[\mathfrak{A}] \to [\mathfrak{A}^{(1)}]$ must be trivial. This completes the proof.

The close connection between the Hilbert class field and the class number can be useful in many situations. We describe here a rather simple one.

13.3 Theorem. Let E be a finite dimensional extension of K such that $E \cap K^{(1)} = K$. Then h_K divides h_E.

PROOF. The extension $EK^{(1)}$ of E is abelian and unramified and so is contained in $E^{(1)}$. Thus $h_K = (K^{(1)} : K) = (EK^{(1)} : E)$ divides $h_E = (E^{(1)} : E)$.

One case where this result can be applied occurs if there is a prime of K which is totally ramified in E. In this case every subfield of E which is strictly larger than K is ramified over K and so $E \cap K^{(1)} = K$ since $E \cap K^{(1)}$ is unramified over K.

If β is a primitive p^nth root of unity for some prime integer p, then the prime p is totally ramified in $Q(\beta)$. If

$$Q \subseteq K \subseteq E \subseteq Q(\beta)$$

then a divisor $\mathfrak{P}$ of p in K is totally ramified in E so h_K divides h_E.

Chapter VI

APPLICATION OF THE GENERAL THEORY TO QUADRATIC FIELDS

This chapter contains examples to illustrate the ideas of the earlier chapters. We are mainly concerned with $K = Q(\sqrt{d})$ and its class number, Hilbert class field, etc. The case $d = -5$ is worked out in detail. We use local Artin maps to make some calculations that could also be made in other ways but our purpose is mainly illustration rather than the end result of the calculation. One example is given with a cubic field.

1. THE CONDUCTOR OF $Q(\sqrt{d})$

Let d be a square free integer and $K = Q(\sqrt{d})$. We shall compute $\mathfrak{f}(K/Q)$ using Corollary 12.6 of Chapter V. The case for $\mathfrak{p}_\infty$ can be settled at once because $\mathfrak{p}_\infty$ is ramified if and only if $d < 0$. Thus $\mathfrak{p}_\infty$ divides the conductor if and only if $d < 0$.

Now let p be a prime integer; Q_p, $\mathbf{U}_p$ as usual denote the completion of Q at p and the group of units in Q_p, respectively.

It is necessary to determine the p-local norms from $K_\mathfrak{p}$ in Q_p. (We write $K_\mathfrak{p}$ for the completion of K at some prime dividing p.) Since we are only interested in those p which ramify in K we have $(K_\mathfrak{p} : Q_p) = 2$. This means that all squares in Q_p are norms. To see what is left we consider $\mathbf{U}_p/\mathbf{U}_p{}^2$. By the calculation in Section 9, Chapter V we have

$$[\mathbf{U}_p : \mathbf{U}_p{}^2] = 2[Z_p : 2Z_p] = \begin{cases} 2 \text{ if } p \text{ odd} \\ 4 \text{ if } p = 2. \end{cases}$$

We consider p *odd* first.

Let u be an integer prime to p and not congruent to a square modulo p. Such a u exists for p odd because the units in $Z/(p)$ have order $p-1$ (even) and so not all elements can be squares. Then the multiplicative group generated by u and $\mathbf{U}_p^2$ is larger than $\mathbf{U}_p^2$ and

$$\mathbf{U}_p = \langle u, \mathbf{U}_p^2 \rangle.$$

Now let $w \in \mathbf{U}_p^{(1)}$. Then either $w = v^2$ or $w = uv^2$ for some v in $\mathbf{U}_p$. In the second case we have

$$1 \equiv w \equiv uv^2 \bmod p$$

and so u is congruent to a square mod p.

This is impossible (by choice) so w is a square. This shows $\mathbf{U}_p^{(1)}$ is contained in the group of p-local norms. Since every ramified prime divides the conductor we have the following:

For odd p, the p-local conductor is $\mathfrak{f}_p(K/Q) = p$.

Now we take $p = 2$. In this case $\mathbf{U}_2^2$ has index 4 in U_2 and we assert

$$\mathbf{U}_2 = \langle 3, 5, \mathbf{U}_2^2 \rangle.$$

To verify this one observes that all squares of units are $\equiv 1 \bmod 8$ and that $3 \not\equiv 5 \bmod 8$ implies $3 \notin \langle 5, \mathbf{U}_2^2 \rangle$.

It is necessary to determine for which $a\mathbf{U}_2^{(a)}$ contained in the group of 2-local norms. If $w \in \mathbf{U}_2^{(3)}$ (so $w \equiv 1 \bmod 8$) then

$$w = 3^i 5^j v^2, \qquad v \in \mathbf{U}_2, \quad i,j = 0 \text{ or } 1.$$

Examine the possible cases with various i,j and read this as a congruence modulo 8. It turns out that $i = j = 0$ is the only possibility. Thus $\mathbf{U}_2^{(3)}$ consists of squares and so these are 2-local norms.

Notice that $\mathbf{U}_2 = \mathbf{U}_2^{(1)}$ so $\mathbf{U}_2^{(1)}$ is contained in the norm group only when 2 is not ramified. We are left to decide in the ramified case whether $\mathbf{U}_2^{(2)}$ is contained in the norm group or not.

We must have

$$\mathbf{U}_2^{(2)} = \langle 5, \mathbf{U}_2^2 \rangle$$

and so the 2-local conductor is 2^2 if and only if 5 is a norm.

The field K_2 consists of all elements $x + y\sqrt{d}$ with x, y in Q_2 so 5 is a norm if and only if

$$\mathrm{N}_2(x + y\sqrt{d}) = x^2 - dy^2 = 5$$

can be solved with $x, y \in Q_2$. Since all squares are norms this is equivalent to solving

$$x^2 - dy^2 = 5z^2.$$

But this implies dy^2 is a norm from $Q_2(\sqrt{5})$; that is:

$$\mathrm{N}(x+z\sqrt{5}) = dy^2.$$

This is equivalent, with the condition that d is a norm from $Q_2(\sqrt{5})$.

The norms from this extension are easily determined because 2 is not ramified. This means that every unit is the norm of a unit in $Q_2(\sqrt{5})$ by Proposition 3.11 of Chapter V. Furthermore the index in $Q_2{}^*$ of the norm group is 2 by Corollary 3.15, Chapter V so the group of norms from $Q_2(\sqrt{5})$ is precisely $\langle 2^2\rangle U_2$.

This means the 2-local conductor of K/Q is 2^2 if and only if $d \in \mathbf{U}_2$ (d is square free). In other words d must be odd. If d is even the 2-local conductor is 2^3. (All of this is under the assumption that 2 is ramified in K.)

Recall (from Chapter I) that 2 is ramified if and only if $d \equiv 2, 3 \bmod 4$ and the odd ramified primes are the divisors of d.

Now combine the above remarks and examine the powers of the prime divisors of d and $\mathfrak{f}(K/Q)$ to get the result in the following form.

1.1 Theorem. The conductor of $Q(\sqrt{d})/Q$ is

$$\begin{aligned} \mathfrak{f} = (d) \qquad & d > 0 \qquad d \equiv 1 \bmod 4,\\ (d)\,\mathfrak{p}_\infty \qquad & d < 0 \qquad d \equiv 1 \bmod 4,\\ (4d) \qquad & d > 0 \qquad d \equiv 2, 3 \bmod 4,\\ (4d)\,\mathfrak{p}_\infty \qquad & d < 0 \qquad d \equiv 2, 3 \bmod 4. \end{aligned}$$

We know by the Kronecker–Weber Theorem that $Q(\sqrt{d})$ must be contained in $Q(\beta_n)$ for some primitive nth root of unity β_n. The least n that will suffice is the one such that $Q(\beta_n)$ has the same conductor as $Q(\sqrt{d})$.

1.2 Corollary. $Q(\sqrt{d}) \subseteq Q(\beta_D)$. Here $D = |d|$ if $d \equiv 1 \bmod 4$ and $D = |4d|$ if $d \equiv 2, 3 \bmod 4$.

PROOF. With the indicated D we have $D > 0$ and the conductor of $Q(\sqrt{d})/Q$ is (D) or $(D)\,\mathfrak{p}_\infty$. By Proposition 3.3, Chapter III, the kernel of the Artin map for $Q(\beta_D)/Q$ is the congruence subgroup for the modulus $(D)\,\mathfrak{p}_\infty$ and so

$$\mathbf{H}(Q(\beta_D)/Q) \subseteq \mathbf{H}(Q(\sqrt{d})/Q).$$

The result follows by the Classification Theorem 9.16, Chapter V.

2. TWO EXAMPLES

EXAMPLE 1. Let $K = Q(\sqrt{-5})$. We shall prove $K^{(1)} = K(\sqrt{-1})$ and $K^{(2)} = K^{(1)}$. The decomposition of rational primes in $K^{(1)}$ is also described by explicitly giving the decomposition groups.

The first step is to show $h_K = 2$. The Minkowski constant is small here and by Theorem 11.19, Chapter I, every ideal class $[\mathfrak{B}]$ in $\mathbf{C}_K$ contains an ideal $\mathfrak{B}_1$ with $\mathrm{N}_{K|Q}(\mathfrak{B}_1) \leqslant 2$. The prime 2 is ramified in K so $(2) = \mathfrak{P}^2$ and $\mathfrak{P}$ is the only nontrivial integral ideal having norm $\leqslant 2$. This implies at once $h_K \leqslant 2$ since $[\mathfrak{P}]$ is the only possible nontrivial class. By Exercise 2, Section 4, Chapter I, h_K is not 1 so $h_K = 2$.

This gives dimension formulas

$$(K^{(1)} : K) = 2, \qquad (K^{(1)} : Q) = 4.$$

Since $K^{(1)}/Q$ is normal it has an abelian Galois group—all groups of order 4 are abelian. Thus $K^{(1)}$ is contained in some cyclotomic extension of Q (by Kronecker–Weber) and to know which one we calculate the conductor. It turns out that $\mathfrak{f}(K^{(1)}/Q) = \mathfrak{f}(K/Q)$ and the following argument proves this in a more general setting.

2.1 Lemma. Let F be an algebraic number field, K/F an abelian extension and E a subfield of $K^{(1)}$ which contains K and is abelian over F. Then $\mathfrak{f}(E/F) = \mathfrak{f}(K/F)$.

PROOF. Let $\mathfrak{p}$ be a prime of F and $\mathfrak{P}$ a divisor of $\mathfrak{p}$ in K. Since E/K is unramified, every $\mathfrak{P}$-adic unit in $K_{\mathfrak{P}}$ is the local norm of a unit in $E_{\mathfrak{P}}$ (the completion of E at some prime over $\mathfrak{P}$) by Proposition 3.11 of Chapter V. Now apply Corollary 13.6, Chapter V to calculate the $\mathfrak{p}$-local conductors. The group $\mathbf{U}_{\mathfrak{p}}^{(b)}$ in $F_{\mathfrak{p}}$ consists of local norms from K if and only if it consists of local norms from E. Thus the same power of $\mathfrak{p}$ divides both conductors and they are equal.

In our example the conductor is $(20)\,\mathfrak{p}_\infty$ so

$$K^{(1)} \subseteq Q(\beta_{20}).$$

This field $Q(\beta_{20})$ has dimension 8 over Q so it would be easy to list all the subfields and find $K^{(1)}$. Rather than do this, we take a more systematic approach that applies in more general circumstances.

The 20th root of unity, β_{20}, is a product $\beta_4\beta_5$ and the Galois group G of $Q(\beta_{20})$ is the direct product of the groups of $Q(\beta_4)$ and $Q(\beta_5)$ over Q. We may write

$$G = \langle\tau\rangle \times \langle\sigma\rangle$$

$$\tau(\beta_4) = \beta_4^{-1} \qquad \sigma(\beta_4) = \beta_4$$

$$\tau(\beta_5) = \beta_5 \qquad \sigma(\beta_5) = \beta_5{}^2.$$

Then $\tau^2 = \sigma^4 = 1$.

We compute the group fixing K. We have

$$\sqrt{-5} = \beta_4 \sqrt{5},$$

where β_4 is a suitable 4th root, and $\sqrt{5} \in Q(\beta_5)$. Thus σ and τ both send $\sqrt{-5}$ to $-\sqrt{-5}$ and $\sigma\tau$ fixes K. Since $\langle \sigma\tau \rangle$ has order 4 it is the full group fixing K. Now $K^{(1)}$ must be left fixed by the subgroup of index 2 in $\langle \sigma\tau \rangle$; this is $\langle \sigma^2\tau^2 \rangle = \langle \sigma^2 \rangle$. Since σ^2 fixes both K and β_4 (which we now write as i) it follows that

$$K^{(1)} = K(i) = Q(\sqrt{-5}, i).$$

It is not difficult to work out the decomposition of primes in $K^{(1)}$. We consider a few cases. The procedure (for unramified primes) is to evaluate the Artin map. This is done most conveniently by the formula

$$\varphi_{K^{(1)}|Q}(\mathfrak{p}) = \operatorname{res} \varphi_{Q(\beta_{20})|Q}(\mathfrak{p})$$

where "res" means restriction to $K^{(1)}$. The Artin map for the extension $Q(\beta_{20})/Q$ is easily evaluated and has been done in Section 3, Chapter III.

A prime p of Q splits completely in $K^{(1)}$ if

$$\operatorname{res} \varphi_{Q(\beta_{20})|Q}(p) = 1;$$

that is, if the automorphism $\beta \to \beta^p$ is the identity on $K^{(1)}$ where $\beta = \beta_{20}$. This occurs precisely when either

$$\beta^p = \beta \quad \text{and} \quad p = 20k + 1, \qquad \text{or} \quad \beta^p = \sigma^2(\beta).$$

This means

$$\beta_4{}^p = \beta_4 \quad \text{and} \quad \beta_5{}^p = \beta_5{}^4.$$

Thus $p \equiv 1 \bmod 4$ and $p \equiv 4 \bmod 5$. Equivalently $p = 20k + 9$.

In a similar way one determines the primes which are unramified and have a specific decomposition group. The full Galois group of $K^{(1)}/Q$ is

$$\langle \tau, \sigma \rangle | K^{(1)} \cong \langle \tau, \sigma \rangle / \langle \sigma^2 \rangle.$$

This has three subgroups of order 2 generated by

$$\tau | K^{(1)}, \qquad \sigma | K^{(1)}, \qquad \tau\sigma | K^{(1)}.$$

A prime p has decomposition group generated by τ if the automorphism $\beta \to \beta^p$ is the same as τ or $\tau\sigma^2$. It must happen that

$$\beta_4{}^p = \beta_4{}^{-1} \qquad \text{and} \quad \beta_5{}^p = \beta_5 \quad \text{or} \quad \beta_5^{-1}.$$

Thus $p \equiv -1 \bmod 4$ and $p \equiv \pm 1 \bmod 5$. Equivalently p has one of the forms $20t + 11$ or $20t + 19$. The remaining cases are summarized in the table below.

The decomposition groups for the ramified primes 2, 5 can be discussed in a similar way except the p-local Artin maps must be used for $p = 2, 5$ in place of the global Artin map. Just as above the p-local Artin map for the extension $K^{(1)}/Q$ is the restriction to $K^{(1)}$ of the p-local map for $Q(\beta)/Q$. Let θ_2, θ_5 denote these maps; we have to compute the restrictions to $K^{(1)}$ of

$$\theta_2(Q_2{}^*) \quad \text{and} \quad \theta_5(Q_5{}^*).$$

Take $p = 2$ first. The 2-conductor for $Q(\beta)/Q$ is 2^2 so

$$\theta_2(\mathbf{U}_2{}^{(2)}) = 1.$$

The group $Q_2{}^*/\mathbf{U}_2{}^{(2)}$ is generated by the cosets containing 2 and 3 so the decomposition group of (2) in $Q(\beta)$ is

$$G(2) = \langle \theta_2(2), \theta_2(3) \rangle.$$

The procedure in Section 12, Chapter V shows how to evaluate these maps. Namely $\theta_2(2) = \varphi ji(y)$ where

$$2^{-1}y \equiv 1 \bmod 4$$
$$y \equiv 1 \bmod 5\mathfrak{p}_\infty.$$

The choice $y = 26$ (or $40t-14$) will do and $\varphi ji(26) = \varphi(13)$ is the map $\beta \to \beta^{13} = \beta_4 \beta_5{}^3$. Thus

$$\theta_2(2) = \sigma^3.$$

Similarly $\theta_2(3) = \varphi ji(51) = \tau$.

Thus $G(2) = \langle \tau, \sigma^3 \rangle$ is the full Galois group and on $K^{(1)}$ the decomposition group of (2) has order 4. We know 2 has ramification number $2 = e$ in K and so also in $K^{(1)}$. Thus (2) has one prime factor in $K^{(1)}$ with relative degree $f = 2$. This holds of course because the decomposition group has order $ef = 4$.

In a similar way $G(5) = \langle \tau\sigma^3 \rangle$. We summarize the decomposition theory of primes in $K^{(1)}$ in the table.

Rational prime	Decomposition group	e, f
$20t+1$ $20t+9$	1	1, 1
$20t+3$ $20t+7$	$\langle \tau \rangle \mid K^{(1)}$	1, 2
$20t+11$ $20t+19$	$\langle \tau \rangle \mid K^{(1)}$	1, 2
$20t+13$ $20t+17$	$\langle \sigma \rangle \mid K^{(1)}$	1, 2
2	$\langle \tau, \sigma \rangle \mid K^{(1)}$	2, 2
5	$\langle \tau\sigma^3 \rangle \mid K^{(1)}$	2, 1

We close this example by showing $K^{(1)}$ has class number one and so $K^{(2)} = K^{(1)}$. The discriminant of $K^{(1)}$ can be estimated. Use the elements

$$1,\ i,\ \frac{1+\sqrt{5}}{2},\ \frac{1+\sqrt{5}}{2}i$$

as a free Z-basis and find the discriminant of this basis is $-2^4 5^2$. The Minkowski constant can be calculated using this for Δ. One finds every ideal class for $K^{(1)}$ contains an ideal whose norm is $\leqslant 3$. It is necessary to look at the factors of (2) and (3). We saw above that both (2) and (3) are divisible by primes with relative degree 2 so their norms are 2^2, 3^2. So there are no ideals with norm $\leqslant 3$ except the unit ideal. Hence the class group is trivial for $K^{(1)}$.

EXAMPLE 2. By way of contrast to the previous example we consider the non-normal extension of Q, $K = Q(\theta)$, $\theta^3 = 11$. In Section 11, Chapter I we saw $h_K = 2$ and that (2) was the product $\mathfrak{P}_2 \mathfrak{P}_2'$ of two primes in K. Moreover $\mathfrak{P}_2$ was not principal but

$$\mathfrak{P}_2{}^2 = (\theta^2 - 5).$$

The group of units in K is $\langle \pm 1 \rangle \times \langle u \rangle$ where we may assume $u > 0$ (and θ is the real cube root).

The problem of describing $K^{(1)}$ is not conveniently solved by using the class field theory of Q because K is not normal over Q. The problem is solved by appealing directly to the technique used in the proof of the existence theorem. Since $K^{(1)}$ has dimension 2 over K it is a Kummer 2-extension and $K^{(1)}$ can be found as a subfield of the extension of K obtained the square roots of all the S-units for a suitable S. The requirements in Section 9, Chapter V, placed upon S are satisfied if we take

$$S = \{\mathfrak{P}_\infty, \mathfrak{P}_\infty', \mathfrak{P}_2, \mathfrak{P}_2'\}$$

with $\mathfrak{P}_\infty, \mathfrak{P}_\infty'$ denoting the two infinite primes. The S units can be found now. Suppose $\alpha \in K^*$ and (α) is divisible only by primes in S. For suitable integers a and b, (α) has the factorization

$$(\alpha) = \mathfrak{P}_2{}^a(\mathfrak{P}_2')^b = (\mathfrak{P}_2 \mathfrak{P}_2')^b \mathfrak{P}_2^{a-b} = (2^b)\, \mathfrak{P}_2^{a-b}.$$

Only even powers of $\mathfrak{P}_2$ are principal, so $a - b = 2k$ and

$$(\alpha) = (2^b)(\theta^2 - 5)^k.$$

This implies α is a unit times $2^b(\theta^2 - 5)^k$. It follows

$$K^S = \langle \pm 1 \rangle \times \langle u \rangle \times \langle 2 \rangle \times \langle \theta^2 - 5 \rangle.$$

If we adjoin to K a root of $X^2-\alpha$ with α a product of distinct generators listed above then $\mathfrak{P}_2$ will ramify whenever 2 or θ^2-5 divides α. To get an unramified extension it is necessary to take $\alpha = \pm u$. But K has a real prime so we select $\alpha = u > 0$ to prevent ramification at infinity. Finally then

$$K^{(1)} = K(\sqrt{u}).$$

The decomposition of primes in $K^{(1)}$ is completely described by their orders in $\mathbf{C}_K$ (as is always the case for any K in $K^{(1)}$). That is, a prime $\mathfrak{P}$ of K is divisible by a prime of $K^{(1)}$ having relative degree 2 if and only if $[\mathfrak{P}]$ has order 2 in $\mathbf{C}_K$. For example $\mathfrak{P}_2$ remains prime in $K^{(1)}$.

3. THE EXTENDED CLASS GROUP

It is possible that an algebraic number field K has class number $h_K = 1$ and still has an abelian extension that is unramified at all finite primes. For example take $K = Q(\sqrt{3})$ and consider the extension $K(\sqrt{-1})$. This is easily seen to be unramified except at the real primes of K.

We consider for any algebraic number field K, the modulus $\mathfrak{m}$ = product of all the real primes of K. Set $\mathfrak{m} = 1$ if there are no real primes. In the real case the elements of $K_{\mathfrak{m},1}$ are called *totally positive* because they map onto positive elements in every real imbedding of K. The *extended class group* is the quotient

$$\mathbf{C}_K{}^+ = \mathbf{I}_K/i(K_{\mathfrak{m},1});$$

the ideals modulo the principal ideals with a totally positive generator.

The congruence subgroup $i(K_{\mathfrak{m},1})$ belongs to an ideal group with conductor $\mathfrak{m}$. The class field to this ideal group is called the *extended Hilbert class field* and is denoted by $K^{(+)}$.

The inclusion $i(K^*) \supseteq (K_{\mathfrak{m},1})$ implies $K^{(1)} \subseteq K^{(+)}$. The difference between these two fields can be measured in terms of K.

3.1 Theorem. Let r be the number of real primes of K; $\mathbf{U}_K, \mathbf{U}_K{}^+$ the group of units and totally positive units of K respectively. Then

$$(K^{(+)} : K^{(1)}) = 2^r[\mathbf{U}_K : \mathbf{U}_K{}^+]^{-1}.$$

PROOF. The Artin maps induce isomorphisms

$$G(K^{(+)}/K) \cong \mathbf{I}_K/i(K_{\mathfrak{m},1}),$$

$$G(K^{(1)}/K) \cong \mathbf{I}_K/i(K^*)$$

and so $G(K^{(+)}/K^{(1)}) \cong i(K^*)/i(K_{\mathfrak{m},1})$.

Now consider the diagram below to compute the order of this group.

$$\begin{array}{ccccccccc}
1 & \to & \mathbf{U}_K{}^+ & \to & K_{\mathfrak{m},1} & \to & i(K_{\mathfrak{m},1}) & \to & 1 \\
 & & \downarrow & & \downarrow & & \downarrow & & \\
1 & \to & \mathbf{U}_K & \to & K^* & \to & i(K^*) & \to & 1 \\
 & & \downarrow & & \downarrow & & \downarrow & & \\
1 & \to & \dfrac{\mathbf{U}_K}{\mathbf{U}_K{}^+} & \to & \dfrac{K^*}{K_{\mathfrak{m},1}} & \to & \dfrac{i(K^*)}{i(K_{\mathfrak{m},1})} & \to & 1
\end{array}$$

The bottom row is exact so it follows

$$(K^{(+)} : K^{(1)}) = [K^* : K_{\mathfrak{m},1}][\mathbf{U}_K : \mathbf{U}_K{}^+]^{-1}.$$

To finish the proof we need to compute $[K^* : K_{\mathfrak{m},1}]$. Let $e_i(\alpha)$ denote the sign of α at the ith real imbedding of K. Then

$$\alpha \to (e_1(\alpha), \ldots, e_r(\alpha))$$

is a homomorphism of K^* to a group of order 2^r. The map is onto by the approximation theorem and the kernel is precisely $K_{\mathfrak{m},1}$ so $[K^* : K_{\mathfrak{m},1}] = 2^r$ as required.

In case $K = Q(\sqrt{d})$ there is a criterion that is quite easy to state although in practice it may be difficult to apply.

If $d > 0$ then $\mathbf{U}_K$ has the form

$$\mathbf{U}_K = \langle \pm 1 \rangle \times \langle u \rangle$$

where we may assume $u > 0$.

3.2 Theorem. Let $K = Q(\sqrt{d})$, d square free. Then

$$\begin{aligned}
K^{(+)} &= K^{(1)} && \text{if } d < 0 \text{ or} \\
& && \phantom{\text{if }} d > 0 \text{ and } \mathrm{N}_{K|Q}(u) = -1 \\
(K^{(+)} : K^{(1)}) &= 2 && \text{if } d > 0 \text{ and } \mathrm{N}_{K|Q}(u) = 1.
\end{aligned}$$

Proof. The case $d < 0$ requires no argument. For $d > 0$ the dimension $(K^{(+)} : K^{(1)})$ is 1 or 2 depending upon the two cases $\mathbf{U}_K{}^+ = \langle u^2 \rangle$ or $\mathbf{U}_K{}^+ = \langle u \rangle$. Evidently u is totally positive if both u and its conjugate $\bar{u}$ are positive. This is equivalent to

$$\mathrm{N}_{K|Q}(u) = u\bar{u} = +1.$$

The alternative $u\bar{u} = -1$ occurs if $\bar{u} < 0$; in other words u^2 generates $U_K{}^+$.

In practice this result may be difficult to apply because there is no general formula to tell us (in terms of d) when -1 is the norm of a unit. It is not difficult to determine when -1 is a norm from $Q(\sqrt{d})$ but this is not sufficient.

For example with $d = 34$ the element $(5+3\sqrt{34})/3$ has norm -1 but no unit has norm -1.

There is a procedure by which u can be found. A method using the continued fraction expression of $\sqrt{d}$ can be found in the book by Borevich–Shafarevich, Number Theory, Chapter 2, Section 7. Also see the exercises following Section 11, Chapter I.

We shall use the close connection between K^+ and $K^{(1)}$ to study the class group of K. In the first example of Section 2 it was very convenient to have $K^{(1)}/Q$ normal and abelian. We always have $K^{(1)}$ and $K^{(+)}$ normal over Q but not necessarily abelian. We consider the next best case—look at the largest subfield which is abelian over Q.

Definition. For any abelian extension K/Q, the *genus field* of K over Q is the largest abelian extension E of Q contained in $K^{(1)}$. The *extended genus field* is the largest abelian extension $E^{(+)}$ of Q contained in $K^{(+)}$.

The main object we want to study is $\mathbf{C}_K$. This is isomorphic to $G(K^{(1)}/K)$ so we look at $K^{(1)}$. The genus field is introduced to describe at least a part of $K^{(1)}$ by using the class field theory of Q. The remaining problem is to describe the part of the extension from the genus field to $K^{(1)}$. The field $K^{(+)}$ is introduced only for technical reasons. It turns out to be slightly easier to describe the extended genus field and from it obtain the genus field.

It is necessary to look closely at a number of groups, subgroups, fields and subfields so we fix some notation that will be used for the rest of this section.

$K = Q(\sqrt{d})$ with d a square free integer and σ is an element of $G^+ = G(K^{(+)}/Q)$ which is not the identity on K; that is, $\sigma|K$ generates $G(K/Q)$. We first describe the extension $K^{(1)}/E$ in terms of K.

3.3 Theorem. (a) $G(K^{(1)}/E) \cong$ subgroup of $\mathbf{C}_K$ generated by the ideal classes of the form $[\sigma(\mathfrak{A})\mathfrak{A}^{-1}]$, $\mathfrak{A} \in \mathbf{I}_K$.

(b) $G(K^{(1)}/E) \cong (\mathbf{C}_K)^2$.

Proof. Let $G = G(K^{(1)}/Q)$ so that E is the field left fixed by the commutator subgroup (G, G) generated by all commutators

$$(u, v) = uvu^{-1}v^{-1}.$$

The Artin map gives an isomorphism

$$\varphi : \mathbf{C}_K \to C \subseteq G$$

in such a way that for each ideal $\mathfrak{A}$ of K we have

$$\varphi[\sigma\mathfrak{A}] = \sigma\varphi[\mathfrak{A}]\sigma^{-1}.$$

It follows that

$$\varphi[\sigma(\mathfrak{A})\mathfrak{A}^{-1}] = (\sigma, a), \qquad a = \varphi[\mathfrak{A}].$$

Thus φ maps the group in part (a) into (G, G) and we must show every commutator in (G, G) has the form (σ, a) with $a \in C$.

We know C is a normal subgroup of G of index 2 and $G/C \cong G(K/Q)$. It follows that every element in G has the form a or σa with $a \in C$. If we use the abelian property of C the following commutator identities can be proved:

$$(\sigma a, a_1) = (\sigma, a_1)$$

$$(\sigma a, \sigma a_1) = (\sigma, a_2), \quad a_2 = \sigma a^{-1} a_1 \sigma^{-1}.$$

It follows that every commutator is in the image of φ and part (a) holds.

To prove (b) we begin with a prime ideal $\mathfrak{P}$ of K and suppose $\mathfrak{p} = \mathfrak{P} \cap Q$. There are three cases to consider:

(i) $\mathfrak{p} = \mathfrak{P}$ so $\mathfrak{P} = \sigma(\mathfrak{P})$ and $[\sigma(\mathfrak{P})\mathfrak{P}^{-1}] = 1$;
(ii) $\mathfrak{p} = \mathfrak{P}^2$ same as above;
(iii) $\mathfrak{p} = \mathfrak{P}\mathfrak{P}_1$, $\mathfrak{P}_1 = \sigma(\mathfrak{P})$, $[\sigma(\mathfrak{P})\mathfrak{P}^{-1}] = [\mathfrak{P}_1 \mathfrak{P}^{-1}] = [\mathfrak{p}\mathfrak{P}^{-2}] = [\mathfrak{P}^{-1}]^2$.

This shows that every class $[\sigma(\mathfrak{A})\mathfrak{A}^{-1}]$ is a square in $\mathbf{C}_K$. Conversely if $[\mathfrak{A}] = [\mathfrak{B}^2]$ then $\sigma(\mathfrak{B})\mathfrak{B}^{-1} = \mathfrak{B}\sigma(\mathfrak{B})\mathfrak{B}^{-2} = \mathrm{N}_{K|Q}(\mathfrak{B})\mathfrak{B}^{-2}$. Since $\mathrm{N}_{K|Q}(\mathfrak{B})$ is principal it follows $[\mathfrak{A}] = [\sigma(\mathfrak{B}^{-1})\mathfrak{B}]$ proving (b). This provides two descriptions of $G(K^{(1)}/E)$. Next we clarify the relation between E and $E^{(+)}$.

3.4 Theorem. $(K^{(+)} : K^{(1)}) = (E^{(+)} : E)$.

Proof. This is proved by purely group theoretical tricks. It is clear that $E = E^{(+)}$ when $K^{(1)} = K^{(+)}$ so we may suppose $(K^{(+)} : K^{(1)}) = 2$. Let

$$G^+ = G(K^{(+)}/Q)$$

$$C^+ = G(K^{(+)}/K)$$

so that $G^+ = \langle \sigma, C^+ \rangle$ and $G = \langle \sigma, C \rangle$. Let τ be the automorphism (of order 2) which leaves $K^{(1)}$ fixed. Then $\tau \in C^+$ and the group $\langle \tau \rangle$ is normal in G^+. This forces τ to commute with every element in G^+. Also the natural restriction map to $K^{(1)}$ establishes

$$G^+/\langle \tau \rangle \cong G$$

$$C^+/\langle \tau \rangle \cong C.$$

If we show $(G^+, G^+) \cong (G, G)$ then it follows that

$$(K^{(+)} : E^{(+)}) = (K^{(1)} : E),$$

which is enough to prove the result.

The arguments above show $(G^+, G^+) = (\sigma, C^+)$ using only the fact that C^+ is abelian with index 2. One can verify the identity $(\sigma, a_1 a_2) = (\sigma, a_1)(\sigma, a_2)$ for a_1, a_2 in C^+ so the function $a \to (\sigma, a)$ is a homomorphism of C^+ onto

(G^+, G^+). Under this map the image of τ is $(\sigma, \tau) = 1$ because τ is in the center and so the image can be identified with the image $C^+/\langle\tau\rangle \cong C \to (\sigma, C) = (G, G)$ and the result follows.

Now we show how to construct $E^{(+)}$.

3.5 Lemma. Except possibly for infinite prime factors we have $\mathfrak{f}(E^{(+)}/Q) = \mathfrak{f}(K/Q)$.

This follows from the proof of Lemma 2.1.

Let D be the positive integer such that the "finite" part of $\mathfrak{f}(K/Q)$ is (D). Thus

$$D = |d| \quad d \equiv 1 \bmod 4$$

$$D = 4|d| \quad d \equiv 2, 3 \bmod 4.$$

Let β be a primitive Dth root of unity.

3.6 Lemma. $E^{(+)}$ is the largest subfield of $Q(\beta)$ which contains K and has the property that for each prime p of Q the ramification number of p is the same in K as it is in $E^{(+)}$.

Proof. By the conductor calculation and the classification theorem we find $E^{(+)} \subseteq Q(\beta)$. On the other hand a field F having the same ramification for each p as K must be unramified over K at every finite prime. This implies $\mathfrak{f}(F/K)$ contains only real primes and so $\mathfrak{f}(F/K) | \mathfrak{f}(E^{(+)}/K)$. It follows $F \subseteq E^{(+)}$.

Let us factor D as

$$D = p_1 p_2 \cdots p_t$$

where each p_i is an odd prime when D is odd and when D is even p_1 is either 2^2 or 2^3 and $p_2, \ldots, p_t$ are odd primes. Let β_{p_i} denote a primitive p_ith root of unity. We may select these roots so that

$$\beta = \prod_{p|D} \beta_p.$$

Let Γ denote the Galois group of $Q(\beta)/Q$ and Γ_p the Galois group of $Q(\beta_p)$ over Q. There is a natural isomorphism

$$\Gamma \cong \prod \Gamma_p$$

which may be described explicitly as follows. Each element in Γ is described by an automorphism

$$\beta \to \beta^r \quad (r, D) = 1$$

and we let this correspond to the same map on each $Q(\beta_p)$. This is then a map onto the direct product because for any map

$$\tau: \beta_p \to \beta_p^{\,r}$$

in Γ_p there is an integer y such that

$$y \equiv r \bmod p$$

$$y \equiv 1 \bmod q \qquad q|D,\ q \neq p.$$

Then $\beta \to \beta^y$ maps onto τ.

We shall regard the isomorphism above as an identification. The key idea of this section is to describe the ramification of a prime p in a subfield of $Q(\beta)$ by using the Γ_p.

3.7 Theorem. Let F be a subfield of $Q(\beta)$ and p a prime divisor of D. The ramification index of p in F is the number

$$[\Gamma_p : \Gamma_p \cap \Gamma_F]$$

where Γ_F is the subgroup of Γ which leaves F fixed.

PROOF. Let θ_p denote the p-local Artin map for the extension $Q(\beta)/Q$ and let

$$\text{res} : \Gamma \to \Gamma/\Gamma_F$$

denote the restriction homomorphism onto $G(F/Q)$. Then $\text{res}\,\theta_p$ is the local Artin map for F/Q and

$$|\text{res}\,\theta_p(Q_p{}^*)| = ef$$

with e and f the ramification index and relative degree, respectively, of a prime divisor of p in F.

We know $Q_p{}^* = \langle p \rangle \times \mathbf{U}_p$ and the order of $\text{res}\,\theta_p(p)$ is f because the smallest power of p that is a local norm from F_p is p^f. It follows that

$$|\text{res}\,\theta_p(\mathbf{U}_p)| = e.$$

Now we prove $\theta_p(\mathbf{U}_p) = \Gamma_p$. Once this is done we then have

$$e = |\text{res}\,\theta_p(\mathbf{U}_p)| = |\text{res}\,(\Gamma_p)| = [\Gamma_p : \Gamma_p \cap \Gamma_F]$$

to complete the proof.

Take $u \in \mathbf{U}_p$ and write $D = pD_0$. We evaluate $\theta_p(u)$ by first selecting an element y (an integer, say) such that

$$y \equiv u \bmod p$$

$$y \equiv 1 \bmod D_0$$

$$y > 0$$

and then $\theta_p(u) = \varphi(y)$, $\varphi =$ global Artin map for $Q(\beta)/Q$. This is the map $\beta \to \beta^y$ and because of the congruences satisfied by y we see $\varphi(y) \in \Gamma_p$. Conversely for any map τ in Γ_p there is an integer y such that $\tau(\beta_p) = \beta_p{}^y$ and

$y \equiv 1 \bmod D_0$. Thus $\theta_p(y) = \varphi(y) = \tau$. This proves $\theta_p(\mathbf{U}_p) = \Gamma_p$ and finishes the proof of the theorem.

This result will help us determine $E^{(+)}$. We have Γ_K, the subgroup of Γ fixing K and the unknown group Γ_0 which is the group fixing $E^{(+)}$. We translate Lemma 3.6 into group theoretic language as

$$[\Gamma_p : \Gamma_p \cap \Gamma_K] = [\Gamma_p : \Gamma_p \cap \Gamma_0].$$

Since $K \subseteq E^{(+)}$, it holds that $\Gamma_0 \subseteq \Gamma_K$ so in fact

$$\Gamma_p \cap \Gamma_K = \Gamma_p \cap \Gamma_0.$$

Now $E^{(+)}$ is the largest field for which this holds so Γ_0 must be the smallest subgroup of Γ for which this holds. Evidently Γ_0 must contain each intersection $\Gamma_p \cap \Gamma_K$ and so the minimal property insures that

$$\Gamma_0 = \prod_{p|D} \Gamma_p \cap \Gamma_K.$$

This is a direct product because $\prod \Gamma_p$ is a direct product. From this we get next

$$G(E^{(+)}/Q) = \Gamma/\Gamma_0 = \prod_{p|D} \frac{\Gamma_p}{\Gamma_p \cap \Gamma_K}.$$

The number $[\Gamma_p : \Gamma_p \cap \Gamma_K]$ is the ramification index of p in K—namely 2. We have proved so far the following:

3.8 Theorem. $(E^{(+)} : Q) = 2^t$, where t is the number of prime divisors of the discriminant of K/Q.

There is a nice consequence of this calculation.

3.9 Theorem. Let $K = Q(\sqrt{d})$ and suppose the discriminant of K/Q has t prime divisors. Then the group $\mathbf{C}_K/(\mathbf{C}_K)^2$ has order 2^{t-1} if $d < 0$ or if $d > 0$ and a unit of K has norm -1; it has order 2^{t-2} if $d > 0$ and all units of K have norm 1.

PROOF. By Theorem 3.3 we find $G(E/K) \cong \mathbf{C}_K/(\mathbf{C}_K)^2$. The order of $G(E/K)$ is easily found using Theorems 3.2, 3.4, 3.8.

We can go a little farther without any additional effort. Explicit generators for $E^{(+)}$ can be given from the knowledge of the Galois group Γ_0 given above. Since $\Gamma_0 \cap \Gamma_p$ has index 2 in Γ_p, it follows that $E^{(+)}$ contains a quadratic subfield of $Q(\beta_p)$. When p is odd Γ_p is cyclic so there is only one quadratic subfield—namely

$$Q(\sqrt{p}) \subset E^{(+)} \qquad \text{if} \quad p \equiv 1 \bmod 4$$
$$Q(\sqrt{-p}) \subset E^{(+)} \qquad \text{if} \quad p \equiv -1 \bmod 4.$$

In case D is even but not divisible by 8 then again there is only one quadratic subfield of $Q(\beta_2)$—namely $Q(\sqrt{-1}) = Q(\beta_2)$.

When D is divisible by 8 then $Q(\beta_2)$ contains three quadratic subfields generated by $\sqrt{-1}$, $\sqrt{2}$, $\sqrt{-2}$. One could work out congruence conditions to tell which case arises but it is simpler to just use the following device. Since 8 divides D, d must be even and

$$|d| = 2p_2 \cdots p_t.$$

Suppose $p_2, \ldots, p_r$ are the primes $\equiv 1 \bmod 4$ and $p_{r+1}, \ldots, p_t$ are the primes $\equiv -1 \bmod 4$. Then

$$\sqrt{d} = \sqrt{\pm 2}\sqrt{p_1} \cdots \sqrt{p_r}\sqrt{-p_{r+1}} \cdots \sqrt{-p_t}$$

and so the ambiguous sign is uniquely determined. We can avoid cases by stating the result in the following way.

3.10 Theorem. Let $K = Q(\sqrt{d})$ have discriminant Δ and suppose $|\Delta| = p_1 p_2 \cdots p_t$ with $p_2, \ldots, p_t$ odd primes and p_1 either an odd prime or a power of 2. The extended genus field of K is

$$E^{(+)} = Q(\sqrt{d}, \alpha_2, \ldots, \alpha_t)$$

where

$$\alpha_i = \sqrt{p_i} \qquad \text{if} \quad p_i \equiv 1 \bmod 4,$$

$$\alpha_i = \sqrt{-p_i} \qquad \text{if} \quad p_i \equiv -1 \bmod 4.$$

In the case $d < 0$, $E = E^{(+)}$ is part of $K^{(1)}$. If $d > 0$ and some unit of K has norm -1 in Q then again $E = E^{(+)}$. Notice this case can only occur when d has no prime factor of the shape $4n-1$. This follows because K has no complex prime and so E must not have one either since E/K is unramified. In this case there are no primes for which $\sqrt{-p}$ is in E.

In the case $d > 0$ and all units of K have norm $+1$ then E is the maximal real subfield of $E^{(+)}$. We can list generators for E by inspection of those for $E^{(+)}$ in the theorem.

Suppose $p_2, \ldots, p_r$ are the primes $\equiv 1 \bmod 4$ in case Δ is even and that also $p_1 \equiv 1 \bmod 4$ in case Δ is odd. Then

$$E = Q(\sqrt{d}, \alpha_2', \ldots, \alpha_{t-1}') \quad \text{and}$$

(i) $\alpha_i' = \sqrt{p_i}, \qquad i \leqslant r$

$\alpha_i' = \sqrt{p_i p_t}, \qquad r < i < t \quad \text{if} \quad r \neq t-1;$

(ii) $\alpha_i' = \sqrt{p_i}, \qquad i < t-1$

$\alpha_{t-1}' = \sqrt{2p_t}, \qquad \text{in case} \quad r = t-1.$

We give a few examples in the table below. Notice that the field E determines the order 2^s of $\mathbf{C}_K/(\mathbf{C}_K)^2$. This means that $\mathbf{C}_K$ has exactly s cyclic direct factors of even order appearing in a direct product decomposition. For certain small values of h_K, E determines completely the structure of $\mathbf{C}_K$. For example with $K = Q(\sqrt{-5\cdot 13})$, $h_K = 8$ and $s = 2$ so $\mathbf{C}_K$ is the direct product of a cyclic group of order 4 and one of order 2.

d	h_K	Generators for E
-21	4	$\sqrt{-1}, \sqrt{-3}, \sqrt{-7}$
21	1	$\sqrt{21}$
-42	4	$\sqrt{+2}, \sqrt{-3}, \sqrt{-7}$
42	2	$\sqrt{2}, \sqrt{21}$
-65	8	$\sqrt{-1}, \sqrt{5}, \sqrt{13}$
65	2	$\sqrt{5}, \sqrt{13}$
-130	4	$\sqrt{-2}, \sqrt{5}, \sqrt{13}$
130	4	$\sqrt{2}, \sqrt{5}, \sqrt{13}$
-55	4	$\sqrt{-1}, \sqrt{5}, \sqrt{-11}$
55	2	$\sqrt{5}, \sqrt{11}$
-110	12	$\sqrt{-2}, \sqrt{5}, \sqrt{-11}$
110	2	$\sqrt{5}, \sqrt{22}$

APPENDIX

In the two sections which follow we present some results which were used several times in the text. Since these are not universally taught in courses prerequisite to one in which this text might be used, complete proofs are given. We prove only the versions of the theorems actually required in the main body of the text. More complete results on these matters can be found in several books; for example see "Algebra" by Lang.

APPENDIX A. NORMAL BASIS THEOREM AND HILBERT'S THEOREM 90

Assume L/K is a normal extension with Galois group G. Let $L = K(\theta)$ and $f(X)$ the minimum polynomial of θ and K. Let

$$f(X) = \prod_{i}^{n} (X - \theta_i)$$

be the factorization of $f(X)$ in $L[X]$.

Proposition 1. $L \otimes_K L \cong L_1 \oplus \cdots \oplus L_n$ (ring direct sum) with $L_i \cong L$. If G acts on $L \otimes L$ by the rule $\sigma(a \otimes b) = \sigma(a) \otimes b$, then G permutes the L_i transitively.

PROOF. Let G act on $L[X]$ by operating on the coefficients of a polynomial. Then the ideal $(f(X))$ is left invariant (as a set) because $f(X)$ has coefficients in K and G permutes transitively the ideals $(X-\theta_i)$ containing $f(X)$. (G permutes the roots transitively.) By CRT we find

$$\frac{L[X]}{(f(x))} \cong \sum \frac{L[X]}{(X-\theta_i)} \cong \sum \oplus L_i$$

with $L \cong L_i$. Clearly the action of G permutes the L_i transitively. Now to get the statement of the proposition we observe the isomorphisms

$$L \otimes_K L \cong L \otimes \frac{K[X]}{(f(X))} \cong \frac{L[X]}{(f(X))}$$

are consistent with the action of G.

Theorem. (Normal Basis Theorem in Cyclic Case). Assume $G = \langle \sigma \rangle$ is cyclic of order n. There is an element α in L such that $\alpha, \sigma(\alpha), \ldots, \sigma^{n-1}(\alpha)$ is a K-basis of L.

PROOF. View the automorphism σ as a K-linear transformation on the K-space L. If x_i is a K-basis for L, then $x_i \otimes 1$ is a $(1 \otimes L)$-basis for $L \otimes L$. This means σ has the same minimum equation over K as $\sigma \otimes 1$ has over $1 \otimes L$. By Proposition 1, $\sigma \otimes 1$ is a permutation matrix and so has minimum equation $X^n - 1$. Thus the minimum and characteristic equation for σ coincide. By linear algebra there is a "cyclic vector". That is there exists an element $\alpha \in L$ whose images under powers of σ span L over K.

REMARK. This theorem holds for any G with the conclusion that an element exists in L whose images under G give a basis of L over K. We shall not require this more general result.

We introduce some further notation. Let β be the map $L \otimes L \to L$ defined by $\beta(a \otimes b) = ab$. This is a K-algebra homomorphism onto L. All but one of the L_i in Proposition 1 must be in the $\ker \beta$. Let L_1 be one that is not and let e_1 denote the identity element of L_1. We still assume $G = \langle \sigma \rangle$ is cyclic and set

$$\sigma^{i-1}(L_1) = L_i, \quad \sigma^{i-1}(e_1) = e_i$$

so that e_i is the identity of L_i and

$$\beta(e_1) = 1, \quad \beta(e_j) = 0 \qquad j \neq 1.$$

Proposition 2. If the elements $\lambda_1, \ldots, \lambda_n$ in L are not all zero then the K-linear transformation

$$\alpha \to \sum \lambda_i \sigma^i(\alpha), \quad \alpha \in L$$

is not the zero transformation.

PROOF. Consider maps

$$L \xrightarrow{h} L \otimes L \xrightarrow{f} L \otimes L \xrightarrow{\beta} L$$

where $h(\alpha) = \alpha \otimes 1$, $f(\alpha \otimes \gamma) = \sum \sigma^i(\alpha) \otimes \gamma\lambda_i$ and β as above. The composite of the maps, $\beta f h$ is the transformation we want to show is nonzero. We see that $h(L)$ generates $L \otimes L$ as a vector space over $1 \otimes L$. Since f is a $1 \otimes L$-linear map we see $\beta f h$ is zero if and only if the image of f is in $\ker \beta$. To prove this is not the case first take an index k such that $\lambda_k \neq 0$. Then

$$\begin{aligned} \beta f(e_{n-k+1}) &= \beta \sum \sigma^i \otimes 1 (e_{n-k+1}) 1 \otimes \lambda_i \\ &= \sum \beta(e_{n-k+1+i}) \lambda_i \\ &= \lambda_k \neq 0, \end{aligned}$$

and this proves the result.

Corollary (Hilbert's Theorem 90). Let L/K have cyclic Galois group $G = \langle \sigma \rangle$. If $\alpha \in L$ and $\mathrm{N}_{L/K}(\alpha) = 1$, then $\alpha = \beta/\sigma(\beta)$ for some $\beta \in L$.

PROOF. Let $\lambda_i = \alpha\sigma(\alpha) \cdots \sigma^{i-1}(\alpha)$ for $i = 1, 2, \ldots, n$. Notice $\lambda_n = \mathrm{N}(\alpha) = 1$ and $\alpha\sigma(\lambda_i) = \lambda_{i+1}$. By the last result there is an element $\gamma \in L$ such that

$$\beta = \sum \lambda_i \sigma^i(\gamma) \neq 0.$$

It is easily checked that $\sigma(\beta) = \alpha^{-1}\beta$ and the result follows.

APPENDIX B. MODULES OVER PRINCIPAL IDEAL DOMAINS

Let R be a principal ideal domain. We shall prove some facts about finitely generated modules over R. For the sake of completeness, we shall first record some definitions. Let M be a (left) R-module. M is *torsion free* if x in R, m in M and $xm = 0$ implies either $x = 0$ or $m = 0$. M is *finitely generated* if there exist elements $m_1, \ldots, m_n$ in M such that $M = Rm_1 + \cdots + Rm_n$. The elements $m_1, \ldots, m_n$ are a *generating set*. If the equation $x_1 m_1 + \cdots + x_n m_n = 0$, with x_i in R, implies $x_i = 0$ for each i then M is *free* of *rank* n. In this case M is isomorphic to the direct sum of n copies of R. We shall also write

$$R^{(n)} = R \oplus \cdots \oplus R = \sum_1^n Re_i$$

where $e_i = (0, \ldots, 1, \ldots, 0)$, 1 in the ith coordinate, for the free module of rank n.

Theorem 1. Any R-submodule of $R^{(n)}$ is a finitely generated free module with rank at most n.

PROOF. Let M be a submodule of $R^{(n)}$ and set

$$M_k = M \cap (Re_1 + \cdots + Re_k).$$

We shall prove by induction on k that M_k is (0) or is free of rank at most k.

For $k = 1$ we find $M_1 = \mathfrak{A}e_1$ with $\mathfrak{A}$ some ideal of R. Since $\mathfrak{A}$ is principal, it is either (0) or free of rank 1. The same is then true of M_1. Now suppose $k \geqslant 2$ and M_{k-1} is free of rank at most $k-1$. Let $\mathfrak{A}_k$ denote the set of all elements x in R such that

$$m = b_1 e_1 + \cdots + b_{k-1} e_{k-1} + xe_k \tag{1}$$

is in M_k for some b_i in R. Then $\mathfrak{A}_k = Ra_k$. Now let

$$m_0 = a_1 e_1 + \cdots + a_k e_k$$

be an element in M_k having the generator of $\mathfrak{A}_k$ as the coefficient of e_k. For any m in M_k we have m in the form (1) with $x = ba_k$ for some b in R. Then

$$m = bm_0 + (m - bm_0)$$

shows $M_k = Rm_0 + M_{k-1}$. If m_0 is already in M_{k-1}, so $a_k = 0$, then $M_k = M_{k-1}$ so the inductive step holds. If $a_k \neq 0$, then inspection of the kth coordinates shows M_k is the direct sum of Rm_0 and M_{k-1}. Furthermore Rm_0 is free of rank 1 so it follows that M_k is free of rank $\leqslant k$.

Corollary. If $0 < m < n$ are integers and

$$R^{(n)} \cong R^{(m)} \oplus V$$

for some module V, then V is free of rank $n-m$.

PROOF. By the theorem, $V \cong R^{(t)}$ for some integer t. If R happens to be a field, then $t = n-m$ by dimension counting. When R is not a field, take any maximal ideal $\mathfrak{A}$ of R so that $R/\mathfrak{A} = F$ is a field. Now use the observation

$$R^{(n)}/\mathfrak{A}R^{(n)} \cong F^{(n)}$$

to count dimension again and complete the proof.

The result about modules most frequently used in the text is the following.

Theorem 2. Let M be a finitely generated, torsion free R-module which can be generated by n elements but no fewer. Then M is free of rank n and any generating set with n elements gives a free basis for M.

PROOF. The proof is by induction on n. Let $m_1, \ldots, m_n$ be a generating set. If $n = 1$, then $M = Rm_1 \cong R$ so the result holds. Assume the theorem is correct for torsion free modules on $n-1$ generators. Define an R-homomorphism ϕ from $R^{(n)}$ to M by

$$\phi(x_1, \ldots, x_n) = x_1 m_1 + \cdots + x_n m_n.$$

This is onto M because the m_i generate M. The theorem will be proved if we show ϕ is one-to-one. We shall reach a contradiction by assuming now $W = \ker\phi \neq 0$.

Assertion 1. $W = R(y_1, \dots, y_n)$ for some $y_i \in R$.

To prove this we consider the projection map

$$\pi : (w_1, \dots, w_n) \to w_n$$

defined on W with values in R. This map is an isomorphism of W with a nonzero ideal of R because any element in the kernel of π has the form $(w_1, \dots, w_{n-1}, 0)$ and

$$w_1 m_1 + \cdots + w_{n-1} m_{n-1} = 0.$$

However the induction hypothesis can be applied to $Rm_1 + \cdots + Rm_{n-1}$ to obtain this is free on the $n-1$ generators $m_1, \dots, m_{n-1}$. Thus $w_1 = \cdots = w_{n-1} = 0$. Now if $\pi(W) = Ry_n$ then $W = R(y_1, \dots, y_n)$ where $(y_1, \dots, y_n)$ is in W.

Assertion 2. There is a basis $f_1, \dots, f_n$ for $R^{(n)}$ and a nonzero element b in R such that $bf_n = (y_1, \dots, y_n)$.

Notice that the proof of this assertion is independent of the rest of the proof of the theorem and is valid so long as $(y_1, \dots, y_n) \neq 0$.

Let $Ry_1 + \cdots + Ry_n = Rb$ and let z_i be an element of R such that $z_i b = y_i$. Then

$$Rz_1 + \cdots + Rz_n = R$$

and

$$a_1 z_1 + \cdots + a_n z_n = 1$$

for some a_i in R.

Consider the homomorphism θ from $R^{(n)}$ to R defined by

$$\theta(x_1, \dots, x_n) = x_1 a_1 + \cdots + x_n a_n.$$

Then $\theta(z_1, \dots, z_n) = 1$. Write $(z) = (z_1, \dots, z_n)$. For any v in $R^{(n)}$ we have

$$v = \theta(v)(z) + [v - \theta(v)(z)]$$

and this shows $R^{(n)} = R(z) + \ker\theta$. It is easy to see the sum is direct and that $R(z)$ is free of rank 1. Thus $\ker\theta$ is free with a basis $f_1, \dots, f_{n-1}$. If we set $f_n = (z)$ then $bf_n = (y_1, \dots, y_n)$ as required.

Now we complete the proof of the theorem. The element $(y) = (y_1, \dots, y_n)$ is in W so

$$0 = \phi(y) = \phi(bf_n) = b\phi(f_n).$$

Since M is torsion free and $b \neq 0$ it follows that $\phi(f_n) = 0$. But then M is generated by $\phi(f_1), \ldots, \phi(f_{n-1})$ since $R^{(n)} = \sum Rf_i$. This means M can be generated by fewer than n elements contrary to our assumption.

Corollary. Let R have quotient field K and let L be a field containing K. If M is finitely generated R-submodule of L, then M is free with rank at most the dimension of L over K.

PROOF. Since L is a field containing R, M must be torsion free. By the theorem, M is free with rank n for some n. An easy common denominator argument shows that any minimal generating set for M is a linearly independent set over K.

EXERCISE. Prove the Unimodular Row Lemma: If R is a PID, and $y_1, \ldots, y_n$ are elements in R such that $Ry_1 + \cdots + Ry_n = R$ then there exists an $n \times n$ matrix Y with entries in R such that the first row of Y is $(y_1, \ldots, y_n)$ and $\det Y =$ unit in R. If $n \geqslant 2$, then in fact, there is such a Y with $\det Y = 1$.

BIBLIOGRAPHY

The references listed here are a few that would most likely be of help to the student wishing to better understand the results in the text. It is obviously not an attempt at a bibliography for the student interested in the history of the subject. The article by Hasse in the volume edited by Cassels and Frohlich will be helpful in this regard.

E Artin, Theory of Algebraic Numbers. Lecture notes, Gottingen, 1969.

E. Artin and J. Tate, "Class Field Theory." Benjamin, New York, 1967.

Z. I. Borevich and I. R. Shafarevich, "Number Theory." Academic Press, New York, 1967.

J. W. Cassels and A Frohlich, eds., "Algebraic Number Theory." Thompson Publ. Washington, D.C., 1967.

L. J. Goldstein, "Analytic Number Theory." Prentice-Hall, Englewood Cliffs, New Jersey, 1971.

H. Hasse, "Vorlesungen über Klassenkörpertheorie." Physica, Wurzburg, 1967.

S. Lang, "Algebra." Addison-Wesley, Reading, Massachusetts, 1965.

S. Lang, "Algebraic Number Theory," Addison-Wesley, Reading, Massachusetts, 1970.

J. Neukirch, "Klassenkörpertheorie." Universität Bonn, Bonn, 1967.

J. P. Serre, "Corps Locaux." Hermann, Paris, 1962.

A. Speiser, *J. reine angew. Math.* **149** (1919), p. 174–188.

A. Weil, "Basic Number Theory." Springer-Verlag, Berlin and New York, 1967.

E. Weiss, "Algebraic Number Theory," McGraw Hill, New York, 1963.

INDEX

Pure and Applied Mathematics

A Series of Monographs and Textbooks

Editors **Paul A. Smith and Samuel Eilenberg**

Columbia University, New York

1: Arnold Sommerfeld. Partial Differential Equations in Physics. 1949 (Lectures on Theoretical Physics, Volume VI)
2: Reinhold Baer. Linear Algebra and Projective Geometry. 1952
3: Herbert Busemann and Paul Kelly. Projective Geometry and Projective Metrics. 1953
4: Stefan Bergman and M. Schiffer. Kernel Functions and Elliptic Differential Equations in Mathematical Physics. 1953
5: Ralph Philip Boas, Jr. Entire Functions. 1954
6: Herbert Busemann. The Geometry of Geodesics. 1955
7: Claude Chevalley. Fundamental Concepts of Algebra. 1956
8: Sze-Tsen Hu. Homotopy Theory. 1959
9: A. M. Ostrowski. Solution of Equations and Systems of Equations. Third Edition, in preparation
10: J. Dieudonné. Treatise on Analysis: Volume I, Foundations of Modern Analysis, enlarged and corrected printing, 1969. Volume II, 1970. Volume III, 1972
11: S. I. Goldberg. Curvature and Homology. 1962.
12: Sigurdur Helgason. Differential Geometry and Symmetric Spaces. 1962
13: T. H. Hildebrandt. Introduction to the Theory of Integration. 1963.
14: Shreeram Abhyankar. Local Analytic Geometry. 1964
15: Richard L. Bishop and Richard J. Crittenden. Geometry of Manifolds. 1964
16: Steven A. Gaal. Point Set Topology. 1964
17: Barry Mitchell. Theory of Categories. 1965
18: Anthony P. Morse. A Theory of Sets. 1965
19: Gustave Choquet. Topology. 1966
20: Z. I. Borevich and I. R. Shafarevich. Number Theory. 1966
21: José Luis Massera and Juan Jorge Schaffer. Linear Differential Equations and Function Spaces. 1966
22: Richard D. Schafer. An Introduction to Nonassociative Alegbras. 1966
23: Martin Eichler. Introduction to the Theory of Algebraic Numbers and Functions. 1966
24: Shreeram Abhyankar. Resolution of Singularities of Embedded Algebraic Surfaces. 1966
25: François Treves. Topological Vector Spaces, Distributions, and Kernels. 1967
26: Peter D. Lax and Ralph S. Phillips. Scattering Theory. 1967.
27: Oystein Ore. The Four Color Problem. 1967
28: Maurice Heins. Complex Function Theory. 1968

29: R. M. Blumenthal and R. K. Getoor. Markov Processes and Potential Theory. 1968

30: L. J. Mordell. Diophantine Equations. 1969

31: J. Barkley Rosser. Simplified Independence Proofs: Boolean Valued Models of Set Theory. 1969

32: William F. Donoghue, Jr. Distributions and Fourier Transforms. 1969

33: Marston Morse and Stewart S. Cairns. Critical Point Theory in Global Analysis and Differential Topology. 1969

34: Edwin Weiss. Cohomology of Groups. 1969

35: Hans Freudenthal and H. De Vries. Linear Lie Groups. 1969

36: Laszlo Fuchs. Infinite Abelian Groups: Volume I, 1970. Volume II, 1973

37: Keio Nagami. Dimension Theory. 1970

38: Peter L. Duren. Theory of H^p Spaces. 1970

39: Bodo Pareigis. Categories and Functors. 1970

40: Paul L. Butzer and Rolf J. Nessel. Fourier Analysis and Approximation: Volume 1, One-Dimensional Theory. 1971

41: Eduard Prugovečki. Quantum Mechanics in Hilbert Space. 1971

42: D. V. Widder: An Introduction to Transform Theory. 1971

43: Max D. Larsen and Paul J. McCarthy. Multiplicative Theory of Ideals. 1971

44: Ernst-August Behrens. Ring Theory. 1972

45: Morris Newman. Integral Matrices. 1972

46: Glen E. Bredon. Introduction to Compact Transformation Groups. 1972

47: Werner Greub, Stephen Halperin, and Ray Vanstone. Connections, Curvature, and Cohomology: Volume I, De Rham Cohomology of Manifolds and Vector Bundles, 1972. Volume II, Lie Groups, Principal Bundles, and Characteristic Classes, in preparation

48: Xia Dao-xing. Measure and Integration Theory of Infinite-Dimensional Spaces: Abstract Harmonic Analysis. 1972

49: Ronald G. Douglas. Banach Algebra Techniques in Operator Theory. 1972

50: Willard Miller, Jr. Symmetry Groups and Their Applications. 1972

51: Arthur A. Sagle and Ralph E. Walde. Introduction to Lie Groups and Lie Algebras. 1973

52: T. Benny Rushing. Topological Embeddings. 1973

53: James W. Vick. Homology Theory: An Introduction to Algebraic Topology. 1973

54: E. R. Kolchin. Differential Algebra and Algebraic Groups. 1973

55: Gerald J. Janusz. Algebraic Number Fields. 1973

In preparation

Samuel Eilenberg. Automata, Languages, and Machines: Volume A

H. M. Edwards. Riemann's Zeta Function

Wayne Roberts and Dale Varberg. Convex Functions

A. B. S. Holland. Introduction to the Theory of Entire Functions